ELEMENTS OF A SUSTAINABLE WORLD

Elements of a Sustainable World

John Evans

University of Southampton

OXFORD
UNIVERSITY PRESS

Great Clarendon Street, Oxford, OX2 6DP,
United Kingdom

Oxford University Press is a department of the University of Oxford.
It furthers the University's objective of excellence in research, scholarship,
and education by publishing worldwide. Oxford is a registered trade mark of
Oxford University Press in the UK and in certain other countries

Published in the United States of America by Oxford University Press
198 Madison Avenue, New York, NY 10016, United States of America

British Library Cataloguing in Publication Data

Data available

Library of Congress Control Number: 2019954843

ISBN 978–0–19–882783–2 (Hbk.)
ISBN 978–0–19–882784–9 (Pbk.)

DOI: 10.1093/oso/9780198827832.001.0001

Printed and bound by
CPI Group (UK) Ltd, Croydon, CR0 4YY

Preface

We now have 118 known chemical elements as our palette in our context of sustaining our world. Whatever patterns we would like to paint with the elements, we are influenced by the pressures that we all are exerting on spaceship Earth and its inhabitants of all species. In this book our context is considered in terms of the four spheres of the ancient world: Earth, Air, Fire and Water. This book shows how chemical principles can be used to understand the pressures on our world spanning from greenhouse emissions through freshwater supplies to energy generation and storage. The supply of the chemical elements is key to their contribution to alleviating these pressures. Most synthetic and radioactive elements are not available in sufficient supply to contribute in this. Some solutions, such as wind turbines, batteries, fuel cells and automotive exhaust remediation pose questions about sustainable supplies of critical elements. With an eye on the target of the IPCC of capping the temperature anomaly to 1.5 ^{o}C (RCP2.6), options for carbon capture and storage, and the generation of energy and element supply from the sea are assessed. The consequences of the escape of plastics and pharmaceuticals into the wider environment for water integrity are also considered. The final chapter assesses the prospects for planet Earth as viewed in 2019. I have included questions at the end of each chapter. If we consider the elements of learning of learning as being knowledge, understanding, application and evaluation, the bulk of the questions are on a combination of knowing something and understanding it well enough to tackle a new situation. There are no "write what you know about" questions.

This book is a substantial amplification of the 12-lecture slot half unit so that it could provide a course book for a full semester unit. It is based upon students having experience of undergraduate chemistry to the second year. In principle, sections could be extracted for use from year 2, through graduating years to Masters level, depending on the experience of the cohort. The aim has been to marry chemical principles and environmental context. For those mainly experienced in chemical principles, it should provide routes to understand environmental consequences and the demands of sustainability using concepts like embodied energy, carbon and water. For those coming to the topic with a strong interest in sustainability it will provide a chemical basis for the causes of environmental impact and also guidelines for technologies and lifestyles that should be sustainable.

Acknowledgements

The thoughtline to this book began at the Royal Society Theo Murphy Symposium held in July 2010 entitled The sustainable planet: opportunities and challenges for science, technology and society.[1] I am grateful to Judith Howard and Martyn Chamberlain for organising this far-reaching and inspirational symposium. As a result, we initiated a new course in Sustainable Chemistry at Southampton, and I wish to thank Robert Raja for his encouragement in that. I am grateful that Sonke Adlung at OUP saw merit in the proposal that I submitted to him in 2017, and to the Press for agreeing to accept a submission.

Over the last two years this has figured largely for me. I am very grateful for friends and colleagues for their advice and support. Particularly, I am grateful to Phil Gale, Gill Reid, Bill Levason, David Read and Peter Wells at Southampton. In and around Romsey, the support from friends at RMC has been very important to me. The writing has impacted most on my family. Beccy and Simon, and Lisa and Ed and, of course, Hilary have been the bedrock. I know so many things that I have not been able to do over the last two years when Hilary has herself given much time to her own roles. Now I should care for our expended carbon budget and paint the house!

John Evans, 4 November 2019

[1] Ed. J. A. K. Howard, M. Chamberlain, The sustainable planet: opportunities and challenges for science, technology and society, *Phil. Trans. R. Soc A*, 2011, **369**, 1713–1882.

Bibliography

This bibliography comprises a list of some key books and reports that were very in influential in writing this book.

D. J. C. MacKay, *Sustainable Energy - without the hot air*, UIT Cambridge, Cambridge, UK, 2009.

J. M. Allwood, J. M. Cullen, *Sustainable Materials with both eyes open*, UIT Cambridge, Cambridge, 2012.

M. F. Ashby, *Materials and the Environment*, Second Edition, Butterworth-Heinemann, Oxford, 2013.

J. T. Houghton, *Global Warming. The Complete Briefing*, Fifth Edition, Cambridge University Press, 2015.

M. F. Ashby, *Materials and the Environment*, Butterworth-Heinemann, Elsevier, Oxford, 2nd. ed., 2018.

Global Warming of 1.5 °C IPCC report SR15, 2018, https://www.ipcc.ch/sr15/

USGS National Minerals Information Center, Commodity Summaries, 2019.

Contents

Planet Earth

1.1 Planetary Resources

For some years, the WWF organization has published some quantitative assessments of the use of our planet.[1] Fig. 1.1 shows the estimates for the total **ecological footprint** of humanity on this world, and its subdivision into different classes of use.[2,3] The area of global hectare area (the biocapacity of the world) is not a constant, and this value increases with the biological productivity. In 1961 world biocapacity was estimated to be 9.5 Gha, clearly in excess of the ecological footprint for that year. Increases in productivity through to 2013 raised biocapacity to 12.2 Gha, considerably in deficit with the ecological footprint. Recognizing that there is some margin for error in these figures, the pattern is still clear. Overall, the total pressure on our world has increased very markedly over the past 50 years (by a factor of c.4) (Fig. 1.1).

To some measure this might be linked to the change in human population through that time (Fig. 1.2), but the estimated increase in our ecological footprint also suggests there is an increase in the per capita footprint. As the expansion in the figure shows, population has essentially doubled between 1970 and 2020.

However, this need not be an inexorable trend. Public health and educational opportunities can change this pattern to a considerable extent. As life expectancy increases beyond 50 there is a general trend across most countries in the world for the fertility rate to decline. Indeed, the fertility rate has halved over the last 50 years (Fig 1.3) and is now approaching 2.4.[4] In time, this would help stabilize population levels.

1.2 Differential Pressure on Land, Freshwater, and Oceans

The effects of humanity on the components of the world's systems have varied considerably (Fig. 1.1). The overall load on the planet has not increased as rapidly as population. Indeed the meat- and fish-supplying space (fishing and grazing) showed relatively small changes, but cropland a significant rise. This may be a combination of change of land use, increased efficiency, and overuse. However,

[1] http://wwf.panda.org.

[2] http://data.footprintnetwork.org/.

[3] The ecological footprint is defined as the biologically productive area that is needed to provide everything that people use.

[4] Data from the World Bank, http://data.worldbank.org/indicator.

Elements of a Sustainable World. John Evans, Oxford University Press (2020). © John Evans.
DOI: 10.1093/oso/9780198827832.003.0001

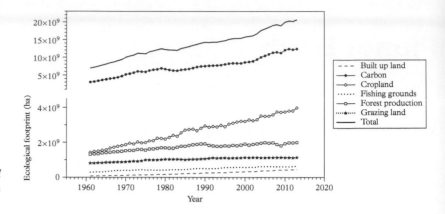

Figure 1.1 *Estimated world ecological footprint (hectares) by land type from 1961 to 2013.*

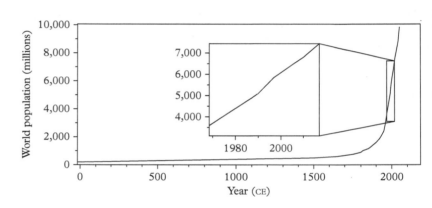

Figure 1.2 *World population from the start of the Christian era (CE) to that estimated for 2050.*

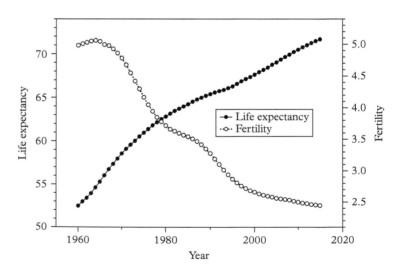

Figure 1.3 *World life expectancies and lifetime fertility rates from 1960 to 2015.*

the dominant change is in the carbon- related part of the system, and that is largely due to energy use.[5] Not surprisingly, a strong theme in this book will be to address carbon- and energy-related factors.

1.3 Effects of CO$_2$ Increases

Over a similar time frame, the mean annual concentration of atmospheric CO$_2$ has shown an increase of *c.*25% with an upward sloping curve (Fig. 1.4).[6] This curve has been extended backwards using the record of atmospheric composition frozen in ice cores (Fig. 1.5).[7] That shows that the trend has been rising since the 1700s from a value of 280 ppm (molecules of CO$_2$/10^6 molecules of air), taken as the pre-industrial level. Exceeding a value of 300 ppm in the 1900s can be seen as a significant departure from the patterns derived from analysis of ice cores dated over the past 800,000 years (Fig. 1.6).[8]

In the main, the sources of atmospheric CO$_2$ attributable to the burning of fossil fuels can be tracked by production and sales records. From a starting level

[5] The carbon footprint is represented by the area of forest land required to sequester the CO$_2$ emissions from fossil fuel use.

[6] P. Tans, NOAA/ESRL and R. Keeling, Scripps Institution of Oceanography (http://scrippsco2.ucsd.edu/) (http://www.esrl.noaa.gov/gmd/ccgg/trends/).

[7] D. M. Etheridge, L. P. Steele, R. L. Langenfelds, R. J. Francey, Division of Atmospheric Research, CSIRO, Victoria, Australia; J.-M. Barnola, Laboratoire de Glaciologie et Geophysique de l'Environnement, Saint Martin d'Heres, France; V. I. Morgan, Antarctic CRC and Australian Antarctic Division, Hobart, Tasmania, Australia.

[8] D. Lüthi, M. Le Floch, B. Bereiter, T. Blunier, J.-M. Barnola, U. Siegenthaler, D. Raynaud, J. Jouzel, H. Fischer, K. Kawamura, and T. F. Stocker. 'High-resolution carbon dioxide concentration record 650,000–800,000 years before present'. *Nature*, 2008, **453**, 379–82.

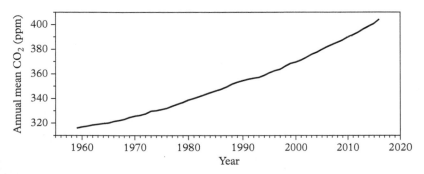

Figure 1.4 *Mean annual concentrations of atmospheric CO$_2$ measured at the Mauna Loa Observatory, Hawaii, since 1959.*

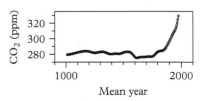

Figure 1.5 *Historical record of atmospheric CO$_2$ from 75 year averages from ice cores.*

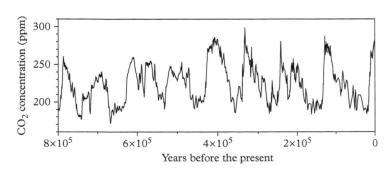

Figure 1.6 *Record of atmospheric CO$_2$ from ice cores over the past 800,000 years.*

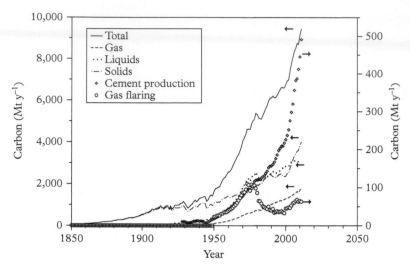

Figure 1.7 *Carbon from fossil fuels burnt annually since 1850.*

[9] T. Boden and B. Andres, 'Global CO_2 emissions from fossil-fuel burning cement manufacturing and gas-flaring: 1751–2014', http://cdiac.ess-dive.lbl.gov/ftp/.ndp030/global.1751_2014.ems.

of $c.3$ Mton y^{-1} in 1751 there was a rise to 8 Mton y^{-1} by 1800 and 50 Mton y^{-1} in 1849; all of this was by the burning of solid fuel.[9] Post 1850, overall use increased substantially and diversified in terms of source and use (Fig. 1.7). Burning of liquid fuels commenced in 1870, but solid fuel predominated until after 1950. The two sources have been largely comparable in their use since 1970, although there has been a significant increase in solid-fuel consumption since 2000. The burning of natural gas was at a low level until 1950, and even by 2011 its use in terms of carbon was about half that of solid and of liquid fuels. The flaring of natural gas peaked at just over 100 Mton y^{-1} in the 1970s (out of a total of $c.9450$ Mton y^{-1}) and has fallen since. The largest source of industrially related production that burns fossil fuels is cement production. This tracked with gas-flaring until the late 1970s, and since then has continued to rise, nearing 500 Mton y^{-1} in 2011, 5% of the total.

These changes in atmospheric CO_2 concentration have been accompanied by other observations that might be expected to be correlated with it. The ability of atmospheric gases to absorb energy and thus increase the temperature of the atmosphere was proposed in the early 1900s by Joseph Fourier and Claude Pouillet. John Tyndall measured the absorption properties of specific gases, notably water vapour and CO_2.[10] These effects were modelled by Svante Arrhenius, who considered the consequence of the burning of coal, the prime fossil fuel of that time.[11] His conclusions included a prediction of a temperature rise of 8–9 °C in the Arctic regions if the level of CO_2 increased by a factor of

[10] J. Tyndall, 'On the absorption and radiation of heat by gases and vapours, and on the physical connexion of radiation, absorption, and conduction. The Bakerian Lecture', *Phil. Mag.*, 1861, **22**, 169–94.

[11] S. Arrhenius, 'On the influence of carbonic acid in the air upon the temperature of the ground', *Phil. Mag.*, 1896, **41**, 237–76.

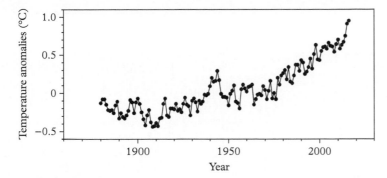

Figure 1.8 *Variations in mean annual atmospheric temperatures compared to the base period of 1901–2000.*

2.3–3 times its level at the time of his writing. That would translate as a change from 295 to 740–890 ppm.

A lengthy debate has transpired since then about the degree to which changes have been due to natural variations or human activity. Improved experimentation and understanding of the interactions between atmosphere, hydrosphere, and biosphere have aided discrimination between contributions, and the bulk of studies and analyses indicate that, qualitatively at least, Arrhenius was onto something. Radiocarbon dating has allowed the characterization of the exchange rates between the atmosphere and oceans, and this resulted in the conclusion that most of the CO_2 added to the atmosphere would be absorbed by the oceans.[12] Hence evidence for trends caused by the increase should be sought in both the oceans and atmosphere at the least.

The assessed annual global land and ocean temperatures[13] have shown a general rise and increase in slope as a baseline to fluctuations of a more short-term nature (Fig. 1.8). The CO_2 can have a direct chemical effect. This can be illustrated by the monitoring of the partial pressure of the gas at an ocean monitoring point with the pH in that region (Fig. 1.9). Again there are short-term variations, but there is a trend in the mean of the pH from *c.*8.12 to *c.*8.06, corresponding to an increase in the concentration of H^+ of *c.*15%.[14] Such a change will cause a shift in the equilibrium concentrations of the carbonate anions ($pK_{a2} = 10.239$ at 25 °C and 0 ion strength) (Eq. 1.1).

$$[HCO_3]^- = [CO_3]^{2-} + H^+. \tag{1.1}$$

This in turn will influence the availability of the aragonite form of calcium carbonate, which is crucial for the formation of shells of marine organisms.

[12] R. Revelle and H. E. Suess, 'Carbon dioxide exchange between atmosphere and ocean and the question of an increase of atmospheric CO_2 during the past decades', *Tellus*, 1957, **9**, 18–27.

[13] GISTEMP Team, 2017: GISS Surface Temperature Analysis (GISTEMP). NASA Goddard Institute for Space Studies. Dataset accessed 2017-11-30 at https://data.giss.nasa.gov/gistemp/. J. Hansen, R. Ruedy, M. Sato, and K. Lo, 2010: 'Global surface temperature change', *Rev. Geophys.*, 2010, **48**, RG4004.

[14] J E. Dore, R Lukas, D. W. Sadler, M. J. Church and D. M. Karl, 'Physical and biogeochemical modulation of ocean acidification in the central North Pacific', *Procl. Natl. Acad. Sci.*, 2009, **106**, 12235–40.

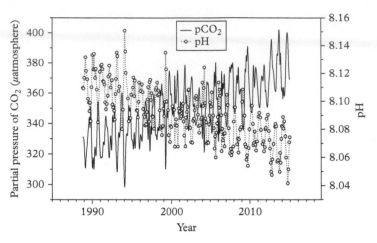

Figure 1.9 *Variations in the seawater CO_2 concentration and pH at Station Aloha near Mauna Loa, Hawaii 1988–2014.*

1.4 Solar Energy Supply

The energy expended using fossil fuels can be set against that supplied by the Sun. This is complicated by the plethora of units used in different communities, so a set of energy definitions is collected in Table 1.1. These relate to materials and processes, and are often described in practical units of energy: kWh or barrels of oil equivalents (BOE). The margin note shows that roughly $\frac{1}{3}$ of the energy consumed in the world is from electricity installations.[15]

[15] World power demand (avg.) 15 TW. Installed electricity capacity 4.5 TW.

The estimate of the solar power received on an imaginary plane normal to the Sun's rays at the edge of the upper atmosphere of Earth averages at 1,367 W m^{-2} (called the solar constant). The maximum value of this is on January 1 when the Earth's elliptical orbit brings it closest to the Sun (by 7%). This impinges upon the Earth as a disc of radius 6,370 km, providing a total input of 1.74×10^{17} W, or 174 PW (petawatts). Spread over the whole surface of the Earth, this affords 342 W m^{-2}, on average (684 W m^{-2} in the day). It is worth noting that the average power demand is 15 TW (terawatts), or less than 10^{-4} of that impinging onto the Earth.

Overall the losses through the atmospheric are described with an **air mass factor** (AM), the ratio of the path length travelled through the atmosphere compared with its depth (Eq. 1.2). The thickness of the most dense region, the **troposphere**, is considerably less at the poles than at the equator (Table 1.2). At the outside of the atmosphere, $AM = 0$. When the Sun is overhead then $\theta_z = 0$, and $AM = 1$. At an incident angle of 60°, $AM = 2$. For solar panel design, an air mass factor of 1.5 is used, corresponding to $\theta_z = 48.2°$. The reduction in

Table 1.1 *Energy definitions and conversions.*

Energy return on investment (EROI)	output/input
Embodied energy of a material	$MJ\,kg^{-1}$
Energy content	$J\,kg^{-1}$
Energy density	$J\,L^{-1}$
Energy concentration	$J\,m^{-2}$
Energy efficiency	J/J
Power density	$W\,m^{-2}$
Energy intensity	J/\$
Tonne of oil equivalent (TOE)	42 GJ
Barrel of oil (*c*.159 L) equivalent (BOE)	6.1 GJ
1eV	$8065.7\ cm^{-1}$
	$1.6921 \times 10^{19}\,J$
	$96.487\ kJ\,mol^{-1}$
1 kWh	3.6 MJ
1 BOE	1694 kWh

Table 1.2 *Height of the troposphere*

At the poles	30,000 ft
At the equator	56,000 ft

Table 1.3 *Solar radiation reaching the surface*

AM value	Proportion	Range (%)
1	0.72	15
1.5	0.64	20
2	0.56	30

solar radiation is not straightforward and depends upon atmospheric conditions (Table 1.3), but at intermediate latitudes the daytime value of 684 $W\,m^{-2}$ will be reduced to *c*.438 $W\,m^{-2}$, which will have a diurnal variation

$$AM = \frac{path}{atmosphere} = \frac{1}{\cos\theta_z}, \quad \theta_z < 70°. \qquad (1.2)$$

1.5 Solar Radiation Spectrum

The light from the Sun can be modelled as a **black body**. In everyday parlance this sounds odd, but a black body is a theoretical material that has no intrinsic spectroscopic properties other than that due to its temperature; approximations to this would be a red-hot poker and a incandescent light bulb. One of the key early successes of quantum theory was the successful calculation of an emission maximum using Planck's equation (Eq. 1.3) [B_ν = spectral radiance at frequency ν]

$$B_\nu(\nu, T) = \frac{2h\nu^3}{c^2} \frac{1}{e^{\frac{h\nu}{k_B T}} - 1}. \qquad (1.3)$$

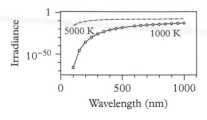

Figure 1.10 *Calculated emission for a black body.*

[16] Laboratory for Atmospheric and Space Physics, University of Colorado, Boulder, Colorado, USA; http://lasp.colorado.edu/home/sorce/instruments/sim/.

As shown in Fig. 1.10, the amplitude of the emission increases greatly with the temperature of the body and also the wavelength of the maximum intensity shortens greatly.

In Wien's law (Eq. 1.4, where $b = 2.8977 \times 10^6$ nm K), the irradiance curves are recognized as having the same shape, and the wavelength of the maximum intensity is inversely rated to the temperature (Fig. 1.11).

$$\lambda_{max} = \frac{b}{T}. \tag{1.4}$$

The energy spectrum of the Sun reaching Earth is measured by the Solar Radiation and Climate Experiment (SORCE), which is a NASA-sponsored satellite mission. One of the spectrometers, SIM (Spectral Irradiance Monitor), has measured the spectrum over the wavelength range 310–2,400 nm (Fig. 1.12), which spans across the irradiance maximum.[16] The energy is mostly in the near

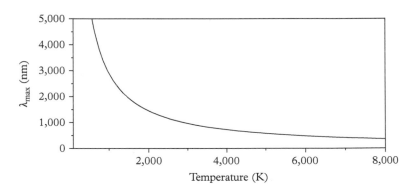

Figure 1.11 *A Wien's law plot of the temperature dependence of the wavelength of maximum intensity for the emission from a black body.*

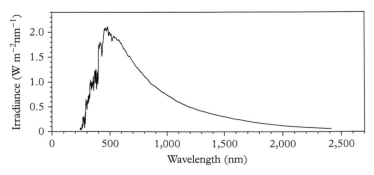

Figure 1.12 *The solar spectrum measured by the SORCE spectrometer SIM at the top of the atmosphere.*

infrared (3,500–700 nm, 51.7%) and the visible (400–700 nm, 38.3%) regions with a minority in the ultra-violet (200–400 nm, 8.7%). The peak radiation is in the visible region (cyan) near 500 nm wavelength, equivalent to a surface temperature of *c.*5,250 °C. The surface regions of the Earth will also afford black body emission. However, the much lower temperatures from those sources will result in much lower intensities and also much longer wavelengths. As shown in Fig. 1.13, the maximum moves from the mid-IR to the far-IR as the temperature is reduced.

The radiation reaching the surface of the Earth is reduced by atmospheric scattering and there are also some specific absorption bands (Fig. 1.14). The strong interactions in the N_2 molecule give a set of **highest occupied molecular orbitals** (HOMOs) of $\pi^4\sigma^2$ which provides the triple bond. There is a large energy gap to the **lowest unoccupied molecular orbitals** (LUMOs), the π^* set. This results in there being no low-lying excited states and it is essentially

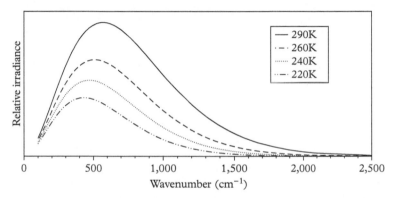

Figure 1.13 *Calculated black body temperatures for temperatures of that at the Earth's surface.*

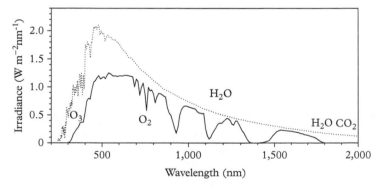

Figure 1.14 *Comparison of the irradiance reaching outer atmosphere (dotted) and at ground level (solid).*

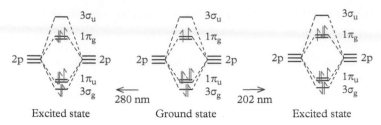

Figure 1.16 *Electronic configurations of the ground state of O$_2$ ($^3\Sigma^-_g$) and first two spin-allowed transitions (left $^3\Sigma^+_u$, right $^3\Sigma^-_u$).*

Figure 1.15 *LUMO (π^*) for the beta spins of ^{3}O$_2$.*

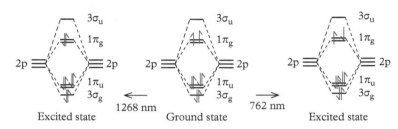

Figure 1.17 *Electronic configurations of the ground state of O$_2$ ($^3\Sigma^-_g$) and first two spin-forbidden transitions (left $^1\Delta_g$, right $^1\Sigma^+_g$).*

Figure 1.18 *Space-filling model of ozone.*

transparent below an energy of 100,000 cm^{-1} (100 nm). The situation for O$_2$ is more complicated (Fig. 1.15) and the ground state of the molecule is a spin triplet ($S = 1$) with two unpaired α spins in the π^* orbitals (Fig. 1.16).

The absorptions due to atmospheric oxygen in the UV region arise from very low cross-sections at wavelengths below 270 nm to 200 nm (the Herzberg continuum largely in the stratosphere). These are due to these spin allowed π to π^* transitions of the beta spin electron and associated vibrational fine structure. These are spin-allowed bands; in other words; there is no change in the spin quantum number, S:

$$\Delta S = 0. \tag{1.5}$$

Extension 1.1 *An oddity: liquid oxygen is blue*

One of the characteristics of oxygen is that the liquid is blue in colour, as well as being paramagnetic. The spin-allowed absorption bands in the ultraviolet cannot account for the blue colour, which is expected to result from absorption bands in the red. The electronic configurations of the two lowest-energy spin-forbidden transitions ($\Delta S = -1$ by spin flips within the π^* set) are shown in Fig. 1.17. The lowest-energy absorption from ^{3}O$_2$ to ^{1}O$_2$ is in the near IR (1,268 nm), as is the second transition (762 nm). There are two strong absorptions at 630 and 577 nm which arise from the harmonic of the first excited state (Fig. 1.14). These are much weaker bands than those dominant in the gas phase but do give observable dips in the solar spectrum at the Earth's surface.

In the ultraviolet much of the atmospheric absorption of the solar spectrum is due to electronic transitions in ozone. There is an electronic transition of ozone centred close to 600 nm that gives rise to the Chappuis bands in the troposphere; there are weaker absorptions near 1,000 nm called the Wulff bands. The main effect though is due to the absorption near 260 nm which has a cross-section 10^4 higher than the Wulff bands (called Hartley and Huggins bands). Based on the energy level diagram in Fig. 1.19, the Chappuis transition can be assigned to a spin-allowed transition from the HOMO to a π^* LUMO state (Fig. 1.20); the spin-forbidden transition will give rise to the Wulff bands. The intense bands near 260 nm emanate mainly from the deeper-lying HOMO-2 orbital with π non-bonding character.

At the other end of the irradiance spectrum (Fig. 1.14), above 2500 nm (4000 cm^{-1}), the absorption is due to the stretching modes of water with their associated rotational fine structure. In the near infra red, the absorptions are mainly due to higher harmonics and combination bands of the of the fundamentals found in the mid-infrared region. There are also spin-forbidden transitions to $^1O_2$ states in the visible and NIR (See Extension 1.1).

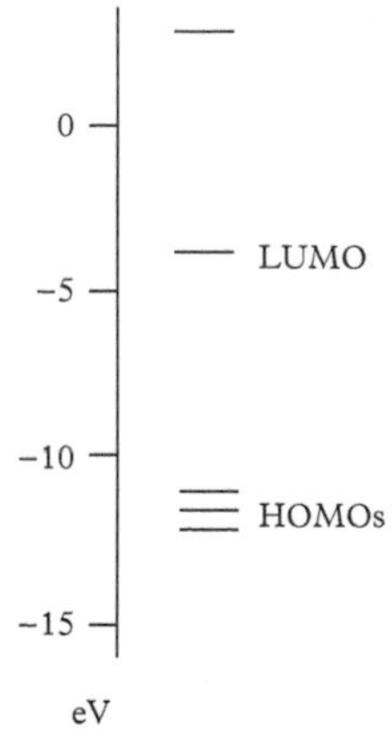

Figure 1.19 *Frontier orbitals of ozone.*

1.6 Absorption of the Energy Emission from Earth

The surface of the Earth itself behaves approximately as a black body, with its spectrum fitting a temperature of 290 K, although the distribution of temperatures will vary with latitude and season. The emission maximum is in the mid infrared

Figure 1.20 *Transitions from the HOMO (left) and HOMO-2 (right) levels to the LUMO (π^*) of O_3.*

near 600 cm^{-1} (Fig 1.13). Above the surface and the boundary layer is the region of the atmosphere called the troposphere. This is the region of most weather events and stretches to about 9 km at the poles and about 17 km at the equator (Table 1.2). Separating that from the **stratosphere** is a region called the **tropopause**. This is a quiet region and movement across the tropopause is relatively small. Hence this can also be considered another black body, with a temperature of 220 K and an emission maximum near 400 cm^{-1} (Fig. 1.13). The measured emission spectra (Fig. 1.21) show maxima following the tracks followed by the black-body curves at 299 and 257 K for tropical and sub-Arctic

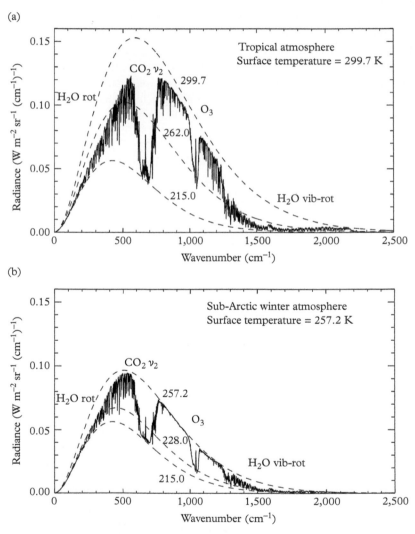

Figure 1.21 *Emission spectra from the Earth measured at the top of the atmosphere for tropical and sub-Arctic winter regions.*

conditions.[17] The curve for the minimum emission tracks that of a black-body at 215 K, taken to be the temperature at the tropopause. The area between those curves represents the energy that could be dissipated by a clear atmosphere if there were no absorbers in the troposphere.

However, this is clearly not the case. In the sub-Arctic region the observed emissions in the 800–900 and 1,100–1,200 cm^{-1} regions of the spectrum are close to that of the Earth's surface, consistent with a virtually transparent troposphere.

However, below 150 cm^{-1} and between 600 and 700 cm^{-1} the emission level is essentially that of the tropopause. In other words, the IR radiation in those regions has been absorbed and retained in the troposphere.

The vast majority of atmosphere components are IR transparent. The third most abundant gas is argon, which being mono-atomic can have no vibrations. The two major components N_2 and O_2 are obviously homonuclear diatomics with a centre of symmetry.[18] Thus their sole vibration creates no dipole change and is IR inactive.

The fundamental energy of a vibration, v, is dependent upon the reduced mass (μ) of a diatomic molecule with masses m_1 and m_2, and the force constant of the bond (k) (Eq. 1.6):

$$v = \frac{1}{2\pi}\sqrt{\frac{k}{\mu}}, \mu = \frac{m_1 m_2}{m_1 + m_2}. \tag{1.6}$$

For a harmonic oscillator, the selection rule is that the quantum number v provides a single transition energy by

$$\Delta v = \pm 1. \tag{1.7}$$

However, the energy profile of a vibration is not symmetrical, flattening out as the interatomic distance approaches the dissociation distance for the bond. This **anharmonicity** allows other transitions to occur:

$$\Delta v = \pm 1, \pm 2, \ldots. \tag{1.8}$$

The result of this is a suite of harmonic and combination bands, as was observed in the solar radiance spectra observed on the surface of Earth (Fig 1.14). Absorption due to water and carbon dioxide was evident in regions above 1,100 nm in wavelength (c.9,000 cm^{-1}).

The three major contributors to the absorptions are water vapour, CO_2 (Fig. 1.22), and O_3 (Fig. 1.18). This is discussed in more detail in Extension 1.3, but we can assign the molecular motions that are most significant in the absorption of IR radiation (Fig. 1.21). For CO_2, the absorption features near 650 cm^{-1} are due to the bending of CO_2; the complex features at low energy are due to rotations of vapour phase water. Particularly in the tropical regions there is evidently a higher level of water vapour in the troposphere, and in that region the absorption due to the rotational fine structure of gas phase water associated with the bending

[17] J. Harries, B. Carli, R. Rizzi, C. Serio, M. Mlynczak, L. Palchetti, T. Maestri, H. Brindley and G. Masiello, 'The Far-Infrared Earth', *Rev. Geophys.* 2008, **26**, RG4004.

[18] It is helpful in this context that N_2O, NO_2, and NO are endothermic and endoergic (Table 1.4).

Table 1.4 *Formation energies (kJ mol^{-1}).*

NO_x	ΔH^o	ΔG^o
N_2O	81.6	103.8
NO_2	33.9	51.9
NO	90.4	86.6

Figure 1.22 *Space-filling models of CO_2 and H_2O.*

[19] J. H. Butler and S. A. Montzka, 'The NOAA Annual Greenhouse Gas Index (AGGI)', 2019 in https://www.esrl.noaa.gov.

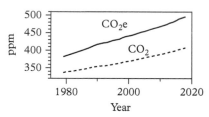

Figure 1.23 *Estimation of the effective concentration of IR absorbing gases as CO₂ equivalents.*

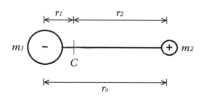

Figure 1.24 *A rigid dipolar diatomic molecule of masses m_1 and m_2 separated by $r_0 = r_1 + r_2$.*

vibration is also apparent. The other obvious feature can be assigned to modes of ozone. These and other gases add to the absorption of energy in the troposphere. Their effects are cumulated relative to CO_2, and thus are represented as CO_2 equivalents (CO_2e). Estimates are available from 1979 (Fig. 1.23).[19] The rate of increase of CO_2e is significantly higher than that of CO_2 alone, indicating that there are other contributors with marked effects.

Water has a permanent dipole moment and so rotation of the molecule will provide a dynamic dipole that can interact with electromagnetic radiation and thus transitions are allowed. This is not the case for the homonuclear diatomics N_2 and O_2, nor for linear CO_2.

If we begin thinking about rotational spectra with a diatomic molecule, we will need to consider a heteronuclear molecule such as HCl or CO, depicted in general in Fig. 1.24. The dipole can be envisaged as rotating about the centre of gravity, C, the position of which will conform to the balance of moments (Eq. 1.9):

$$m_1 r_1 = m_2 r_2 = m_2 (r_0 - r_1). \tag{1.9}$$

The moment of inertia, I, is defined by Eq. 1.10:

$$I = m_1 r_1^2 + m_2 r_2^2. \tag{1.10}$$

Using the balance of moments equations, the definition of the moment of inertia can be rewritten in terms of the reduced mass, μ, as Eq. 1.11:

$$I = \frac{m_1 m_2}{m_1 + m_2} r_0^2 = \mu r_0^2. \tag{1.11}$$

The energy of the rotational levels, $\epsilon_{\mathcal{J}}$ in cm^{-1}, has been shown to be Eq. 1.12:

$$\epsilon_{\mathcal{J}} = B\mathcal{J}(\mathcal{J} + 1), \mathcal{J} = 0, 1, 2, \ldots. \tag{1.12}$$

where B, the rotational constant, is given by Eq. 1.13:

$$B = \frac{h}{8\pi^2 I c}. \tag{1.13}$$

The transition energies follow the selection rule (Eq. 1.14):

$$\Delta \mathcal{J} = \pm 1, \tag{1.14}$$

and differ depending upon the $\mathcal{J}$ in question. For example, we can compare the transition energies from 0 and from 1 (Eqs. 1.15, 1.16):

$$\Delta\epsilon_{0\text{-}1} = \epsilon_1 - \epsilon_0 = B(1+1) - 0 = 2B \tag{1.15}$$

$$\Delta\epsilon_{1\text{-}2} = \epsilon_2 - \epsilon_1 = B(2)(2+1) - B(1+1) = 4B. \tag{1.16}$$

Rotational splittings are comparable to, or less than, the Boltzmann energy under atmospheric conditions and so there will be significant populations of a number of ground states, giving rise to a sequence of lines separated by $2B$. These series will be sensitive to isotopic substitution due to the relationship between energy and the moment of inertia. As an example, we compare $^1H^{35}Cl$ (B_1) with $^2H^{35}Cl$ (B_2) (Eq. 1.17):

$$\frac{B_1}{B_2} = \frac{I_2}{I_1} = \frac{\mu_2}{\mu_1} = \frac{(2)(35)(1+35)}{(1)(35)(2+35)} = \frac{(2)(36)}{(1)(37)} = 1.94. \tag{1.17}$$

Hence the spectra of atmospheric gases will include sequences for the isotopomers. This also shows that the spacing between the rotational lines will tend to reduce with increasing mass. That leaves water displaying a rotational spectrum over a very wide energy range.

The complexity of the spectra increases with less symmetrical structures. The linear molecule HCl has two rotational degrees of freedom, but they are degenerate and can be defined by one rotational constant. For non-linear, pyramidal molecules like ammonia (NH_3), or CFC-11 ($CFCl_3$), there are three rotational degrees of freedom. The threefold axis of symmetry (labelled z or the parallel axis, $\parallel$) contains one axis and also means that two of the Cartesian axes (x and y, the perpendicular axes, $\perp$) are rendered equivalent. This is called a **symmetric-top** molecule, which needs two moments of inertia to describe the rigid shape, each with its own quantum number to relate the energy levels (Eq. 1.18).

$$\epsilon_{\mathcal{J},K} = B\mathcal{J}(\mathcal{J}+1) + (A-B)K^2, \mathcal{J} = 0, 1, 2, \ldots K = \mathcal{J}, \mathcal{J}-1 \ldots -\mathcal{J}, \tag{1.18}$$

where B and A are the rotational constants given by Eq. 1.19:

$$B = \frac{h}{8\pi^2 I_\perp c}, A = \frac{h}{8\pi^2 I_\parallel c} \tag{1.19}$$

and the selection rules is (Eq. 1.20):

$$\Delta\mathcal{J} = \pm 1, \Delta K = 0. \tag{1.20}$$

From these relationships the transition energies can be calculated (Eq. 1.21). The spectrum is shown to be independent of the rotation about the z-axis (the $\parallel$

axis) since rotation around it will not move the molecule dipole (Eq. 1.22); this is the same as for the rotation about the internuclear axis of the linear molecule.

$$\epsilon_{J+1,K} - \epsilon_{J,K} = [B(\mathcal{J}+2)(\mathcal{J}+1) + (A-B)K^2] - [B\mathcal{J}(\mathcal{J}+1) + (A-B)K^2] \tag{1.21}$$

$$= 2B(\mathcal{J}+1). \tag{1.22}$$

For most other molecules like water, called an **asymmetric top**, the situation is more complex. There will be three rotational degrees of freedom, each with its own moment of inertia. With additional structural complexity, the approximation of a rigid molecule becomes less valid. The motion itself creates molecular distortions.

In the water vapour spectrum the fine structure around 1,600 cm^{-1} can be assigned to the rotational-vibrational structure. The vibrational frequency is labelled the Q branch of the spectrum ($\Delta \mathcal{J} = 0$), and straddling it are the P and R branches where the rotational quantum numbers respectively decrease, giving a lower transition energy, and increase, raising the transition energy from that of the vibration alone. The selection rules for simultaneous vibrations and rotations vary with the shape and symmetry of the molecule and the vibration (Eqs 1.23–1.26):

Diatomic linear molecules:vibrations parallel to the principal axis:

$$\Delta v = \pm 1, \Delta \mathcal{J} = \pm 1 (\Delta \mathcal{J} \neq 0). \tag{1.23}$$

Linear molecules: vibrations perpendicular to the principal axis:

$$\Delta v = \pm 1, \Delta \mathcal{J} = 0, \pm 1. \tag{1.24}$$

Symmetric-top molecules: parallel vibrations:

$$\Delta v = \pm 1, \Delta \mathcal{J} = 0, \pm 1, \Delta K = 0. \tag{1.25}$$

Symmetric-top molecules: perpendicular vibrations:

$$\Delta v = \pm 1, \Delta \mathcal{J} = 0, \pm 1, \Delta K = \pm 1. \tag{1.26}$$

In all cases, including the anharmonicity of the vibration allows Δv also to take higher integer values. This pattern shows that the rotational fine structure of a linear molecule associated with vibrational modes along the molecular axis will have no Q branch. All other modes (Fig. 1.25) will, including the bending modes of linear molecules like CO_2 and N_2O. The perpendicular vibrational modes of symmetric-top molecules will be further complicated with series of rotational transitions based on the moments of inertia both $\parallel$ and $\perp$ to the principal axis.

Figure 1.25 *Space-filling model of N_2O.*

Extension 1.2 *Beyond the model of the rigid rotor*

The treatment used so far for rotational transitions has inbuilt inconsistencies. One of these is that we have assumed that there is no change in the bond length during a rotation and even when the motion is combined with a vibration. In reality vibrations occur hundreds of times within the timescale of a molecular rotation, and so the bond length does undergo change, and this will affect the value of the rotational constant (Eq. 1.27). Also, there is a centrifugal effect so that the higher the rotational level, the greater the tendency to increase the bond length. As a result, the rotational spectrum of, for example, a diatomic molecule will show a decrease in line separation and that decrease will increase with J values.

$$B \propto \frac{1}{r_o^2}. \tag{1.27}$$

This effect can be accounted for with a centrifugal distortion constant, D (Eq. 1.28), amounting to a contribution for diatomics of $< 0.01 \ cm^{-1}$. Hence in the rotational-vibrational spectrum, which may be recorded at the order of c.1,000 cm^{-1}, this effect is small.

$$\epsilon_{J+1} - \epsilon_J = 2BJ(J+1) - 4D(J+1)^3. \tag{1.28}$$

However, the effect of the change in the equilibrium interatomic distance of a diatomic with vibrational level is significant. The energy required to effect the ro-vibrational transition, the band origin or centre, is of energy $\bar{\omega}_e$ (Eq. 1.29). Anharmonicity provides a shift to lower energy by the factor x_e; then the energy of the transitions (Eq. 1.30) is further modified by the rotational constant, B, and quantum numbers, J, of the vibrational ground state (J'') and vibrational excited state (J') (Eq. 1.29). For $v = 0$ to 1, this corresponds to Eq. 1.31.

$$\epsilon_v = \left(v + \frac{1}{2}\right)\bar{\omega}_e - \left(v + \frac{1}{2}\right)^2 \bar{\omega}_e x_e \tag{1.29}$$

$$\Delta\epsilon_{J,v} = \epsilon_{J',v=1} - \epsilon_{J'',v=0} \tag{1.30}$$

$$\Delta\epsilon_{J,v} = \bar{\omega}_e(1 - 2x_e) + B(J' - J'')(J' + J'' + 1)cm^{-1}. \tag{1.31}$$

*So far we have considered the vibrational and rotational energies as coming from independent factors—called the **Born–Oppenheimer approximation**. But we have accepted that the bond length, and hence the rotational constant, will be affected by the vibrational energy level (Eq. 1.32):*

$$B_v = B_e - \alpha\left(v + \frac{1}{2}\right). \tag{1.32}$$

Taking this together, we can amend Eq. 1.31 to include the break-down of the Born–Oppenheimer approximation (Eq. 1.33):

$$\Delta\epsilon_{J,v} = \bar{\omega}_e(1 - 2x_e) + B_1 J'(J'+1) - B_o J''(J''+1)cm^{-1}. \tag{1.33}$$

Both anharmonicity and the effect of vibrational energy level on the rotational constant mean that the higher-level transitions become less separated in energy.

Extension 1.3 *The IR absorptions of triatomic molecules*

Four key triatomic molecules in the troposphere are CO_2, O_3, N_2O, and H_2O (Fig. 1.26). CO_2 is linear and centrosymmetric (point group $D_{\infty h}$). Of the nine (3N) degrees of freedom of the triatomic, there are three resulting in translations and two to rotations (the rotation around the interatomic axis is not a valid motion). The remaining four (3N-5) motions comprise two stretching vibrations and two rotations. The symmetric stretch results in no dipole moment change and thus is IR inactive. The antisymmetric stretch is of relatively high energy (2,345 cm^{-1}) compared to the emission spectrum of the surface of the Earth. The bending vibration is doubly degenerate and occurs around 667 cm^{-1}, close to the maximum black-body surface emission.

N_2O is linear and non-centrosymmetic (Point group $C_{\infty v}$). As for carbon dioxide, there are two stretching vibrations and a doubly degenerate pair of rotational modes. Again, the antisymmetric stretch is high in energy (2,224 cm^{-1}), but in this case the symmetric stretch is a very strong allowed IR absorption and has an energy within the sensitive region (1,285 cm^{-1}). The bending vibration occurs around 589 cm^{-1}, also close to the maximum black-body surface emission.

O_3 is non-linear (Point group C_{2v}), and has a third viable rotation axis, leaving three (3N-6) remaining for vibrations: two fundamental stretching modes (1,355 and 1,033 cm^{-1}) and one bend (528 cm^{-1}), all in the sensitive region. All these spectra are

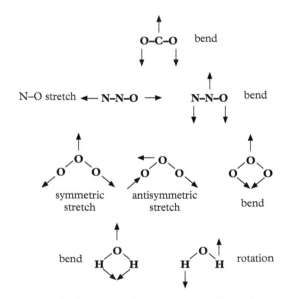

Figure 1.26 *Motions of triatomic molecules that contribute significantly to energy absorption in the troposphere.*

complicated by the presence of isotopes (giving a variety of structural possibilities called isotopomers) by overtone and combination bands to higher energy, and by vibrational-rotational fine structure. Thus the IR spectra provide excellent fingerprints of these gaseous molecules.

H_2O has a structure akin to that of O_3 (Point group C_{2v}), and so the IR fundamentals follow the same pattern. Here the low mass of hydrogen renders the two stretching modes (around $3,700cm^{-1}$) to be outside the intense region of the Earth's surface emission. The bending mode ($1,595\ cm^{-1}$) is towards the top of this energy range, but the absorption spectrum is extended greatly by the highly complex vibrational-rotational pattern as well as additional features from the isotopomers. The water molecule has a permanent dipole moment and low-mass atoms. As a result, absorptions due to the pure rotational spectrum are strongly evident below $1,000\ cm^{-1}$.

The effect of an individual gas on the trapping of energy will be related not only to its concentration, but the intrinsic absorption cross-section of the vibrations and the intensity of the emission of the surface of Earth at the energy of the IR transition. The intensity of an infrared absorption is related to the change in the dipole during the vibration. For stretching vibrations, in which the interatomic distance is changed around the equilibrium bond length, there are two factors which might influence the intensity.

First, a polar bond has a dipole at equilibrium, and so changing the interatomic distance would be expected to change the dynamic dipole (Fig. 1.27). The stretching of a polar bond would change the separation between the partial charges. It may also reduce the covalency and thus tend to localize more charge.

In a second type of effect, the degree of π bonding in a heteronuclear bond can change during the travel of a stretching mode, and this can also change the dipole. This can be illustrated using resonance forms of CO_2 and N_2O (Fig. 1.28). The asymmetric stretch in CO_2 would tend towards the right-hand resonance form with a large induced dipole. Accordingly, the intensity of this mode in the IR is very strong. But the relatively high energy of this vibration means that its contribution to IR energy storage is not as high as it might be. Stretching of the N–O bond would reduce the N–O π bonding, which will also induce an increased dipole for that bond and change the overall molecular dipole. This band is much closer in energy to the maximum for energy emission from the surface of the Earth.

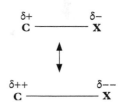

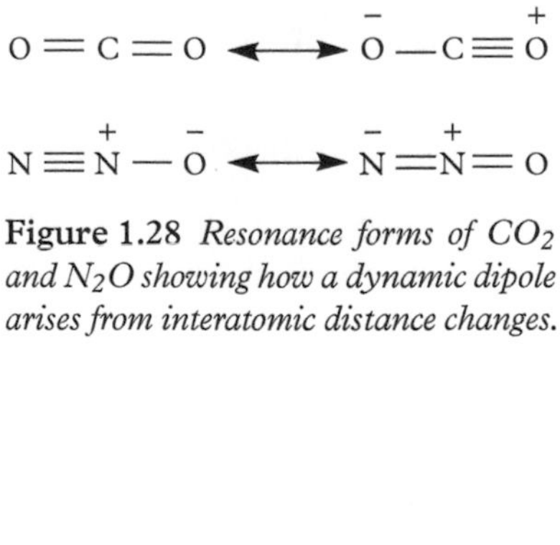

Figure 1.27 *A dynamic dipole from interatomic distance changes.*

$$O = C = O \ \longleftrightarrow \ \overset{-}{O} - C \equiv \overset{+}{O}$$

$$N \equiv \overset{+}{N} - \overset{-}{O} \ \longleftrightarrow \ \overset{-}{N} = \overset{+}{N} = O$$

Figure 1.28 *Resonance forms of CO_2 and N_2O showing how a dynamic dipole arises from interatomic distance changes.*

Extension 1.4 *The IR absorptions of five-atom molecules*

There are four C-X bonds and six X-C-X angles in a tetrahedron, one more than the number of degrees of freedom. It is impossible to increase or decrease all the angles simultaneously as would happen in a symmetric bend, hence only five bending modes are valid. To be IR active a vibrational mode must have the same symmetry as a Cartesian axis. The point group of CH_4 is T_d and x, y, and z all fall within the triply degenerate t_2 representation. The antisymmetric stretches and a set of bending modes

match this symmetry. The symmetric stretch (a_1) is IR inactive, and the remaining C–H bending modes are doubly degenerate (e), and also cannot create a change in dipole moment and thus are IR inactive. The lower symmetry of $CFCl_3$ (C_{3v}) splits the t_2 into two modes: the symmetric stretch (a_1) and the doubly degenerate antisymmetric stretches (e). For all lower-symmetry methane derivatives the number of IR active stretching modes = the number of bonds (four), and similarly the number of bending absorptions will be five.

The IR properties of some halomethane derivatives (Fig. 1.29) are highlighted in Table 1.5. The pentanuclear molecules possess $3 \times 5 - 6 =$ nine vibrational modes: four stretches and five bends. Near the emission maximum of Earth methane exhibits one IR mode, a H–C–H bend. It is a medium-intensity mode, but the absorption is increased because it is triply degenerate (t_2) due to the cubic symmetry of CH_4. There is one IR allowed stretching mode for CX_4 molecules. For CCl_4, particularly due the higher mass of chlorine as opposed to hydrogen, this is considerably lowered in energy from that typical of C–H

Table 1.5 *Calculated parameters for methane derivatives of strong IR bands near the energy of Earth's emission maximum (from Spartan-16).*

Compound	Vibration	ν (cm^{-1})	Relative intensity
CH_4	HCH bend	1,328	15 ($\times$ 3)
CCl_4	C–Cl	811	153 ($\times$ 3)
$CFCl_3$	C–F	1,114	14
	C–Cl	865	253 ($\times$ 2)
CF_2Cl_2	C–F	1,181	217
		1,124	304
	C–Cl	928	396
		676	12
$CHFCl_2$	HCF bend	1,336	25
	HCCl bend	1,263	57
	C–F	1,111	193
	C–Cl	828	236
		748	41

Figure 1.29 *Space-filling models of $CFCl_3$ (CFC-11), CF_2F_2 (CFC-12), $CHFCl_2$ (CFC-21), $CF_2ClCFCl_2$ (CFC-13), and CF_3CH_2F (HCF-134a).*

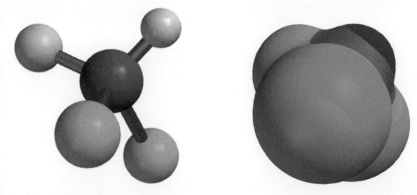

Figure 1.30 *A ball-and-stick model of CHFCl$_2$ (left), with the calculated electrostatic potential surface (right).*

stretches (c.3,000 cm^{-1}) and occurs between the absorptions due to O$_3$ and CO$_2$ (Fig. 1.21) and is of very high intensity. CFC molecules like CFCl$_3$ and CF$_2$Cl$_2$ also have intense IR absorptions due to C–F stretches which are higher in energy than the O$_3$ stretches (c.1,200–1,300 cm^{-1}), and thus have a high radiative forcing potential. A beneficial secondary effect of the Montreal Protocol to protect ozone in the stratosphere has been to reduce the concentration of chloro-organics in the troposphere. The effect is then to reduce the degree of IR absorption around 800 cm^{-1}. The substitutes that have been used are HFC molecules like CF$_3$CH$_2$F. These retain the strong absorptions due to C–F stretches though. The effect of the electron withdrawing fluorines is to render the C–H group more electropositive, as shown for CHFCl$_2$ in Fig. 1.30. Thus motions of the CH unit can have much higher intensity associated with them. In C$_2$ HFC molecules, an unsymmetrical distribution of fluorines can also polarize the C–C bond, as shown for CF$_3$CH$_2$F in Fig. 1.31, and thus also increase the intensity of these modes in the IR.

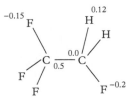

Figure 1.31 *Calculated atomic charges in CF$_3$CH$_2$F (from Spartan-16).*

1.7 Radiative Forcing

This energy absorption in the IR and other regions is quantified as the **radiative forcing** effect (ΔF in W m^{-2}). It is a change in energy balance between energy coming onto the Earth and that radiated back to space, expressed as power (energy s^{-1}) per unit area. Positive values imply the balance is in favour of incoming energy (i.e. warming). The change is set against a reference level taken as that estimated for the year 1750. The general expression for a gas, G, would be compared to its concentration in 1750, G_o:

Table 1.6 *Expressions for calculating radiative forcing, ΔF (W m^{-2}).*

Gas	ΔF(W m^{-2})	Constant
CO_2	$\alpha \ln(C/C_o)$	α, 5.35
CH_4	Eq. 1.37	β, 0.036
N_2O	Eq. 1.38	ϵ, 0.12
$CFCl_3$	$\lambda(X-X_o)$	λ, 0.25
CF_2Cl_2	$\omega(X-X_o)$	ω, 0.32

$$\Delta F = f(G, G_o). \tag{1.34}$$

[20] K. P. Shine, R. G. Derwent, D. J. Wuebbles and J.-J. Morcrette, 'Radiative Forcing of Climate', 41–68, in *Climate Change: The IPCC Scientific Assessment*, ed. J. T. Houghton, G. J. Jenkins and J. J. Ephraumus, Cambridge University Press, Cambridge, 1990.

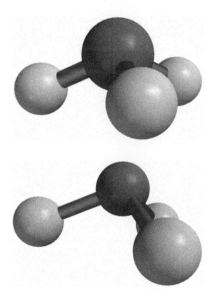

Figure 1.32 *Ball-and-stick models of NH₃ (top) and NF₃ (bottom).*

[21] A. Totterdill, T. Kovács, W. Feng, S. Dhomse, C. J. Smith, J. C. Gómez-Martín, M. P. Chipperfield, P. M. Forster and J. M. C. Plane, 'Atmospheric lifetimes, infrared absorption spectra, radiative forcings and global warming potentials of NF₃ and CF₃CF₂Cl (CFC-115)', *Atmos. Chem. Phys.*, 2016, **16**, 11451–63.

Simplified expressions have been derived from curve fitting of very complex modelling (Table 1.6).[20] These formulae have quite different appearances due to the different circumstances. The simplest expressions are those pertaining to the trace components like the CFCs. These are not naturally occurring and thus the concentration at 1750 was zero. At ppb levels the absorption of IR energy due to these molecules is essentially linear with concentration. There are no saturation effects. Also, the CFCs have absorptions in regions of the IR which are distinct for them, so effects of band overlap are relatively minor.

We can see the chemical basis of radiative forcing with the example of nitrogen trifluoride, NF_3, which is being used as a replacement for the inert perfluorocarbons (PFCs). Like ammonia, NF_3 is a pyramidal molecule (Fig. 1.32) (point group C_{3v}), but the effect of the larger, more electronegative fluorines is to reduce the bond angle from 107 to 102°. It will have allowed rotational transitions, being a symmetric-top molecule with a dipole moment. This tetranuclear molecule will have six degrees of vibrational freedom: three stretches and three bends. In the C_{3v} point group, the z-axis is unique, and both the symmetric stretch and bend are polarized along that axis under the a_1 representation. The other two axes, x and y, are degenerate, and so the antisymmetric stretches and bends are each doubly degenerate (e modes). The vibrational spectrum therefore should consist of four fundamentals: two stretching and two bending.

The most intense vibrational mode is the antisymmetric stretch e mode near 900 cm^{-1}, with the symmetric stretch near 1,030 cm^{-1} (Table 1.7).[21] The bending modes are much weaker and in the region of 500–600 cm^{-1}. By measuring the spectra in absorbance of known gas pressures, (c in mole. cm^{-3}), the absorption cross-sections σ (cm^2 $molec^{-1}$ cm^{-1} bandwidth) could be derived using the Beer–Lambert Law (Eq. 1.35), where A is the absorbance at wavelength λ, and l = the path length (cm).

$$A(\lambda) = \sigma(\lambda)cl \tag{1.35}$$

Table 1.7 *Integrated absorption cross-sections, σ, of NF$_3$ at 296 $\pm$ 2 K.*

Band limits (cm^{-1})	σ (10^{-18} cm^2 molec^{-1} cm^{-1})	Mode
600–700	0.41	bend (sym)
840–860	65.03	stretch (asym)
970–1,085	5.88	stretch (sym)
1,085–1,200	0.1	
1,330–1,400	0.08	
1,460–1,580	0.21	combination
1,720–1,870	0.71	harmonic + combination
1,890–1,970	0.65	harmonic
600–1,970	73.5	

Each of the observed bands shows unresolved rotational structure as envelopes. The intensity of each spectral region was integrated to give the intensity of absorption in those energy bands (Table 1.7). A high proportion of the cross-section is due to the two stretching bands, which occur close to the maximum of the IR emission through the troposphere. As a result, there is potential for this to be a significant greenhouse gas. Note too that the weaker modes and the harmonics and combination bands do have a measurable contribution and so their effects should be included when estimating the forcing effect of trace gases. These contribute to the overall **radiative efficiency**, A_x, of a gas x. The forcing effect is a product of this efficiency and the properties of the atmosphere (impulse) (Eq. 1.36):

$$\Delta F = A_x.\text{impulse}_x. \qquad (1.36)$$

To assess the overall forcing effect of a component like NF$_3$ is more complex than applying the Beer–Lambert law of a well-mixed sample held by fixed windows investigated with a consistent light source. The atmospheric 'pathlength' is effectively the height of the troposphere and varies with latitude. Obviously, the atmosphere is not bounded by fixed, identical walls. The global figures utilize atmospheric modelling computation, measurements of gas concentrations, and the spectroscopic properties. The three-dimensional maps are averaged to provide a global mean of an atmospheric column. But this can be done with different scenarios. As a result there are a variety of measures of radiative forcing:[22]

Instantaneous radiative forcing (IRF): the change in the downward radiative flux at the tropopause.

(Adjusted) radiative forcing (RF): the change in the downward radiative flux at the tropopause after allowing stratospheric temperature to readjust.

Effective Radiative forcing (ERF): the change in the downward radiative flux at the tropopause after allowing atmospheric temperatures, water vapour, and

[22] G. Myhre, D. Shindell, F.-M. Bréon, W. Collins, J. Fuglestvedt, J. Huang, D. Koch, J.-F. Lamarque, D. Lee, B. Mendoza, T. Nakajima, A. Robock, G. Stephens, T. Takemura and H. Zhang, 'Anthropogenic and natural radiative forcing', 2013, in T. F. Stocker, D. Qin, G.-K. Plattner, M. Tignor, S. K. Allen, J. Boschung, A. Nauels, Y. Xia, V. Bex and P. M. Midgley (eds), *Climate Change 2013: The physical science basis. Contributions from Working Group 1 to the Fifth Assessment Report of the Intergovernmental Panel on Climate Change*, Cambridge University Press, Cambridge, UK and New York, NY, USA.

Table 1.8 *Radiative forcings and efficiencies estimated for NF$_3$.*

RF model	Conditions	Radiative forcing (10^{-4} W m^{-2})	Radiative efficiency (W m^{-2} ppb^{-1})
Instantaneous	Clear	3.66	0.35
Instantaneous	All-sky	2.30	0.22
Adjusted	Clear	4.18	0.40
Adjusted	All-sky	2.61	0.25

clouds to readjust. There are additional enhancements made by also allowing rapid-response surface properties to adjust.

In the example of NF$_3$, the adjustment of stratospheric temperature is considered to increase the forcing effect (Table 1.8). The values are very dependent upon the sky conditions. The positions of the main absorptions in the IR means that clouds considerably mitigate the effect of this gas. Clouds effectively act as black-body radiators, and thus screen the effect of greenhouse gases below them. Generally the values tabulated use the (adjusted) radiative forcing model and incorporate a cloudy-sky correction.

The much higher concentration of CO$_2$ means that the dependence of energy absorbance on concentration is much lower. This can be understood from the emission spectra in Fig. 1.21. It is evident from the figure that the absorption of IR near 600 cm^{-1} is very high, so doubling the CO$_2$ concentration cannot double the amount of light absorbed in that region. The logarithmic relationship in Table 1.6 provides a good estimation. Each gas can cross the tropopause into the stratosphere. In that higher atmospheric band it can also exert warming or cooling effects, and thus have an effect on the temperature at the top of the troposphere. The constants given in the table are estimates of the effects of emission from above as well as below the tropopause. It has been estimated that *c*.40% of the additional warming in the troposphere by CO$_2$ may be due to downward emission from the stratosphere.[23]

The formulae in Table 1.6 for N$_2$O and methane show strong interrelationships (Eqs. 1.37–1.39), where the concentrations in ppb are given as M for CH$_4$ and N for N$_2$O, with M_o 722 ppb, and N_o 270 ppb.

$$\Delta F = \beta \left(M^{\frac{1}{2}} - M_o^{\frac{1}{2}} \right) - [f(M, N_o) - f(M_0, N_o)] \tag{1.37}$$

$$\Delta F = \epsilon \left(N^{\frac{1}{2}} - N_o^{\frac{1}{2}} \right) - [f(M_o, N) - f(M_0, N_o)] \tag{1.38}$$

$$f(M, N) = 0.47 ln[1 + 2.01.10^{-5}(MN)^{0.75} + 5.31.10^{-15}M(MN)^{1.52}]. \tag{1.39}$$

There are several ways in which CH$_4$ and N$_2$O interfere by virtue of the IR spectra and atmospheric chemistry. Particularly important here is the overlap

[23] V. Ramanathan, L. Callis, R. Cess, J. Hansen, I. Isaksen, W. Kuhn, A. Lacis, F. Luther, J. Mahlman, R. Reck and M. Schlesinger, 'Climate–chemical interactions and effect of changing tropospheric trace gases', *Rev. Geophys.*, 1987, **25**, 1441–82.

between the N–O stretching mode of N_2O and the bending mode of CH_4 near 1,300cm^{-1}. In this case, absorption of IR light by one gas will render it unavailable for the other. In other words, one gas reduces the effective incident radiation available for the other. In principle this could also compromise on measurement of concentrations, but there are other bands that can be used which do not overlap.[24]

Radiative forcing effects have been tabulated on a wide spectrum of materials, a selection which which are presented in Table 1.9.[22,25] There is about 200 times the concentration of CO_2 (399.6 ppm) in the troposphere as methane (1.8 ppm), but the increased IR absorption properties of each methane molecule mean that the ΔF value is about 25% that of carbon dioxide. This is continued through nitrous oxide (328 ppb) and ozone (337 ppb). The halocarbons are all below 1 ppb in concentration, but their radiative forcing effects are globally significant.

Radiative forcing values have been estimated for the most important components, like CO_2 also since 1979 (Fig. 1.33). After CO_2, methane and nitrous oxide are the two principal contributors. Both are increasing with time, but the percentage change is much larger for N_2O. The other two individualized plots are for CF_2Cl_2 (CFC-12) and $CFCl_3$ (CFC-11). Both have been released as industrial gases (refrigerants and propellants). Both of these gases were identified as contributors to the depletion of the ozone layer in the stratosphere (height 10 km), with evidence of chlorine radicals playing a role. As a result they were included in the Montreal Protocol, which was signed in September 1987 and came into effect at the start of 1989. Chlorofluorocarbons, including CFC-11 and CFC-12,

[24] W. C. Wang, Y. L. Yung, A. A. Lacis, T. Mo and J. E. Hansen, 'Greenhouse effects due to man-made perturbations of trace gases', *Science*, 1976, **194**, 685–90.

[25] T. J. Blasing, *Recent Greenhouse Gas Concentrations*, 2016, DOI:10.3334/CDIAC/atg.032.

Table 1.9 *Changes in radiative forcing of gases in the atmosphere, 1750–2016.*

Gas	Radiative forcing ΔF (W m^{-2})
CO_2	1.94
CH_4	0.50
N_2O	0.20
O_3	0.40
CF_2Cl_2	0.166
$CFCl_3$	0.060
CHF_2Cl	0.049
CCl_4	0.0140
$CF_2ClCFCl_2$	0.022
CF_3CH_2F	0.0140
NF_3	0.00026
SF_6	0.004
CF_4	0.004

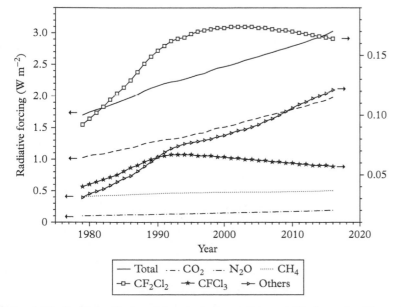

Figure 1.33 *Radiative forcing of the principal contributors to IR absorption. The total, CO_2, N_2O, CH_4 using the left-hand y-axis, and the remainder the right-hand y-axis.*

were to be limited from 1991 and phased out from production and use by major industrial countries from 1996. A reduction in the atmospheric concentration of $CFCl_3$ (CFC-11) has been evident since the early 1990s, and the radiative forcing attributable to CF_2Cl_2, which has a much longer residence time in the atmosphere (100 compared to 45 y), peaked about ten years later.

The basket of fifteen more minor components (the category of 'others') also displayed a levelling off in the 1990s, but has been rising consistently since 2005. The most abundant of these, albeit in parts per trillion (ppt), are carbonyl sulphide (OCS), CHF_2Cl (HCFC-22), CCl_4, $CF_2ClCFCl_2$ (CFC-113), and CF_3CH_2F (HFC-134a). The last of these is a replacement for CFCs and is not considered to contribute to the depletion of stratospheric ozone.

1.8   Atmospheric Lifetime

The long-term effects of a gas will also be dependent upon its lifetime in the atmosphere. The residence times, τ, are defined according to an exponential decay (Eq. 1.40):

$$\frac{N(t)}{N_o} = e^{-t/\tau} \tag{1.40}$$

where N is the number of particles at time t and time zero. The **residence time** (**lifetime**) is that required for the concentration to reduce to $1/e$ of the initial value, that is 36.8%.

Residence times of gases vary enormously (Table 1.10). As we know, water comes and goes from the atmosphere, and ozone too has a relatively short lifetime in the troposphere (c.22 days). This is due to the photolysis of O_3, in the stratosphere and troposphere, by UV light (λ 300–320 nm) to form singlet O_2 and atomic oxygen:

$$O_3 + h\nu = O_2(^1\Delta_g) + O(^1D). \tag{1.41}$$

The Russell Saunders term for the ground state of the oxygen atom is ^3P. This label of 'triplet-P' indicates that there are two unpaired electrons. Russell Saunders terms are explained in Section 2.8. The ^1D ('singlet-D') term indicates the excited state has all electrons paired. Mostly this decays to the triplet ground state by collisions with N_2 or O_2. The remainder mostly reacts with water vapour to form hydroxyl radicals, key reagents in the troposphere via:

$$H_2O + O(^1D) = 2OH. \tag{1.42}$$

The predominant mechanism of N_2O decay is also by a photolytic route ($\lambda < 337$ nm) in the stratosphere, with the reaction rate peaking at a height of about 30 km:

$$N_2O + h\nu = N_2 + O(^1D). \tag{1.43}$$

Table 1.10 *Residence times of gases in the atmosphere.*

Gas	Residence time (y)
CO_2	100–300
CH_4	12
N_2O	121
O_3	c.0.02
CF_2Cl_2	100
$CFCl_3$	45
CHF_2Cl	12
CCl_4	26
$CF_2ClCFCl_2$	85
CF_3CH_2F	13
CF_4	50,000
C_2F_6	10,000
C_3F_8	2,600
NF_3	510
SF_6	1,280

Hence N_2O is also a source of hydroxyl radicals due to the singlet atomic oxygen liberated in its photolysis. This reaction can also occur in the troposphere. Nitrous oxide itself will react with ^1O forming nitric oxide initially which decomposes:

$$N_2O + O(^1D) = 2NO = N_2 + O_2. \tag{1.44}$$

The hydroxyl radicals are mostly consumed by oxidation of CO to CO_2, and provide an efficient means of limiting the concentration of CO in the troposphere:

$$CO + OH = CO_2 + H \tag{1.45}$$

and then

$$H + O_2 + M = HO_2 + M. \tag{1.46}$$

The hydrogen radical reacts with O_2 mediated by a third molecule (M) to afford the hydroperoxyl radical, HO_2.

Hydroxyl radicals also provide a mechanism for the decay of methane, with c.30% of the radicals consumed by this hydrogen-transfer process:

$$CH_4 + OH = CH_3 + H_2O. \tag{1.47}$$

Methyl radicals can also be generated from CH_4 by reaction with ^1O and by photolysis ($\lambda < 140$ nm):

$$CH_4 + O(^1D) = CH_3 + OH \tag{1.48}$$
$$CH_4 + h\nu = CH_3 + H \tag{1.49}$$

and

$$CH_4 + h\nu = CH_2 + H_2. \tag{1.50}$$

The combination of UV light, O_2, and the radicals OH and HO_2 provides a mechanism for the stepwise oxidation of methane to carbon dioxide (Fig. 1.34).

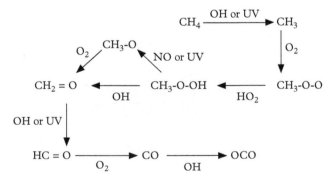

Figure 1.34 *Scheme of the oxidation of CH_4 to CO_2 in the troposphere.*

The route includes two relatively stable molecular entities, methyl hydroperoxide, CH_3OOH, and methanal, CH_2O, before reaching CO. There is a pathway between these and a parallel one via the methoxyl radical, CH_3O, which can be mediated by nitric oxide, NO. Other natural processes operate to reduce the methane concentration, including the bacterial enzyme *methane monooxygenase*, MMO. Overall a mean residence time approximating to observed concentration profiles for these parallel mechanisms is the one tabulated. For methane, this is $c.12$ years. For N_2O, which does does not have a reaction pathway involving OH, the lifetime is much longer, $c.120$ years.

For a range of organochlorine compounds, the prime decomposition route is photolysis of the C–Cl bond, as in CFC-12, CF_2Cl_2. This is essentially inert in the troposphere, but shorter-wavelength UV ($\lambda < 190$ nm) in the stratosphere can cause photolysis, with the loss of one or two Cl atoms, depending upon the excitation energy:

$$CF_2Cl_2 + h\nu = CF_2Cl + Cl \tag{1.51}$$

and

$$CF_2Cl_2 + h\nu = CF_2 + 2Cl. \tag{1.52}$$

In the stratosphere, the halogen radicals cause chain reactions which decompose O_3, which rather offsets the gain in providing a mechanism of loss for the CFCs from the troposphere!

Table 1.10 shows that the longest residence times are for perfluoro compounds, with those of carbon, nitrogen, and sulphur occurring as significant trace atmospheric components. Such long lifetimes present the possibility of diffusion across several layers of the atmosphere around the Earth (Fig. 1.35). In addition to crossing the tropopause (8–16 km) into the stratosphere (in which the temperature rises with increasing height), they may cross the **stratopause**

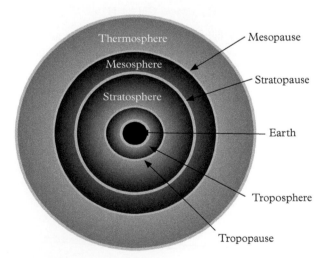

Figure 1.35 *Atmospheric layers and interfaces.*

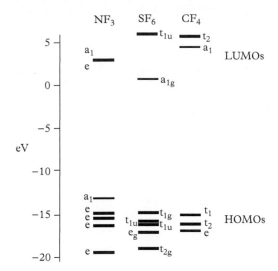

Figure 1.36 *Frontier orbitals calculated for NF$_3$, SF$_6$, and CF$_4$ (Spartan 16).*

(*c*.50 km altitude.) Above that there is the **mesosphere** (temperatures fall with altitude) to a **mesopause** at *c*.80 km above which is the lower **thermosphere** (temperature rises with altitude again). There is evidence of such long-lived species entering both the mesosphere and thermosphere.

NF$_3$ is thought to have a relatively long lifetime of *c*.500 years. Calculations indicate that there is a large energy gap of 16 eV between the HOMO (Fig. 1.36), which is a mixture of the nitrogen lone pair as an antibonding interaction with the 2*p* orbitals on the fluorines (Fig. 1.37), and the LUMOs, a degenerate pair of antibonding orbitals with high $2p_{x,y}$ character.

Photochemical studies indicate that there is no removal in the troposphere as reactions with OH and O$_3$ are relatively slow.[26] There is a very weak onset of absorption for λ < 250 nm, with a rising curve certainly down to < 96 nm. The main channel for decomposition is UV photolysis in the stratosphere (again with a weak onset λ < 250 nm). It estimated that peak of NF$_3$ loss is at a height of *c*.39 km.[27]

$$NF_3 + h\nu = NF_2 + F \qquad (1.53)$$

and then:

$$F + H_2O = HF + OH. \qquad (1.54)$$

Reaction with singlet atomic oxygen is considered to play a finite but minor role:

$$NF_3 + O(^1D) = NF_2(?) + F. \qquad (1.55)$$

As might be anticipated from the effects of halogens on the IR spectra of methane derivatives (Table 1.5), the dynamic dipole of the stretching modes of the polar N–F bonds are high (Table 1.11), particularly those for the asymmetric

Figure 1.37 *One of the LUMOs (top) and HOMO (bottom) calculated for NF$_3$.*

[26] T. J. Dillon, L. Vereecken, A. Horowitz, V. Khamaganov, J. N. Crowley and J. Lelieveld, 'Removal of the potent greenhouse gas NF$_3$ by reactions with the atmospheric oxidants O(^{1}D), OH and O$_3$', *Phys. Chem. Chem. Phys.*, 2011, **13**, 18600–8.

[27] V. C. Papadimitriou, M. R. McGillen, E. L. Fleming, C. H. Jackman and J. B. Burkholder, 'NF$_3$: UV absorption spectrum temperature dependence and the atmospheric and climate forcing implications', *Geophys. Res. Lett.*, 2013, **40**, 440–5.

Figure 1.38 *Space-filling model of SF₆.*

Table 1.11 *Calculated IR parameters for perfluoro gases (from Spartan-16).*

Compound	Vibration	ν (cm^{-1})	Relative intensity
NF₃	N–F	1,093	42
	N–F	963	240 ($\times$ 2)
	FNF bend	678	3
	FNF bend	511	0.5 ($\times$ 2)
SF₆	S–F	961	443 ($\times$ 3)
	FSF bend	609	30 ($\times$ 3)
CF₄	C–F	1,289	419 ($\times$ 3)
	FCF bend	627	4 ($\times$ 3)

Figure 1.39 *The LUMO (top) and and one of the HOMOs (bottom) calculated for SF₆.*

[28] This will have a transition symmetry of t_{1g}, which does not match that of the Cartesian coordinates (t_{1u}). Transitions from the next two lower energy levels will be allowed.

[29] T. Kovács, W. Feng, A. Totterdill, J. M. C. Plane, S. Dhomse, J. C. Gómez-Martín, G. P. Stiller, F. J. Haenel, C. Smith, P. M. Forster, R. R. García, D. R. Marsh and M. P. Chipperfield, 'Determination of the atmospheric lifetime and global warming potential of sulfur hexafluoride using a three-dimensional model', *Atmos. Chem. Phys.*, 2017, **17**, 883–98.

stretches. These occur near 900 cm^{-1}, in between the absorptions of O₃ and CO₂ (Table 1.7), and so the high integrated absorption cross-sections (Table 1.7) and radiative efficiency (Table 1.8) of NF₃, combined with the absence of removal below the middle of the stratosphere, indicates that it can have a significant long-term effect.

The IR properties of octahedral SF₆ (Fig. 1.38) (point group O_h) are broadly similar to those of NF₃: the band positions of the asymmetric stretches are similar and these have a high relative intensity. However the mode is triply degenerate and emanates from the motion of six S–F bonds, resulting in a much higher intensity (Table 1.11). As a result, it is estimated to have a higher radiative efficiency of 0.59 W m^{-2} ppb^{-1}.

The frontier orbitals have rather different properties to those of NF₃. There is no lone pair. The triply degenerate HOMOs are S–F non-bonding (Fig 1.39), and are lower in energy than those of NF₃ (Fig. 1.37). The LUMO has sulphur 3s character and is S–F antibonding. This also is lower in energy those of nitrogen trifluoride and the result is that the HOMO–LUMO gap is little changed. However, the transition from the HOMO (t_{1g}) to the HOMO (a_{1g}) is not dipole allowed.[28] The energy gap between the t_{1u} and the empty a_{1g} is calculated as 17.3 eV. The electronic absorption spectrum exhibits bands that are deep in the vacuum UV (95,000–120,000 cm^{-1}, or λ 105–83 nm).

There is considerable uncertainty about the lifetime of SF₆ in the atmosphere, but is in the order of a few thousand years.[29] As for NF₃, the oxidizing agents in the troposphere do not effect its decomposition. The high energy of the absorption indicates that photochemical decomposition is likely to occur in the region of an altitude of above 80 km into the thermosphere. However, little SF₆ is found there. In part it is because of its high molecular mass, but also considered due to another decomposition mechanism called **electron attachment**. This can occur in two steps:

associative:

$$SF_6 + e = SF_6^-$$ (1.56)

and then dissociation

$$SF_6^- = SF_5^- + F^-.$$ (1.57)

The attachment process would create a very short-lived transient, which may react directly with other reagents or liberate the unsaturated anion which can travel further and collide with a reactive molecule. In the altitude range of 65–80 km there are significant concentrations of anions, such as O_2^-, O_3^-, NO_3^- and CO_3^-, which sequester most free electrons. Hence the electron-attachment mechanism seems viable. The altitude range lowers near the poles as the electron concentration is estimated to be five times higher in those regions than around the equator. By quantifying the expected effect of this mechanism of decomposition, the estimate of the residence time of SF_6 has been lowered from 3,200 to 1,280 years.

Tetrafluoromethane, CF_4 (Fig. 1.40), is the lowest member of the perfluoro-carbon series (PFCs). Its IR spectrum also exhibits intense, triply degenerate asymmetric stretching modes, but these occur at $c.1,290$ cm^{-1} (Table 1.11), higher in energy that the ozone absorptions and above the peak of the emission from the surface of Earth. As a result its radiative efficiency value (0.09 W m^{-2}ppb^{-1}) is considerably lower than than those of NF_3 and SF_6.

It is known to be a very stable molecule due to the strength of the C–F bonds. As a result it is also too inert to react with the oxygenating reagents in the troposphere. The energy of the HOMO (Fig. 1.42) is similar to that of SF_6. Like that of SF_6, it is a non-bonding orbital entirely of fluorine $2p$ character (Fig. 1.42). The LUMO is largely of carbon $2s$ character and is C–F antibonding. In this case too, then transition from HOMO to LUMO is not dipole allowed, being of t_1 symmetry; transition from the next set lower in energy, the t_2 group, is allowed. That energy gap to the LUMO is over 21 eV, considerably higher than that of SF_6.

We would therefore anticipate that photodegradation would require even higher energy excitation. Indeed, CF_4 has been shown to display a peak in its absorption spectrum at $c.22$ eV (λ 56 nm).[30] Such energies are not supplied below the thermosphere, and evidently the gas is also resistant to electron attachment. The estimate placed on its residence time is 50,000 years, but there is a considerable error margin. For many purposes it can be considered as persistent.[31] Other fluoro-organics do undergo photolysis and can be oxidized to $CF_3C(O)F$. This itself can undergo a photoinduced reaction to yield CF_4 (Fig. 1.41). At present this route is dwarfed by the emission of CF_4 from activities on Earth's surface, but it illustrates that CF_4 is a reaction sink.

CO_2 is also considered to be a reaction sink, but there are many more ways for its removal from the troposphere. Most of them are not irreversible chemical change, but involve sequestration as the hydrate in solution or as carbonates. The processes include absorption by land surfaces, dissolution into waters followed by transport in the body of water, and photosynthesis. Photosynthesis does effect

Figure 1.40 *Space-filling model of CF₄.*

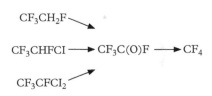

Figure 1.41 *Atmospheric routes to CF₄ from other fluorocarbons.*

[30] J. W. Au, G. R. Burton and C. E. Brion, 'Quantitative spectroscopic studies of the valence-shell electronic excitation of Freons (CFCl₃, CF₂Cl₂, CF₃Cl and CF₄) in the VUV and soft X-ray regions', *Chem. Phys.*, 1997, **221**, 151–68.

[31] A. M. Jubb, M. R. McGillen, R. W. Portmann, J. S. Daniel and J. B. Burkholder, 'An atmospheric photochemical source of the persistent greenhouse gas CF₄', *Geophys. Res. Lett.*, 2015, **42**, 9505–11.

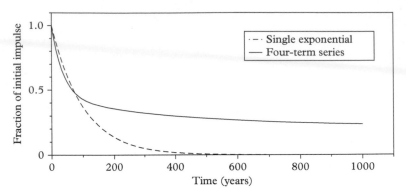

Figure 1.43 *Comparison of the decay of an injection of CO_2 modelled by a four-term function and a single lifetime of 100 years (parameters from note 32).*

Figure 1.42 *The LUMO (top) and one of the HOMOs (bottom) calculated for CF_4.*

[32] F. Joos et al., 'Carbon dioxide and climate impulse response functions for the computation of greenhouse gas metrics: a multimodal analyis', *Atmos. Chem. Phys.*, 2013, **13**, 2793–825.

chemical change, but much of it is reversed rapidly by respiration. Carbon dioxide has been ascribed a residence time of *c*.100 years. The exponential decay of an injection of CO_2 with a 100-year lifetime is shown in Fig. 1.43. The impulse response factor (Eq. 1.58) was refined as a four-term series to provide the best fit of a wide range of experimental data; this takes the form of:

$$\text{IRF}(t) = a_0 + \sum_{i=1}^{3} a_i . e^{-t/\tau_1}. \tag{1.58}$$

The four-term series, derived for a period of 1 to 1,000 years, contains amplitude terms which are similar (0.21–0.28) and includes a time-independent constant (a_o), and three time constants of 394, 37, and 4 years.[32] As can be seen, the single-term exponential bears a similarity to the refined function over the first 100 years, but then fails to account for the much longer residence of a fraction of the gas.

1.9   Global Warming Potential

From a knowledge of the response efficiency of a gas, *A*, and its concentration profile with time (the **impulse response factor**, IRF), the overall **absolute global warming potential** (AGWP) up to a time horizon, *TH*, can be estimated by Eq. 1.59:

$$\text{AGWP}_x = \int_0^{TH} A(t)\text{IRF}_x(t)dt = \int_0^{TH} \Delta F_x(t)dt. \tag{1.59}$$

This is the integration of the radiative forcing of gas (x) ΔF_x. For the majority of atmospheric species that do follow an exponential decay in concentration with time then Eq. 1.60 is valid:

$$\text{AGWP}_x(TH) = A_x \tau_x (1 - e^{-TH/\tau_x}). \qquad (1.60)$$

Tabulations are generally presented for a release per kg of sample, giving units of $W\ m^{-2}\ y\ kg^{-1}$. The time horizon chosen varies with 20 and 100 years often chosen. Values are often compared to those of CO_2 as the reference material using Eq. 1.61 and the result is called the **global warming potential** (GWP) of a gas (x):

$$\text{GWP}_x(TH) = \frac{\text{AGWP}_x(TH)}{\text{AGWP}_{CO_2}(TH)}. \qquad (1.61)$$

What is of particular interest is the change in surface temperature (ΔT_s in K) that these changes in energy flux create. The change in surface temperature can be said to be related to the cumulated radiative forcing effect from the constituent gases and also the climate sensitivity factor, λ, which is estimated to be $0.8\ K\ (W\ m^{-2})^{-1}$, according to Eq. 1.62.

$$\Delta T_s = \lambda \sum_i \Delta F_i. \qquad (1.62)$$

However, this does not account for any timescale for the transfer of energy to a recipient body and the equilibration necessary to achieve a new temperature. There is always a time delay, d, in temperature rise following a radiative forcing event (impulse) at an instant in time. This needs to be tensioned against the residence time of each component in the atmosphere. In this way, each component can usefully be given an **absolute global temperature change potential** (AGTP), which is the temperature change induced by this component after a time horizon (Eq. 1.63)

$$\text{AGTP}_x(TH) = A_x \sum_{j=1}^{2} \frac{\tau_x c_j}{\tau_x - d_j} (e^{-TH/\tau_x} - e^{TH/d_j}). \qquad (1.63)$$

The first term, $j = 1$, models the temperature change of the ocean mixed layer. Accordingly d_1 is relatively short (8.4 years, with a c_1 value of 0.631 K $(W\ m^{-2})^{-1}$). The second term models the response of the deep ocean with a much longer delay time ($d_2 = 409.5$ years with a c_2 value of $0.429\ K\ (W\ m^{-2})^{-1}$).

In parallel to definition of GWP, the **global temperature potential** (GTP) is scaled relative to CO_2 (value of 1) (Eq. 1.64):

$$\text{GTP}_x(TH) = \frac{\text{AGTP}_x(TH)}{\text{AGTP}_{CO_2}(TH)} = \frac{\Delta T(TH)_x}{\Delta T(TH)_{CO_2}}. \qquad (1.64)$$

Some GWP and GTP values are collected in Table 1.12. It shows that the effect of the rises in concentrations of methane and nitrous oxide in the atmosphere is amplified by their high GWP and GTP values. The lower lifetime for methane

Table 1.12 *Global warming (GWP) and temperature (GTP) potentials (over 20 and 100 years).*[22]

Gas	GWP(20)	GWP(100)	GTP(20)	GTP(100)
CO_2	1	1	1	1
CH_4	84	28	67	4
N_2O	264	265	277	234
CF_2Cl_2	10,800	10,200	11,300	8,450
$CFCl_3$	6,900	4,660	6,890	2,340
CHF_2Cl	5,280	1,760	4,200	262
CCl_4	3,480	1,730	3,280	479
$CF_2ClCFCl_2$	6,490	5,280	6,730	4,470
CF_3CH_2F	3,710	1,300	3,050	201
CF_4	4,880	6,630	5,270	8,040
C_2F_6	8,210	11,100	8,880	13,500
C_3F_8	6,640	8,900	7,180	10,700
NF_3	12,800	16,100	13,700	18,100
SF_6	17,500	23,500	18,900	28,200

indicates that it could be muted over a timescale of tens of years. The future effects of N_2O, however, are more likely to grow and be maintained for hundreds of years. It is evident that the halocarbons and perfluoro- gases all have very high potential effects, and their long lifetimes means that their relative warming and temperature potentials increase with the length of the time horizon.

1.10 The Energy Balance

The global balance of power density is summarized in Fig. 1.44. The primary source of energy is solar radiation (340 W m^{-2}). Of this, a mean of 100 W m^{-2} is reflected from the atmosphere and the surface, and a smaller value absorbed by the atmosphere (79 W m^{-2}). The other transfers involve heat. Some involve physical motion from the surface, including the latent heat required for evaporation from the surface and convection eddies (total *c.*100 W m^{-2}). The remainder of the processes are largely in the IR. The emission from the surface is estimated as 398 W m^{-2}, and that returned to the atmosphere from the greenhouse effect (*c.*342 W m^{-2}) effects the near balance shown in Fig. 1.44. That also gives a net IR transfer into the atmosphere of 56 W m^{-2}. Taken with the other energy transfers to the atmosphere, by the absorption of solar radiation, evaporation, and

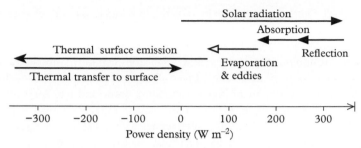

Figure 1.44 *Global power density balance at the Earth's surface. Positive values are energy input, negative values are output. Values from note 22.*

convection processes, which also result in IR radiation, the thermal outgoing from the top of the atmosphere (239 W m^{-2}) essentially balances the non-reflected solar radiation.

The effects of the additional gases on the atmosphere will mainly be on the factor of the thermal transfer to the surface. The estimated effects on the power density balance derived since industrialization[22] (taken as 1750) to 2011 are summarized in Fig. 1.45. The major increase in the **effective radiative forcing (ERF)** (2.83 W m^{-2}) has been ascribed to the **well-mixed greenhouse gases (WMGHG)**. The next-largest increase is due to ozone in the troposphere. Although relatively short-lived, the quantity (called the burden) of O_3 in the troposphere is estimated to have increased from *c.*250 Tg (250 Mt) in 1850 to 337±23 Tg in 2013.

This increase can be traced back to the increased concentrations of two of the main greenhouse gases, methane and nitrous oxide, and also from the lower-concentration, reactive nitrogen oxides: NO and NO_2. Fig. 1.34 in Section 1.8 showed that in an early stage in the oxidation of CH_4 to CO, the radical CH_3O_2 was formed. Also, the reaction of the CO so formed with the hydroxyl radical (Eq. 1.45) starts the sequence to the hydroperoxo and peroxo radical, HO_2 (Eq. 1.46). Both of these peroxyl radicals will react with NO by:

$$HO_2 + NO = NO_2 + OH \qquad (1.65)$$
$$CH_3O_2 + NO = NO_2 + OCH_3. \qquad (1.66)$$

NO itself is formed by reaction of N_2O with $O(^1D)$ (Eq. 1.44), but this supply will be greatly augmented by any releases of NO_x. Nitrogen dioxide will then photolyse in the troposphere ($\lambda < 420$ nm) and the resulting oxygen atom in the electronic round state can add to the oxygen molecule to form ozone by:

$$NO_2 + h\nu = NO + O(^3P) \qquad (1.67)$$
$$O_2 + O(^3P) = O_3. \qquad (1.68)$$

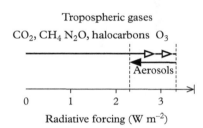

Figure 1.45 *Overall global mean effective radiative forcing (ERF)(W m^{-2}) from 1750 to 2011.*

The small remaining increase in radiative forcing is the net change of a combination of factors: composition changes in the stratosphere due to ozone (-0.05 W m^{-2}) are offset by those of methane and water vapour (0.07 W m^{-2}), and the additional positive values come from contrails (vapour trails) in the atmosphere (0.01 W m^{-2}) and surface albedo effects (0.04 W m^{-2}) due to black carbon aerosols reducing the reflectivity of snow and ice. Atmospheric-borne aerosols are the main source of the effects reducing radiative forcing (-0.15 W m^{-2}), caused by both their interactions with radiation directly and also with clouds. The other smaller reduction is that due to a change in surface albedo due to the alterations in land use. Overall this gives a net change of $+2.3$ W m^{-2}, with values for 5–95% confidence range limits being 1.13–3.33 W m^{-2}.

The overall result indicates an increase in the global mean power density input that is outside the margin of error, and thus is expected to afford an increase in surface temperature. The changes are much larger than those estimated for variations in solar radiance over the same time frame ($+0.05\pm0.05$ W m^{-2}) They can be correlated with human activity and thus called anthropogenic. Volcanic eruptions do not follow the pattern of a continuous year-on-year trend. They can cause reductions in ERF of several W m^{-2}, which can completely offset the anthropogenic background changes. Their effects though are generally relatively short-lived (1–2 years), and form a noise pattern onto a smoothly changing background.

1.11 A Sustainable Approach

There is much evidence that human activity is causing a change in the distribution of elements in and around the Earth. Those elements are predominantly carbon, nitrogen, and oxygen, with fluorine and chlorine also having significant effects. The major cause of potential climate change is due to the change in concentrations of compounds of these elements in the troposphere. The largest factor is that since the eighteenth century we have been releasing the energy of ancient sunlight trapped by photosynthesis and subsequently converted into carbonaceous fuels. The CO_2 that had been fixed by the photosynthesis is regenerated with the combustion of these fossil fuels, mostly for heating, manufacturing, and transport. Methane and nitrous oxide have also increased in the atmosphere due to these activities, and in turn they are significant factors in the increased ozone in the troposphere.

Manufacturing, electrical systems, and refrigeration have resulted in the emission of halogen-containing gases. Many fluorine-containing gases have extremely long life times in the atmosphere and have very large IR cross-sections. As a result relatively low concentrations can have long-term effects.

The remaining chapters will examine the situation across the periodic table. Chemical principles will be applied in the context of an element's availability, to assess if there are other potential limiting factors on human activity or whether that element can provide a means of alleviating a problem. For example, there is indeed much energy available from the Sun which can be exploited. Thus the substitution of fossil fuel combustion by solar-derived energy could reduce emissions of compounds carbon, nitrogen, and oxygen into the troposphere and provide a more sustainable future. The approach of the stabilization triangle (Fig. 1.46) provides a viable approach. Changing the emission slope from upward to downward is unlikely to be achieved in a single step, but will require an accumulation of contributions, some of which may prove to be intermediate technologies. For example, switching from coal to natural gas as a fuel source can be seen as an early stage on the pathway to zero-carbon fuel. In the following two chapters, we can examine the abundance and availability of most elements to look for any pinch points, and then look at the characteristics of their use in valuable commodities.

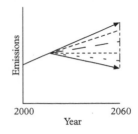

Figure 1.46 *Stabilization triangle*

1.12 Questions

1. (i)   Draw an energy level diagram for the N_2.
    (ii) Derive the lowest energy allowed electronic transition of N_2.
    (iii) Modify the energy level diagram to be appropriate for nitric oxide, NO.
    (iv) Derive the lowest energy allowed electronic transition of NO.

2. The species $[NO_2]^n$, where $n = 1, 0$, and -1 have all been identified in the atmosphere.
    (i)   By considering isoelectronic neutral molecules, propose structures or the cation and anion.
    (ii) Draw qualitative molecular orbital energy level diagrams for these two structures.
    (iii) Propose a structure for NO_2, giving your reasons.
    (iv) What are the differences between the vibrational and rotational properties of the three species?

3. Look up the exact masses of the isotopes of carbon and oxygen. From these values, calculate the reduced masses (μ), moments of inertia (I), and the separation of the rotational transitions of $^{12}C^{16}O$, $^{13}C^{16}O$, and $^{12}C^{18}O$. (Treat CO as a rigid rotor with r_0 113.1 pm).
    Extension It has been estimated that for $^{12}C^{16}O$, $\bar{\omega}_e = 2{,}169.74$ cm^{-1}, $x_e = 0.0061$, $B_e = 1.924$, and $\alpha = 0.018$. Calculate the energies of the first member of the P and R branches, P_1 and R_0. Determine the separation between these lines.

4. (i)   How many IR modes would you predict for $CHFCl_2$ (HFC-21)?
    (ii) How many stretches and bends do you predict?
    (iii) Which of these modes are likely to contribute very significantly to the radiative efficiency of HFC-21?

5. (i)   The principal contributions to the radiative efficiency of chloroform, $CHCl_3$, have been reported to be in two infrared regions: 720–810 cm^{-1} and 1,175–1,255 cm^{-1}. The absorption cross-sections in these two regions are reported to be 6.13×10^{-17} and 0.561×10^{-17} cm^2 $molec^{-1}$ cm^{-1}, respectively. Assign the vibrational modes that are responsible for these absorptions.

    (ii) The radiative efficiency of $CHCl_3$ was estimated to be 0.070 W m^{-2} ppb^{-1}. Predict the order of the radiative efficiencies for CH_3Cl, CH_2Cl_2, $CHCl_3$, and CCl_4.

    (iii) The lifetimes in the atmosphere of the same chlorocarbons are reported to be 1.0, 0.4, 0.4, and 26 years, respectively. Propose the order of their global warming potentials.

6. (i)   The electronic absorption spectrum of ozone, O_3, in the visible and near IR region exhibits bands peaking near 16,500 cm^{-1} (600 nm) and much weaker bands peaking over 10,000–12,000 cm^{-1}. Propose transition types that are responsible for these absorption bands.

    (ii) In the UV region, above 30,200 cm^{-1}, other transitions occur which create new chemical species. What type of process occurs? Identify the products.

    (iii) Propose two allowed conversions that could occur, and indicate which would be expected to be the major channel.

    (iv) What will be the products of collisions of the new species with water molecules?

7. From the data in question 5, the atmospheric residence times of $CHCl_3$ and CCl_4 are estimated to be 0.4 and 26 years, respectively.

    (i)   Propose a mechanism for the degradation of $CHCl_3$ in the troposphere.

    (ii) Account for the longer atmospheric lifetime of CCl_4 and indicate a mechanism for its degradation.

The Palette of Elements

2

The key questions about resources can be defined as:

1. How much is there?
2. What does it do?
3. What does it take to make it?

In this chapter we examine the background information about the availability of the elements on Earth and the variation of their properties as a function of their position in the periodic table. We will also discuss the chemical principles which help us understand how materials are made and the reasons behind them having their function.

2.1 Natural Abundance of the Elements

2.1.1 Universal Abundance

The abundances of the elements in the universe vary very widely (Fig. 2.1). The vast majority of atoms are either of hydrogen (H, 74%) and helium (He, 23%) and are considered to emanate from the period shortly after the big bang. The rest of the periodic table, comprising the remaining 3%, has been formed by subsequent events in the life and demise of stars, with the exception of some lithium. From $Z = c.42$ (molybdenum, Mo) elements are mostly from the merging of neutron stars. There is a saw-tooth pattern due to the tendency for elements with even numbers to be more stable. Within that, the more stable isotopes tend to have even numbers of neutrons, for example ^{12}C for carbon, ^{16}O for oxygen, ^{56}Fe for iron, and ^{208}Pb for lead.

2.1.2 Crustal Abundance

The estimated natural abundances of the elements in the Earth's crust also show enormous variations (Fig. 2.2), even after removing the four absences of technetium (Tc, $Z = 45$), promethium (Pm, $Z = 61$), astatine (At, $Z = 89$) and francium (Fr, $Z = 87$). These are not immutable values, and are estimates based upon best efforts to measure across the world's surface layer. The three most

Elements of a Sustainable World. John Evans, Oxford University Press (2020). © John Evans.
DOI: 10.1093/oso/9780198827832.003.0002

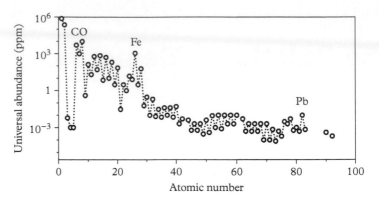

Figure 2.1 *Abundance of the elements in the universe.*

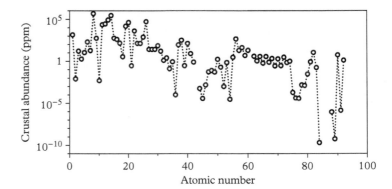

Figure 2.2 *Crustal abundance of the elements.*

abundant elements literally form most of the bedrock of the surface of the planet: oxygen, silicon, and aluminium. Three groups of elements which are estimated to be in low concentration are:

1. the inert gases: helium (He, $Z = 2$), neon (Ne, $Z = 10$), argon (Ar, $Z = 18$), krypton (Kr, $Z = 36$), xenon (Xe, $Z = 54$), radon (Rn, $Z = 86$);

2. the most radioactive elements: actinium (Ac, $Z = 98$), polonium (Po, $Z = 84$), and, again, radon ($Z = 86$);

3. the platinum group of metals (PGMs): ruthenium (Ru, $Z = 44$), rhodium (Rh, $Z = 45$), palladium (Pd, $Z = 46$), osmium (Os, $Z = 76$), iridium (Ir, $Z = 77$), platinum (Pt, $Z = 78$).

An example of changed values is afforded by the lanthanides ($Z = 57$ to 71); newer estimates have differentiated the rare earth elements in terms of their abundances and, generally, have increased them significantly.

2.2 Mineral Availability

The short- and long-term availability of minerals is of considerable interest for a wide spectrum of technological purposes and thus data is collected by public and private sector organizations. These include the US[1] and British[2] Geological Surveys. Both provide information online about mineral resources. As example is the Risk List[3] published by the BGS (Table 2.1). This considered the known reserves

[1] US Geological Survey(USGS): https://www.usgs.gov.

[2] British Geological Survey (BGS): http://www.bgs.ac.uk.

[3] Minerals UK Risk List 2015: https://www.bgs.ac.uk/mineralsuk/statistics/riskList.html.

Table 2.1 *Supply risk index and crustal abundance of selected elements.*

Element(s)	Symbol	Relative supply	Crustal abundance (ppm)
Rare earth elements	REE	9.5	0.28–43
Antimony	Sb	9.0	0.2
Bismuth	Bi	8.8	0.18
Germanium	Ge	8.6	1.3
Vanadium	V	8.6	138
Gallium	Ga	8.6	16
Strontium	Sr	8.3	320
Tungsten	W	8.1	1
Molybdenum	Mo	8.1	0.8
Cobalt	Co	8.1	26.6
Indium	In	8.1	0.052
Arsenic	As	7.9	2.5
Magnesium	Mg	7.6	28,104
Platinum group elements	PGE	7.6	0.000037–0.0015
Lithium	Li	7.6	16
Rhenium	Re	7.1	0.000188
Selenium	Se	6.9	0.13
Tin	Sn	6.0	1.7
Nickel	Ni	5.7	26.6
Uranium	U	5.5	1.3
Iron	Fe	5.2	52,157
Titanium	Ti	4.8	4,136
Copper	Cu	4.8	27
Aluminium	Al	4.8	84,149

[4] A ppm is expressed in mass terms, that is, mg/kg.

(fully surveyed), resources (potential reserves), reserve distributions, recycling, substitutability (where a mineral can substitute for another), production metal fraction, companion metal fraction (where obtained as a by-product), and the political stability of the supplier and producer countries. The rare earth elements and heavy main-group elements with crustal abundances of 1 ppm or less are assessed to have limited supply options. At the other end of the scale, elements with abundances in the range of 10 to over 10^3 ppm[4] do have very low supply risk. Evidently, the crustal natural abundance is a factor in the risk of supply, but demand, location, and processing are also key factors and give rise to some marked deviations from a correlation between the two parameters, with vanadium and uranium being evident exceptions.

2.3　Element Transmutation

The presence of radioactive elements in the Earth's crust indicates that there is a natural change which may alter the atomic number of a particular atom; in other words it becomes a different element. The distribution of elements is also changed by human activity.

2.3.1　Natural Transmutation

Isotopes which are unstable follow a variety of pathways of relaxation (Fig. 2.3). All processes can be a considerable source of energy as heat or high-energy (γ) radiation. A γ emission alone will not change the atomic or mass numbers, but will allow the nucleus to relax to a more stable state.

A β^- emission is the loss of an electron from the nucleus caused by the transformation of a neutron into a proton. Thus the overall charge and formal mass balance (A in Fig. 2.3) is maintained, but the nuclear charge (Z) will

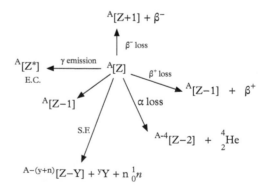

Figure 2.3 *Mechanisms of natural transmutation.*

increase by 1. The obverse of this is β^+ (positron) emission. In this case a proton is converted into a neutron and so the nuclear charge decreases by 1. A similar transformation is effected by **electron capture** (EC). Here an extra-nuclear electron is captured by the nucleus as an alternative mechanism for the conversion of a proton into a neutron. A common decay mechanism of heavy elements is α emission. In this case the nucleus fragments by expelling a ^{4}He nucleus. Thus the atomic number will decrease by 2 and the atomic mass by 4. A more major fragmentation occurs under **spontaneous fission** (SF). Here, two massive components of the original nuclcus are generated, with an accompanying evolution of neutrons. By these mechanisms some elements are continually formed from the feedstock of radioactive isotopes.

2.3.2 Induced Transmutation

There are other transmutation processes which require induction and are not known on the Earth's crust.[5] **Induced fission** (Fig. 2.4) is similar to spontaneous fission but requires neutron activation, as would be available in a nuclear reactor. The opposite type of transmutation is fusion, in which two nuclei merge, resulting generally in the loss of a particle, such as a neutron. The capture of a neutron by a nucleus may promote another decay process other than fission. For example β^- loss will give an increase in the atomic number by 1. With sufficient flux this can happen in a string of neutron-capture and β^- loss processes and give a ladder of increasing atomic numbers. However, as the ladder extends this pathway becomes less probable, and so the alternative synthesis method is by bombardment of a target by a beam of ions of a second element. Initially the

[5] Evidence from isotopic ratios was found in Oklo in Gabon in 1972 which indicated there had been a self-sustaining nuclear reactor at the site about 1.7 billion years ago.

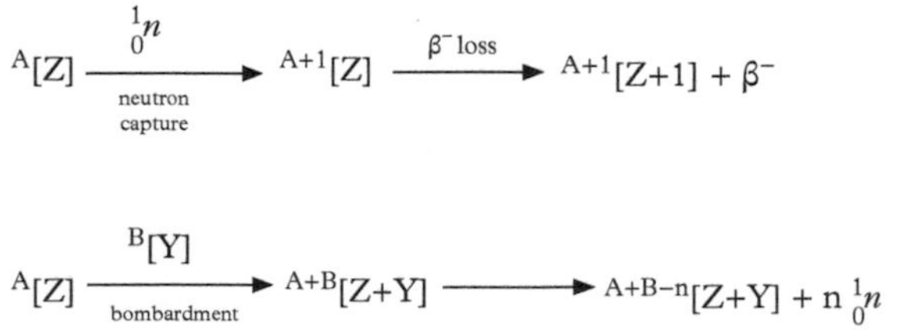

$$^A[Z] \xrightarrow[\text{induced fission}]{{}^1_0 n} {}^{A-(y+n-1)}[Z-Y] + {}^y[Y] + n\,{}^1_0 n$$

$$^A[Z] \xrightarrow[\text{fusion}]{{}^B[Z]} {}^{A+B-1}[2Z] + {}^1_0 n$$

$$^A[Z] \xrightarrow[\text{neutron capture}]{{}^1_0 n} {}^{A+1}[Z] \xrightarrow{\beta^-\text{loss}} {}^{A+1}[Z+1] + \beta^-$$

$$^A[Z] \xrightarrow[\text{bombardment}]{{}^B[Y]} {}^{A+B}[Z+Y] \longrightarrow {}^{A+B-n}[Z+Y] + n\,{}^1_0 n$$

Figure 2.4 *Mechanisms of induced transmutation.*

two nuclei may merge, but will probably require the loss of a number of neutrons to afford an observable isotope.

2.4   Element Supply

6 http://www.rsc.org/periodic-table.

7 G. Audi, O. Bersillon, J. Blachot and A. H. Wapstra, 'The NuBASE evaluation of nuclear and decay properties', *Nucl. Phys.* A, 2003, **729**, 3–128.

In the periodic table there have been several different numbering schemes for the columns. In this book the IUPAC numbering of groups 1–18 is adopted.[6] Period 1 just consists of hydrogen in group 1 and helium in the group number of all the inert gases, 18. As gases, those two elements will be considered mainly in Chapter 4 (Air).

2.4.1   s-Elements

The bulk of the *s*-elements are in group 1 (having ns^1 valence electrons) and group 2 (having ns^2 valence electrons) in periods 2–7; these are called the alkali and alkaline earth elements, respectively (Fig. 2.5). The two elements in Period 7 are radioactive.[7] The most stable isotope of radium has a mass of 226. It follows a decay path through the loss of an α particle from the nucleus. Thus the atomic number is reduced by 2 and the mass by 4 units, according to Eq. 2.1:

$$\ce{^{226}_{88}Ra} \rightarrow \alpha + \ce{^{222}_{86}Rn}. \tag{2.1}$$

The half-life of $\ce{^{226}Ra}$ is 1,600 y (Table 2.2). This is essentially the only naturally occurring isotope and its supply as a trace element (1 part per trillion, ppt) is maintained by the decay train from uranium-238. The naturally occurring $\ce{^{223}Ra}$ ($t_{1/2}$ 11.4 d) has been approved for use as an α emitter radiopharmaceutical.[8] Pathways to enhance the supply of the isotope are being developed, including microgeneration from $\ce{^{227}Ac}$, the major route of which is shown in Eq. 2.2:

$$\ce{^{227}_{89}Ac} \rightarrow \beta^- + \ce{^{227}_{90}Th} \rightarrow \ce{^{223}_{88}Ra} + \alpha. \tag{2.2}$$

Figure 2.5 *Elements in groups 1 and 2*

Group

Period	1	2
2	Li	Be
3	Na	Mg
4	K	Ca
5	Rb	Sr
6	Cs	Ba
7	Fr	Ra

8 Trade name Xofigo—used in treatment of late-stage prostrate cancer.

Table 2.2 *Longest-lived isotopes of groups 1 and 2.*

Isotope	Half-life	Decay	Product
$\ce{^{223}Fr}$	22 min	β^-	$\ce{^{223}Ra}$
		α	$\ce{^{219}At}$
$\ce{^{226}Ra}$	1600 y	α	$\ce{^{222}Rn}$

Francium is available in even lower traces; it is estimated that there is only *c*.30 g of the element in the Earth's crust at any moment. The most long-lived isotope is $\ce{^{223}Fr}$. It is formed by α decay of $\ce{^{227}Ac}$, itself a trace component of uranium minerals. The half-life of $\ce{^{223}Fr}$ is 22 min. It also can follow a α decay route, to astatine-219, but it also has a parallel decay process, by loss of a β^- particle (electron) from the nucleus thus increasing *Z* by 1 (Fig. 2.3, Eq. 2.3):

$$\ce{^{223}_{87}Fr} \rightarrow \beta^- + \ce{^{223}_{88}Ra}. \tag{2.3}$$

Francium synthesis has been achieved by bombarding a target of $\ce{^{197}Au}$ with beam of $\ce{^{18}O}$ using a linear accelerator (Eq. 2.4), albeit to a level of 10^{-18} mol of an isotope with a $t_{1/2}$ of 3.2 min.

$$^{197}_{79}\text{Au} + {}^{18}_{8}\text{O} \rightarrow {}^{210}_{87}\text{Fr} + 5n. \tag{2.4}$$

As a result of their low availability and high radioactivity, the chemistry of the two period 7 elements is excluded from further discussion, indicated by the shading in Fig. 2.5. They are not available in sufficient quantities to make much impact on the context of Planet Earth; they should be considered as rare, but not endangered, species.

2.4.2 *p*-Elements

For the purposes of this book, volatile *p* block elements (Fig. 2.6) are discussed in Chapter 4 (Air), and these are indicated in the figure by mid-shading. Those elements which only have radioactive isotopes are also indicated in Fig 2.6 (dark shading). Technically bismuth is also radioactive. Naturally, it has a single isotope which is an α emitter. However, the half-life is about 10^9 times the age of the Universe, and its decay has not been observed.

Of the effectively radioactive elements, four are synthetic elements all with half-life of seconds or less (Table 2.3): nihonium (Nh, Z = 113), flerovium (Fl, Z = 114), moscovium (Mc, Z = 115) and livermorium (Lv, Z = 116). Production is by high-energy ion-atom collisions. For example, collisions of ^{48}Ca ions with ^{248}Cm afforded a transient isotope of livermorium, and loss of neutrons generated a few atoms of ^{193}Lv (Eq. 2.5):

$$^{248}_{96}\text{Cm} + {}^{48}_{20}\text{Ca} \rightarrow {}^{296}_{116}\text{Lv}^{*} \rightarrow {}^{293}_{116}\text{Lv} + 3n. \tag{2.5}$$

Table 2.3 *Long lived isotopes of the p elements.*

Isotope	Half-life	Decay	Product
^{209}Bi	10^{19} y	α	^{205}Tl
^{209}Po	125 y	α	^{205}Pb
		β^+	^{209}Bi
^{210}Po	138 d	α	^{206}Pb
^{286}Nh	9.5 s	α	^{282}Rg
^{289}Fl	1.9 s	α	^{285}Cn
^{290}Mc	650 ms	α	^{286}Nh
^{293}Lv	57 ms	α	^{289}Fl

Thus these synthetic elements too are unlikely prospects for the amelioration of Planet Earth. All decay via α emission resulting in the formation of the Z-2 element (cf. Eq. 2.1).

The remaining radioactive element in Table 2.3 is polonium (Po, Z = 84). The longest-lived isotope, ^{209}Po, also decays by loss of an α particle to form ^{205}Pb

Period \ Group	13	14	15	16	17	18
2	B	C	N	O	F	Ne
3	Al	Si	P	S	Cl	Ar
4	Ga	Ge	As	Se	Br	Kr
5	In	Sn	Sb	Te	I	Xe
6	Tl	Pb	Bi	Po	At	Rn
7	Nh	Fl	Mc	Lv	Ts	Og

Figure 2.6 *The p-element block.*

(Eq. 2.6). But in addition, it can emit a positron as a proton converts to a neutron, resulting in a decrease in Z of 1 (Eq. 2.7). ^{209}Po, and the other isotope with a $t_{1/2}$ of years, ^{208}Po, are also synthetic and are created by bombardment of lead or bismuth in a cyclotron.

$$^{209}_{84}\text{Po} \rightarrow \alpha + {}^{205}_{82}\text{Pb} \tag{2.6}$$

$$^{209}_{84}\text{Po} \rightarrow \beta^+ + {}^{209}_{83}\text{Bi}. \tag{2.7}$$

An isotope that is on the radioactive decay chain from ^{238}U, like ^{226}Ra, is ^{210}Po, and this is the most abundant isotope. It is purely an α emitter (Eq. 2.8). Its half-life is 138 d, and so is present in much lower concentrations than even radium-226 ($c.1/500^{th}$):

$$^{210}_{84}\text{Po} \rightarrow \alpha + {}^{206}_{82}\text{Pb}. \tag{2.8}$$

It is synthesized by neutron capture by natural bismuth; this is followed by a β emission giving the increase in Z of 1 (Eq. 2.9):

$$^{209}_{83}\text{Bi} + n \rightarrow {}^{210}_{83}\text{Bi} \rightarrow \beta^- + {}^{210}_{84}\text{Po}. \tag{2.9}$$

As an intense emitter of α particles (Eq. 2.8), which are absorbed strongly by living tissue, ^{210}Po is exceptionally toxic with an LD$_{50}$ of $c.1$ μg. It too then, of the elements in Fig. 2.6, will not be further discussed as some of its historical applications can be euphemistically described as niche.

2.4.3   *d*-Elements

Two of the elements in the d-element block in periods 4–6 display radioactivity in their natural state, but this does not necessarily preclude them from conventional use, and hence they are lightly shaded in Fig. 2.7. Naturally occurring rhenium (Re, $Z=75$) has two isotopes and is unusual in that it is the minor nucleide, ^{185}Re, which is the stable one (Table 2.4). ^{187}Re is a β emitter, following Eq. 2.10. However the half-life of about 40 billion years, longer than the probable age of the universe, and the energy of the electron emission (2.6 keV) is extremely

Table 2.4 *Long-lived isotopes of the d-elements.*

Isotope	Half-life	Decay	Product
^{97}Tc	2.6×10^6 y	EC	^{97}Mo
^{98}Tc	4.2×10^6 y	β^-, γ	^{98}Ru
^{99}Tc	2.1×10^5 y	β^-	^{99}Ru
^{185}Re	37.4%	stable	
^{187}Re	4×10^{10} y	β^-	^{187}Os
^{227}Ac	21.7 y	β^-	^{227}Th
		α	^{223}Fr
^{267}Rf	1.3 h	SF	
^{268}Db	28 h	SF	
^{269}Sg	3.2 min	α	^{265}Rf
^{270}Bh	1 min	α	^{266}Db
^{270}Hs	10 s	α	^{266}Sg
^{278}Mt	4 s	α	^{274}Bh
^{281}Ds	14 s	SF	
		α	^{277}Hs
^{282}Rg	100 s	α	^{278}Mt
^{285}Cn	29 s	α	^{269}Fl

		Group									
Period	3	4	5	6	7	8	9	10	11	12
4	Sc	Ti	V	Cr	Mn	Fe	Co	Ni	Cu	Zn
5	Y	Zr	Nb	Mo	Tc	Ru	Rh	Pd	Ag	Cd
6	La	Hf	Ta	W	Re	Os	Ir	Pt	Au	Hg
7	Ac	Rf	Db	Sg	Bh	Hs	Mt	Ds	Rg	Cn

Figure 2.7 *The d-element block.*

low—similar to that of the K_α X-ray emissions of chlorine. In essence it may be handled with little consideration of it being formally radioactive.

$$^{187}_{75}\text{Re} \rightarrow \beta^- + {}^{187}_{76}\text{Os}. \tag{2.10}$$

Technetium (Tc) has the lowest atomic number ($Z = 43$) of the elements which only have radioactive isotopes. Two of the nucleides have half-lives of millions of years. The longer-lived of these, ^{98}Tc, decays by β emission, increasing the atomic number by 1 to form ^{98}Ru. ^{97}Tc, however, follows an electron-capture (EC) mechanism in which an extra-nucleus electron converts a proton into a neutron. Thus Z decreases by 1 to form ^{97}Mo (Eq. 2.11):

$$^{97}_{43}\text{Tc} \xrightarrow{\text{EC}} {}^{97}_{42}\text{Mo}. \tag{2.11}$$

With such half-life times, technetium would not be present without it being continually formed, mainly by fission in uranium and thorium ores. ^{99}Tc results from the spontaneous fission (SF) of ^{238}U and neutron-induced fission of ^{235}U; it decays by β emission of energy 293.7 keV, with a half-life of 210,000 y. It is estimated that there are *c.*18,000 t in the Earth's crust at any given time, with 1 kg of uranium containing *c.*1 ng of technetium. However, spent nuclear fuel rods provide much more concentrated 'ores' of ^{99}Tc; for example, fission of 1 g of ^{235}U affords 27 mg of this isotope. Technetium is a near neighbour in the d block to other elements that have a substantial track record in applications in catalysis, but its use would demand appropriate containment.

However, the radioactive nature of technetium has been used to advantage in radiopharmaceutical, particularly diagnostic, imaging.[9] A breadth of ligands have been designed to deliver a γ emitter to potential disease sites which are imaged by computer tomography (CT). The source of the emitter is ^{99}Mo, which is normally created by neutron-activated fission of uranium enriched in ^{235}U. This molybdenum isotope is a β emitter ($t_{1/2} = 66$ h) and the majority (87.5%) of the technetium-99 formed by this process has the nucleus in an excited, metastable state, labelled as ^{99m}Tc (Eq. 2.12). This relaxes to the nuclear ground state, ^{99}Tc, by a γ-ray emission at 140.5 keV. That passes through organic material and can be imaged by X-ray detectors within a CT scanner. The $t_{1/2}$ of ^{99m}Tc is 6 h, and so effectively within a day, the γ source had been spent, and converted into ground state. That is a weak β emitter, and so presents much-reduced handling challenges.

[9] *Technetium-99m radiopharmaceuticals: Manufacture of kits*, IAEA, Tech. Rep. Ser. 499, 2008, Vienna.

$$^{99}_{42}\text{Mo} \rightarrow \beta^- + {}^{99m}_{43}\text{Tc} \rightarrow \gamma(140.5\text{keV}) + {}^{99}_{43}\text{Tc} \rightarrow \beta^- + {}^{99}_{44}\text{Ru}. \tag{2.12}$$

The majority of the period 7 elements of the d block are synthetic. A relatively neutron-rich isotope, for example ^{48}Ca, is often used as the collision partner with a heavy element with the aim of affording a neutron-rich product. It has been calculated that unidentified neutron-rich isotopes of copernicium (Cn, $Z = 112$)

[10] A. V. Karpov, V. I. Zagrebaev, Y. M. Palenzuela and W. Greiner, 'Superheavy Nuclei: Decay and Stability', *Exciting Interdisciplinary Physics*, 2013, 69–79.

like ^{291}Cn and ^{293}Cn could have a half-life of $c.100$ y.[10] However the $t_{1/2}$ of the longest-lived known isotope, ^{285}Cn, is just 29 s. An example of such a synthesis is for hassium-270 in which ions of magnesium-26 are bombarded onto a target of curium-248 (Eq. 2.13). The result is the longest-lived of the known isotopes of element 108.

$$^{248}_{96}\text{Cm} + {}^{26}_{12}\text{Mg} \rightarrow {}^{274}_{108}\text{Hs}^* \rightarrow {}^{270}_{108}\text{Hs} + 4n. \tag{2.13}$$

Rutherfordium (^{267}Rf, $Z = 104$) and dubnium (^{268}Db, $Z = 105$) both have isotopes with half-lives of over 1 h and decay by spontaneous fission (SF). Seaborgium (^{269}Sg, $Z = 106$), bohrium (^{270}Bh, $Z = 107$), and roentgenium (^{282}Rg, $Z = 111$) decay by α loss with $t_{1/2}$ values of minutes. For hassium (Hs, $Z = 108$), meitnerium (Mt, $Z = 109$), darmstadtium (Ds, $Z = 110$), and copernicium (Cn, $Z = 111$), half-life values are all in the order of seconds, with α loss also the dominating decay process.

$$^{235}_{92}\text{U} \rightarrow \alpha + {}^{231}_{90}\text{Th} \rightarrow \beta^- + {}^{231}_{91}\text{Pa} \rightarrow {}^{227}_{89}\text{Ac} + \alpha. \tag{2.14}$$

Actinium (Ac, $Z = 89$) does have one isotope with a $t_{1/2}$ of the order of years. ^{227}Ac is continually formed through the radioactive decay chain of ^{235}U (Eq. 2.14), and thus is present in trace amount in uranium ores (1 t of uranium in ore contains $c.0.2$ mg of ^{227}Ac). Two other isotopes with much shorter half-lives come from other decay trains from thorium-232: ^{225}Ac, $t_{1/2}$ 10 d and ^{228}Ac, $t_{1/2}$ 6.2 h. As for most other elements in period 7, they are interesting species, but are unlikely to participate in ameliorating Planet Earth's situation.

2.4.4　*f*-Elements

In the normal representations of the periodic table, the challenge of arranging the *f*-elements is generally solved by two rows fanning out from group 3. The 4*f*-elements fan out from lanthanum (La) in period 6, and the 5*f*-elements from actinium (Ac) in period 7. Hence the titles of these rows are lanthanides and actinides. In this text, as adopted by the Royal Society of Chemistry (see footnote 6), we have presented La, with an empty 4*f* shell, within the *d* block, rather than lutetium (Lu) with a filled *f* block. This is consistent with our treatment of groups 11 and 12. The *f*-elements are not normally shown with a group number. So instead the columns have been given the sum of all the valence electrons (Fig. 2.8).

In period 6 we find the element with the second-lowest Z which has only radioactive nuclei, promethium (Pm, $Z = 61$) (Table 2.5). Three isotopes have half-life times of the order of years, with the longest-lived being ^{145}Pm (17.7 years). The traces of promethium therefore depend upon generation through radioactive decay processes: from europium-151 (to ^{147}Pm) and from uranium. The estimate for the abundance of the element in the Earth's crust is 500–600 g. These two naturally occurring isotopes decay by different mechanisms: ^{145}Pm

Table 2.5 *Radioactivity properties of the f-elements*

Isotope	Half-life	Decay	Product
^{145}Pm	17.7 y	EC	^{145}Nd
^{232}Th	1.4×10^{10} y	α	^{228}Ra
^{231}Pa	3.3×10^4 y	α	^{227}Ac
^{238}U	4.5×10^9 y	α	^{234}Th
		SF	
		$\beta^-\beta^-$	^{238}Pu
^{237}Np	2.1×10^6 y	α	^{233}Pa
^{244}Pu	8.1×10^7 y	α	^{240}U
		SF	
^{243}Am	7370 y	SF	
		α	^{239}Np
^{247}Cm	1.6×10^7 y	α	^{243}Pu
^{247}Bk	1380 y	α	^{243}Am
^{251}Cf	898 y	α	^{247}Cm
^{252}Es	472 d	α	^{248}Bk
		EC	^{252}Cf
		β^-	^{252}Fm
^{257}Fm	100.5 d	α	^{253}Cf
		SF	
^{258}Md	51.5 d	EC	^{258}Fm
^{259}No	58 min	α	^{255}Fm
		EC	^{259}Md
		SF	
^{266}Lr	10h	SF	

$\Sigma\ (nf + (n+1)d + (n+2)s\ \text{electrons})$

Period	4	5	6	7	8	9	10	11	12	13	14	15	16	17
6	Ce	Pr	Nd	Pm	Sm	Eu	Gd	Tb	Dy	Ho	Er	Tm	Yb	Lu
7	Th	Pa	U	Np	Pu	Am	Cm	Bk	Cf	Es	Fm	Md	No	Lr

Figure 2.8 *The f-element block.*

undergoes electron capture (EC) to generate ^{145}Nd (Eq. 2.15), whilst ^{147}Pm loses an electron from the nucleus resulting in ^{147}Sm (Eq. 2.16). Promethium-147 can be synthesized by neutron-promoted fission of ^{235}U in enriched uranium, but uses are specialized.

$$^{145}_{61}Pm \ \overrightarrow{EC}\ ^{145}_{60}Nd \tag{2.15}$$

$$^{147}_{61}Pm \ \rightarrow\ ^{147}_{62}Sm + \beta^-. \tag{2.16}$$

There is a wide spectrum of element supply across period 7 of the f-element block. All are radioactive, which creates a duty of care for their applications. At one end of the extreme, thorium (Th, $Z=90$) exhibits an isotope, ^{232}Th, that is an α emitter with a $t_{1/2}$ value similar to that of the age of the Universe. Thorium is considered to have a relatively high abundance in the Earth's crust (5.6 ppm). For uranium (U, $Z=92$), the situation is similar, ^{238}U having a $t_{1/2}$ value comparable to the age of the Earth and a crustal abundance estimated at 1.3 ppm (Table 2.1). Both elements are in sufficient abundance to have uses of global significance.

Several other actinides can be found in trace amounts. The longest-lived isotope of protactinium (Pa, $Z = 91$), ^{231}Pa , is also an α emitter, but its $t_{1/2}$ is shorter than its neighbours by a ratio of 10^{-5} or more. Hence protactinium is only found associated with uranium, as ^{231}Pa is on the decay chain of ^{235}U (Eq. 2.17). It occurs in the uranium ore *uraninite* (pitchblende) at a concentration of c.0.3 ppm in most samples. It is generated in nuclear reactors, and ^{231}Pa is problematical as it forms a long-lived radioactive component of nuclear waste.

$$^{235}_{92}U \rightarrow \alpha + ^{231}_{90}Th \rightarrow ^{231}_{91}Pa + \beta^-. \tag{2.17}$$

Neptunium (Np, $Z=93$) is the first of the transuranic elements. Its longest-lived isotope, ^{237}Np, is also an α emitter, forming ^{233}Pa, with a lifetime much shorter that the age of Earth. Neptunium, plutonium, (Pu, $Z=94$) and americium (Am, $Z=95$) are three transuranics which are found naturally in trace amounts associated with uranium ores. The three elements are formed in sequence by neutron capture followed by β emission. Thus, neutron capture by ^{238}U affords ^{239}U, which decays with a $t_{1/2}$ of 24 min. The isotope of neptunium so formed has a half-life of 2.5 d transmuting into ^{239}Pu (Eq. 2.18). That has a half-life of 24,000 y. A further capture of two neutrons to generate ^{241}Pu leads to a β emission and the formation of ^{231}Am, with a half-life of 432 y. This affords the isotope of neptunium with the longest half-life by α emission (Eq. 2.19). Neptunium-237 and plutonium-239 have been found to be present in uranium ores to ppt levels. Naturally found americium is even rarer, if one avoids areas of nuclear incidents and nuclear weapons tests!

$$^{238}_{92}U + \overrightarrow{n} \ ^{239}_{92}U \ \overrightarrow{24\text{min}} \ \beta^- \ | \ ^{239}_{93}Np \ \overrightarrow{2.5d} \ \beta^- + \ ^{239}_{94}Pu \tag{2.18}$$

$$^{239}_{94}Pu \ \overrightarrow{n} \ ^{240}_{94}Pu \ \overrightarrow{n} \ ^{241}_{94}Pu \ \overrightarrow{14y} \ \beta^- + \ ^{241}_{95}Am \ \overrightarrow{432y} \ \alpha + \ ^{237}_{93}Np. \tag{2.19}$$

These elements are extracted from spent nuclear fuel. Neptunium is extracted from fuel rods in kg quantities. From 1 t of such fuel, there is $c.8$ kg of plutonium and 100 g of americium. It is estimated that the annual production of plutonium is in the region of 70 t.

The next three elements in the row are curium (Cm, $Z = 96$), berkelium (Bk, $Z = 97$), and californium (Cf, $Z = 98$). All are synthetic, and are extractable from nuclear reactors. The neutron flux in a reactor can generate the sequence of nuclear reactions to afford ^{241}Am (Eqs 2.18, 2.19). This americium isotope can also capture a neutron and the resulting ^{242}Am decays by a β emission to form ^{242}Cm (Eq. 2.20); this is an α emitter with a $t_{1/2}$ of 162.8 d. The longer-lived isotope, ^{244}Cm, $t_{1/2}$ 18.1 y, is accessible with higher neutron fluxes, as in a plutonium-based reactor (Eq. 2.21). The production of the longest-lived isotope, ^{247}Cm, is relatively low as it is prone to neutron-activated fission. The total production of all isotopes of curium has been in the order of kg.

$$^{241}_{95}Am \ \overrightarrow{n} \ ^{242}_{95}Am \ \overrightarrow{24\text{min}} \ \beta^- + \ ^{242}_{96}Cm \tag{2.20}$$

$$^{239}_{94}Pu \ \overrightarrow{4n} \ ^{243}_{94}Pu \ \rightarrow \ \beta^- + \ ^{243}_{95}Am \ \overrightarrow{n} \ ^{244}_{95}Am \ \overrightarrow{10h} \ \beta^- + \ ^{244}_{96}Cm \ \overrightarrow{18y} \ ^{240}_{94}Pu + \alpha. \tag{2.21}$$

Synthesis of berkelium can start from ^{244}Cm formed, as in Eq. 2.21 in a high neutron flux reactor. Capture of 5 neutrons affords ^{249}Cm, which decays by β loss to ^{249}Bk, $t_{1/2}$ 64.2 min (Eq. 2.22). This isotope of berkelium has a half-life of the order of a year (330 d). Synthesis of the longest-lived isotope, ^{247}Bk, is again very inefficient. Even the most accessible isotope, ^{249}Bk, has been synthesized only to a total production of grammes.

$$^{244}_{96}Cm \ \overrightarrow{5n} \ ^{249}_{96}Cm \ \overrightarrow{64\text{min}} \ \beta^- + \ ^{249}_{97}Bk \ \overrightarrow{330d} \ ^{249}_{98}Cf + \beta^-. \tag{2.22}$$

Californium-250, $t_{1/2}$ 13 y, can be synthesized in a nuclear reactors, or particle accelerator by neutron bombardment of ^{249}Bk, which undergoes β emission to ^{250}Cf (Eq. 2.23). The longest-lived isotope, ^{251}Cf, requires further neutron capture. The annual production of californium is estimated as less than 1 g.

$$^{249}_{97}Bk \ \overrightarrow{n} \ ^{250}_{97}Bk \ \overrightarrow{3.2h} \ \beta^- + \ ^{249}_{98}Cf \ \overrightarrow{n} \ ^{250}_{98}Cf \ \overrightarrow{n} \ ^{251}_{98}Cf. \tag{2.23}$$

Einsteinium and fermium are the last elements that can be synthesized in a nuclear reactor. A sequence of 15 neutron-capture steps and 6 β emissions affords ^{253}Cf. This which itself undergoes a further β loss to generate ^{253}Es, which largely decays via α loss ($t_{1/2}$ 20.5 d) (Eq. 2.24):

$$^{238}_{92}\text{U} \rightarrow\rightarrow\rightarrow\ ^{253}_{98}\text{Cf} \xrightarrow{18\text{d}} \beta^- +\ ^{253}_{99}\text{Es} \xrightarrow{20\text{d}}\ ^{249}_{97}\text{Bk} + \alpha \tag{2.24}$$

The alternative method of synthesis is by intense neutron irradiation of plutonium, which, following on from ^{253}Es (as formed in Eq. 2.24) can yield ^{254}Es (Eq. 2.25). This decays by a β loss to provide ^{254}Fm. Annual production of einsteinium is of the order of mg, and, perhaps, μg for fermium.

$$^{253}_{99}\text{Es} \xrightarrow{n}\ ^{254}_{99}\text{Es} \xrightarrow{276\text{d}}\ ^{254}_{100}\text{Fm} + \beta \xrightarrow{3.2\text{h}}\ ^{250}_{98}\text{Cf} + \alpha \tag{2.25}$$

The remaining three elements in row 7 of the f block are all synthesized by high-energy bombardment and all are formed in much lower quantities than elements 98–100. Indeed, einsteinium is used as a source for the longest-lived isotope of mendelevium (Md, $Z = 101$), via bombardment of ^{255}Es with α particles (Eq. 2.26). Overall production is probably less than 1 pg per year.

$$^{255}_{99}\text{Es} \xrightarrow{\alpha}\ n +\ ^{258}_{101}\text{Md} \xrightarrow{\text{EC, 52d}}\ ^{258}_{100}\text{Fm.} \tag{2.26}$$

All of the isotopes of nobelium (No, $Z = 102$) and lawrencium (Lr, $Z = 103$) have half-lives of considerably less than 1 day. ^{259}No is the longest-lived isotope ($t_{1/2}$ 58 min), but the most commonly used isotope is ^{255}No ($t_{1/2}$ 3.1 min). It can be formed by bombarding curium-248 or californium-249 with carbon-12 (e.g. Eq. 2.27), yielding $c.100$ atoms per minute of bombardment:

$$^{249}_{98}\text{Cf} +\ ^{12}_{6}\text{C} \rightarrow \alpha + 2n +\ ^{255}_{102}\text{No.} \tag{2.27}$$

The longest lived isotope of lawrencium, ^{266}Lr, has a half-life of $c.1$ hour. However, it has only been identified as a decay product from ^{294}Ts. The longer-lived of the two most accessible isotopes is ^{260}Lr ($t_{1/2}$ 3 min). It can be synthesized by bombarding ^{249}Bk with oxygen-18 (Eq. 2.28), and the quantities involved are probably similar to those of nobelium:

$$^{249}_{97}\text{Bk} +\ ^{18}_{8}\text{O} \rightarrow \alpha + 3n +\ ^{260}_{103}\text{Lr.} \tag{2.28}$$

The supply of the f-elements is depicted in Fig. 2.8. In row 6, promethium is heavily shaded to indicate that its radioactive nature reduces its availability by a high degree, and it will be excluded from the chemical discussion in this work. In row 7 the elements Es, Fm, Md, and Lr similarly have a supply of < 1 g, accompanied with high intrinsic radioactivity, and thus are also shaded out. At the other extreme, thorium and uranium are relatively abundant radioactive elements in the Earth's crust and thus require consideration with regards to a sustainable world, and so are highlighted. The much lower natural supplies of the radioactive elements Pa, Np, Pu, and Am are considerably augmented by the nuclear industry. The remaining three elements, Cm, Bk, and Cf, are available in much smaller quantities than the previous set (g to kg), but it may be that production could be increased if an important application were identified, and this is also indicated by heavy shading.

Table 2.6 *Slater's effective quantum numbers.*

Principal (n)	Effective ($n*$)
1	1
2	2
3	3
4	3.7
5	4.0
6	4.2

2.5 Elemental Properties

Two valuable concepts in comparing properties of elements are **effective nuclear charge**, Z_{eff}, and **electronegativity**, χ. The effective nuclear charge is the net positive charge that is experienced by an electron once the shielding of the nuclear charge, Z, by the other electrons is taken into account, represented by a shielding constant, σ (Eq. 2.29):

$$Z_{eff} = Z - \sigma. \tag{2.29}$$

$$\psi(r) = r^{n*-1} exp\left(\frac{-(Z-\sigma)r}{n*}\right). \tag{2.30}$$

This approach envisages electrons as being hydrogen-like orbitals with the principal quantum number (Table 2.6) and nuclear charge adjusted by the interactions with all the other electrons (Eq. 2.30).[11] The values of the factors were derived by changing empirical adjustments to match the atomic energies, E, from Eq. 2.31 to the experimental values of the atomic spectral energies like ionization energies and electronic transitions. The rules for estimating the shielding constants are given in Table 2.7. There are more complicated refinements to the Slater model which reduce the error,[12] but these serve to give semi-quantitative estimates.

$$E = -k \sum_{i=1}^{N} \left(\frac{Z-\sigma}{n_i*}\right)^2. \tag{2.31}$$

[11] J. C. Slater, 'Atomic shielding constants', *Phys. Rev.* 1930, **36**, 57–64.

[12] E. Clementi and D. L. Raimondi, 'Atomic screening constants from SCF function', *J. Chem. Phys.*, 1963, **38**, 2686–9.

Table 2.7 *Slater screening constants.*

Subshell	Same subshell	Same n, lower l	$n-1$	$< n-1$
1s	0.30	–	–	–
ns,np	0.35	–	0.85	1
nd,nf	0.35	1	1	1

As an example, for lithium Z_{eff} for the *2s* electron will have the atomic number of 3 reduced by the shielding of the two *1s* electrons by 0.35 for each one, resulting in Z_{eff} value of 1.3. Given that the ionization potential of hydrogen is 13.6 eV (1,312 kJ mol^{-1}), we can estimate that for Li by substituting into Eq. 2.31 to give Eq. 2.32:

$$IE = 13.6 \left(\frac{Z - \sigma_{2s}}{n_{2s}*}\right)^2 = 13.6 \left(\frac{1.3}{2}\right)^2 = 5.75 \, eV. \qquad (2.32)$$

The experimental value is 5.4 eV, but the model is close enough to be the basis of considering the variations across the periodic table in a qualitative way.

If we consider the first short period, Li to F, then we can evaluate the likely changes in Z_{eff} and the consequences on other properties. Moving right from Li to Be, there is an increase in the atomic number from 3 to 4, and an extra *2s* electron. This extra electron will only very partially compensate for the increase in nuclear charge since it is in the outermost occupied subshell. Hence Z_{eff} will increase from Li to Be, the value of which can be estimated using Table 2.7. This increase will attract the electrons more strongly and thus reduce the atomic radius. The effect will be to lower the energy of the occupied valence orbitals and thus increase the ionization energy of the atom (Fig. 2.9). Adding a third valence electron will require the electron to occupy a *2p* orbital which is in a higher level subshell. Although not predicted by the Slater parameters (Table 2.8), the ionization energy of boron is reduced and the atomic radius increased. For carbon and nitrogen the extra electrons also occupy *2p* oribitals with a low screening constant and thus Z_{eff} will again increase; the trend in ionization energy and atomic radius will be as for Li to Be. This trend continues to Ne with one exception: between N and O.

Table 2.8 Z_{eff} *values for the highest-energy electron for Li-Ne.*

Element	Z_{eff}
Li	1.30
Be	1.95
B	2.60
C	3.25
N	3.90
O	4.55
F	5.20
Ne	5.85

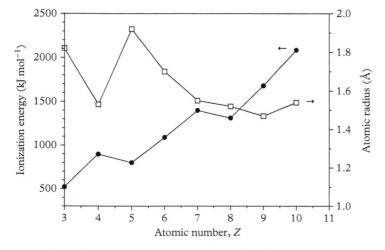

Figure 2.9 *Ionization energies and atomic radii of Li to Ne.*

In this case the fourth electron entering the $2p$ subshell must pair with another. This increased repulsive energy (or screening) somewhat increases the atomic radius and the ionization energy drops slightly.

Electronegativity is considered to be the ability of an atom, in an environment, to attract electrons to itself. The opposite effect is **electropositivity**, which is the ability to release electrons, as displayed by alkali metals. Electronegativity as a concept dates to Berzelius about 200 years ago, but estimating it is still a current issue. Tabulations are generally for an element using parameters with experimental values. For Mulliken (in 1934) these were values of the ionization energy (IE) and electron affinity (EA) of the gaseous atom; the former is a direct measure of the attraction of an electron to the atom, and the latter a measure of its ability to attract a further electron. The Mulliken electronegativity, χ_M, is given by Eq. 2.33:

$$\chi_M = \frac{(IE + EA)}{2}.$$

(2.33)

This was called the **absolute electronegativity** as it was based upon fundamental properties of the atom. In fact this scale post-dated Pauling's scale (1932), χ_P, and so the values of χ_M are generally rescaled to correlate to the Pauling scale. If the units of energy are eV, this is achieved by Eq. 2.34:

$$\chi_M = 0.187(IE + EA) + 0.17.$$

(2.34)

Much later (1989), Allen adapted this approach to use the spectroscopic properties of atoms to provide energies of the n_s and n_p valence electrons, ϵ_s and ϵ_p, to derive an average energy; this new electronegativity measure, χ_A, may be matched to the Pauling scale with a single factor (0.169) (Eq. 2.35). This was also extended to the d block elements, but not to the f-elements.[13] In Eq. 2.36, $p+q$ is the maximum known oxidation state; however, due to the mixture of s and d occupancies p takes values of 1.5–2.0.

[13] J. B. Mann, T. L. Meek, E. T. Knight, J. F. Capitani and L. C. Allen, 'Configuration energies of the d-block elements', *J. Am. Chem. Soc.*, 2000, **122**, 5132–7.

$$\chi_A = 0.169 \left(\frac{n_s \epsilon_s + n_p \epsilon_p}{n_s + n_p} \right)$$

(2.35)

$$\chi_A = 0.169 \left(\frac{p \epsilon_s + q \epsilon_d}{p + q} \right).$$

(2.36)

[14] A. L. Allred and E. G. Rochow, 'A scale of electronegativity based on electrostatic force', *J. Inorg. Nucl. Chem.*, 1958, **5**, 264–8.

Allred and Rochow[14] have utilized the electrostatic potential on the surface of an atom as a principle for estimating electronegativity, on the scale χ_{AR} (Eq. 2.37, r_{cov} in Å). The numerical factors are again to optimize the alignment to the Pauling parameters. Here the Z_{eff} values calculated employing the Slater shielding constants are used directly, and the covalent radii, r_{cov}, are the experimental values taken from bondlengths. This, then, is a measure of the attraction of electrons by an atom at the edge of its covalent bond. As such, there is a relationship to the original Pauling basis of χ (Eq. 2.38).

$$\chi_{AR} = 0.359\frac{Z_{eff}}{r_{cov}^2} + 0.744 \tag{2.37}$$

$$|\chi_A - \chi_B| = \sqrt{D_{AB} - \frac{D_{AA} + D_{BB}}{2}}. \tag{2.38}$$

Pauling ascribed an expectation value for the covalent contribution to the dissociation energy of a heteronuclear bond, *A–B* (in units of eV), as being the arithmetic mean of those of the two homonuclear bonds, *A–A* and *B–B*. The experimental value, D_{AB}, would also contain an ionic contribution due to the polarization of the bond. Hence the difference between the experimental value of D_{AB} and the covalent contribution could be used as an assessment of the electronegativity difference between the two elements. The scale of individual values of χ requires a reference point, and that is now normally taken as hydrogen being 2.20.[15] Pauling also proposed an empirical formula based upon geometric mean, which will afford the relationship in Eq. 2.39:

$$|\chi_A - \chi_B| = \sqrt{\frac{D_{AB} - \sqrt{D_{AA}D_{BB}}}{1.3}}. \tag{2.39}$$

[15] A. L. Allred, 'Electronegativity values from thermochemical data', *J. Inorg. Nucl. Chem.*, 1961, **17**, 215–21.

The relationships between the covalent radii of the elements and also the Pauling electronegativities are presented in Fig. 2.10. There are gaps in the Pauling electronegativities for He, Ne, and Ar due to the absence of bond energy data. The periodicity with atomic number is evident in both datasets, which roughly follow an inverse relationship with each other. There is an abrupt change at the end of a period, followed by a steep variation across the short *s* block. The slope is reduced over the *p*-elements, and a more gradual variation over the transition-

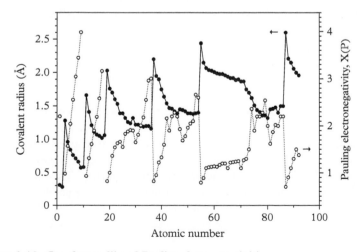

Figure 2.10 *Covalent radii and Pauling electronegativities.*

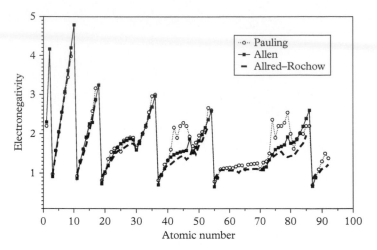

Figure 2.11 *Pauling, Allen, and Allred–Rochow electronegativities.*

element (d and f) blocks. It is also evident that running down the periodic table, there is an increase in r_{cov} with period (row) and a corresponding decrease in χ. A comparison of the values from the Pauling (bond energy based), Allred–Rochow (electrostatic potential based), and Allen (atomic energy level based) methods (Fig. 2.11) shows that over much of the main group periods, there is a close correlation. The Allen scale fills in the holes in the lighter inert gases. However, there are very significant differences for the *4d* and *5d* transition series. The Pauling values, using the bond energies available at the time, render many of these metals more electronegative than silicon, which, as we will see, is rather unexpected. The Allred–Rochow values (from covalent radii) track the Allen set but with a tendency to diverge increasingly to lower values at higher Z.

Overall, the relationships between atomic number, covalent radii, and electronegativity can be displayed as a contour map (Fig. 2.12). The locations of the elements can be seen on the R versus Z grid, with Allen electronegativity as the vertical contour. This shows graphically the tendency for electronegativity to reduce with atomic number as the covalent radii increase. The tracks across the rows of the periodic table follow the opposite trend, but the peaks in electronegativity reduce with each row. The filling of the *4f* block elements, which are of relatively low electronegativity, causes a shrinking of the atomic radius to values lower than might be anticipated from the extra shell. This is called the **lanthanide contraction**. The heaviest elements of the *p* block appear somewhat anomalous with larger variations between them.

Further correction procedures to electronegativity, as developed by Sanderson, are described in Extension 2.1.[16]

[16] R. T. Sanderson, 'Electronegativity and bond energy', *J. Am. Chem. Soc.*, 1983, **105**, 2259–61.

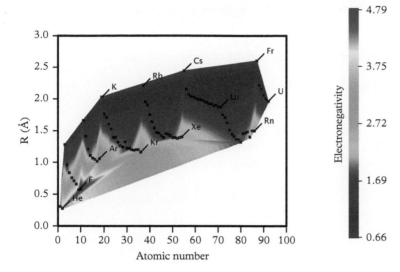

Figure 2.12 *Contour plot of covalent radii and Allen electronegativities against atomic number.*

Extension 2.1 *Further refinements of electronegativity*

Sanderson has used the geometric mean relationship for the covalent bonding in Eq. 2.39 together with the relationship of χ to r in Eq. 2.37 to utilize the covalent radius as a correction factor. Thus the expectation value of the covalent contribution to the dissociation energy of a bond was proportioned by the ratio of R_{cov} (the sum of the covalent radii) to the true distance, R_{obs}, normally implying an increase in energy (Eq. 2.40). The expectation of the ionic contribution to the bond energy follows the normal electrostatic interaction in Eq. 2.41, employing units in eV and Å.

$$D_{cov} = (D_{AA}D_{BB})^{\frac{1}{2}} \frac{R_{cov}}{R_{obs}} \tag{2.40}$$

$$D_{ion} = \frac{1.440}{R_{obs}}. \tag{2.41}$$

The additional concept of Sanderson is that of **electronegative equalization**. Here electronegativity itself is allowed to be sensitive to the bond in the chemical environment. The sensitivity of electronegativity to unit charge, $\Delta\chi_i$ is given by Eq. 2.42:

$$\Delta\chi_i = 1.57\chi^{\frac{1}{2}}. \tag{2.42}$$

The fractional charge on the atom of an element E, δ_E, is estimated by Eq. 2.43, where χ_E is the electronegativity of element E and $\Delta\chi_i$ is its electronegativity sensitivity; χ_m is the molecular electronegativity, calculated as the geometric mean of the values of each atom in the molecule:

$$\delta_E = \frac{\chi_m - \chi_E}{\Delta\chi_i}. \tag{2.43}$$

As a result, the actual expectation of the bond energy is not the sum of the covalent and ionic contributions, but each is reduced by blending coefficients, t (Eq. 2.44), which are derived from Eq. 2.45.

$$D_{AB} = t_{cov}D_{cov} + t_{ion}D_{ion}, \tag{2.44}$$

where:

$$t_{ion} = \frac{D_{obs} - D_{cov}}{D_{ion} - D_{cov}} \text{ and } t_{cov} = 1 - t_{ion}. \tag{2.45}$$

These factors form a sound starting point. There are further refinements to provide corrections due to oxidation state, coordination number, and bond type which allow more useful estimations of the stability of a material, and also give insight into the type of material.

Since van Arkel's work in 1941 the classifications of the types of bonding, that is covalent, ionic, and metallic, have been displayed in a triangular manner. Jensen later quantified this using electronegativity means and differences (Fig. 2.13).[17] As might be expected, covalent bonding predominates when there is little electronegativity difference between the bonded atoms and their electronegativity is high, as in F_2. Moving to CsF, there is now a large difference in electronegativity and so there is a high degree of ionicity to the bonding. Compounds with lesser differences would adopt positions within the triangle indicating a mixture of these two types of bonding. However, if both the electronegativity and the difference are low, as in Cs, the atom can dissociate an electron. The electron enters a conduction band, and has a high degree of mobility, namely metallic behaviour.

This approach has been refined[18] by allowing a range of compositions using the weighted mean of the electronegativities. In a compound of formula A_mB_n, this average, χ_{wt}, is given by Eq. 2.46:

$$\chi_{wt} = \frac{m\chi_A + n\chi_B}{m + n}. \tag{2.46}$$

By considering a large range of binary compounds, regions appropriate to the predominance of a type of bonding could be ascribed (Fig. 2.14). The approach has been exemplified in Fig. 2.15,[19] which includes a number of materials that are highlighted in this book. For example, the perfluoro-compounds CF_4, NF_3, and

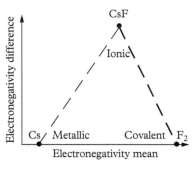

Figure 2.13 *The electronegativity triangle.*

[17] W. B. Jensen, 'A quantitative van Arkel diagram', *J. Chem. Educ.*, 1995, **72**, 395–8.

[18] T. L. Meek and L. D. Garner, 'Electronegativity and the bond triangle', *J. Chem. Educ.*, 2005, **82**, 325–33.

[19] Allen electronegativities have been used generally with Allred–Rochow χ values for f block elements.

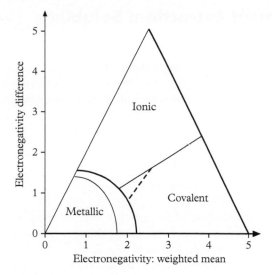

Figure 2.14 *Relationship of material type to electronegativity.*

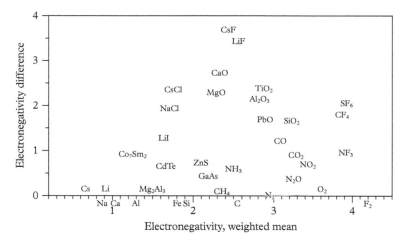

Figure 2.15 *Electronegativity triangle (Allen values) for selected materials.*

SF_6 (Chapters 1 and 7) are all shown in the covalent region, albeit with differing degrees of ionic character.

Two transition zones are identified. One of these transition zones was between covalent and ionic, and included compounds with very borderline characteristics such as Be_2C, Al_4C_3, and CdSe. The other transition zone with curved boundaries (Fig. 2.14) borders onto the metallic region. As can also be seen from Fig. 2.15, this region encompasses materials like silicon, CdTe, and InSb. Hence there are a wide variety of functional materials, such as semiconductors, which fall near metal–insulator transitions.

2.6 Element Extraction: Solubility

Table 2.9 *Standard enthalpies, $\Delta_f H^o$, of a selection of gaseous ions (kJ mol^{-1}).*

Cation	$\Delta_f H^o$
H$^+$	1,536
Li$^+$	686
Na$^+$	610
K$^+$	514
Rb$^+$	490
Cs$^+$	459
Be^{2+}	2,987
Mg^{2+}	2,342
Ca^{2+}	1,919
Sr^{2+}	1,784
Ba^{2+}	1,655
F$^-$	−255
Cl$^-$	−234

[20] The exception is that atomization of Mg is of lower energy than for Ca, but this is a small difference (30 kJ mol^{-1}) compared to the overall change.

The group of elements which are unshaded in Fig. 2.5 comprise ten elements of low electronegativity and high atomic radius (Fig. 2.12). As expected from the electronegativity triangle, the majority of their common compounds have a high degree of ionicity (Fig. 2.15) and the elements themselves are metallic. Formation of the gaseous ions involves two steps: vapourization (Eq. 2.47) and ionization (Eq. 2.48):

$$M_s \rightarrow M_g \tag{2.47}$$
$$M_g \rightarrow M_g^+ + e. \tag{2.48}$$

The enthalpy of formation of the gaseous ions decreases down the group (Table 2.9). As the atomic radius increases, the energy required for both vapourization and ionization generally decreases (Fig. 2.16).[20] The large difference in energies between the two curves is due to the second ionization (Eq. 2.49) that is important for group 2. Removing a second electron from a positive ion is a much higher-energy process. The entropy values (Table 2.10) are far more similar, and of smaller contribution to ΔG^o (≈ 50 kJ mol^{-1}). As a result, the changes in ΔH^o dominate the trends in gaseous ion formation.

$$M_g^+ \rightarrow M_g^{2+} + e. \tag{2.49}$$

The occurrence of these elements on Earth is as ionic salts with small anions of highly electronegative (hard) atoms, generally oxygen or chlorine. The balance

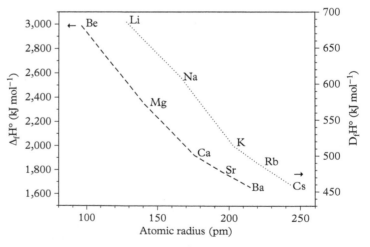

Figure 2.16 *Experimental $\Delta_f H^o$ (kJ mol^{-1}) versus atomic radii of groups 1 and 2.*

between deposited and dissolved materials will depend upon an equilibrium of the type in Eq. 2.50:

$$MX_{(s)} = M^+_{(aq)} + X^-_{(aq)}. \tag{2.50}$$

The dissolution process can be split conceptually into vapourizing the solid (Eq. 2.51) and the solvation of the gaseous ions (Eq. 2.52):

$$MX_{(s)} \rightarrow M^+_{(g)} + X^-_{(g)} \tag{2.51}$$

$$M^+_{(g)} + X^-_{(g)} \rightarrow M^+_{(aq)} + X^-_{(aq)}. \tag{2.52}$$

Interatomic distances can be readily derived on crystal samples, and from that it is evident that the ionic radius varies with coordination number. In solution there is a dynamic interconversion of differing coordination sites. For example in water, Li^+ is understood to be predominantly four-coordinate, Na^+ and Mg^{2+} six coordination, whilst K^+ and Ca^{2+} have a mean near-seven coordinate. Studies[21] though have been able to make useful estimates of ionic radii and the attendant hydration free energies.

As explained in Extension 2.2, the thermodynamic values for aquated ions are conventionally listed against the reference zero of the aquation of the proton. The main reason for this is that the relative values are firm but estimations for the enthalpy of aquation of the proton vary. Values in the region of $-1,050$ to $-1,130$ kJ mol^{-1} have been proposed, and in this work we adopt the values of $-1,110$ kJ mol^{-1}.[22]

The variations in $\Delta_{hyd}H^\circ$ with element (Table 2.11) are shown in Fig. 2.17, and clearly both groups follow a similar shaped curve, with the hydration enthalpy becoming increasingly exothermic with decreasing ionic radius; the values and the changes are much larger for the dications. The absolute entropies of hydration

Table 2.10 *Absolute standard entropies, S°, of a selection of gaseous ions (J mol^{-1} K^{-1}).*

Cation	S°
H^+	108.9
Li^+	133.1
Na^+	148.0
K^+	154.6
Rb^+	164.4
Cs^+	169.9
Be^{2+}	136.3
Mg^{2+}	148.7
Ca^{2+}	154.9
Sr^{2+}	164.7
Ba^{2+}	170.3
F^-	145.6
Cl^-	153.4

[21] Y. Marcus, 'Thermodynamics of Solvation of Ions. Part 5—Gibbs Free Energy of Hydration at 298.15K', *J. Chem. Soc. Faraday Trans.*, 1991, **87**, 2995–9.

[22] J. Barrett, 'Inorganic Chemistry in Aqueous Solution', Royal Society of Chemistry, Cambridge, 2003.

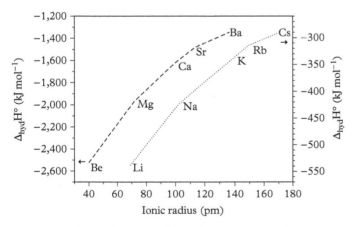

Figure 2.17 *Experimental absolute $\Delta_{hyd}H^\circ$ versus ionic radii of groups 1 and 2 ($\Delta_{hyd}H^\circ$ of H^+ taken as $-1,110$ kJ mol^{-1}).*

Table 2.11 *Absolute standard enthalpies of hydration, $\Delta_{hyd}H^o$, of a selection of aqueous ions ($J\ mol^{-1}\ K^{-1}$).*

Cation	$\Delta_{hyd}H^o$
H^+	$-1,110$
Li^+	-538
Na^+	-424
K^+	-340
Rb^+	-315
Cs^+	-291
Be^{2+}	$-2,524$
Mg^{2+}	$-1,963$
Ca^{2+}	$-1,616$
Sr^{2+}	$-1,483$
Ba^{2+}	$-1,346$
F^-	-504
Cl^-	-359

[23] The permittivity is the property of a material to affect the Coulombic force between point charges. Relative permittivities are generally tabulated against that of ϵ_o and will reduce the Coulombic force.

(Table 2.12) are all negative, indicating that an ordering process is occuring. This would be expected as the gaseous ion dissolves into the liquid, but the values for the dications indicate a further level of ordering of the solvent structure. For group 2, the variation in $T\Delta_{hyd}S^o$ amounts to ≈ 30 kJ mol^{-1} as opposed to $\approx 2,000$ kJ mol^{-1} for $\Delta_{hyd}H^o$.

The electrostatic model for solvation was developed by Max Born into Eq. 2.58. The Born equation treats the solvent as a uniform medium and predicts the change in Gibbs free energy when a mole (Avogadro's number, N_A) of ions of radius r_i moves from a vacuum of permittivity ϵ_o to a medium of permittivity ϵ (Eq. 2.58).[23] Incorporating the values of the constants, then the Born equation becomes Eq. 2.59, when r_i is in pm. The values increase with the square of the ionic charge and are inversely proportional to the ionic radius, accounting for the relative magnitudes shown in Fig. 2.17.

Extension 2.2 *Thermodynamic values for ions*

For transformations at constant pressure it is the Gibbs free energy, ΔG, which is the factor that affords the equilibrium constant for a reaction, K_p. However the ionization energy (as for Eq. 2.48) and the lattice energy, U, that might be calculated to understand Eq. 2.51, are defined as a change at constant volume and at 0 K $\Delta U(0)$. At 298 K, the energy $\Delta U(298)$ will differ due to there being a heat capacity of a material, C_v. For an ideal gas, the variation due to temperature is given by 0.5 RT for each degree of freedom ($RT = 2.477$ kJ mol^{-1} at 298 K). Thus for the reaction in Eq. 2.28, one mole of $M^+{}_g$ is converted into a mole of $M^{2+}{}_g$ and a mole of electron gas. This extra mole of ideal, monatomic gas will have three translational degrees of freedom. Hence Eq. 2.53 provides a value for $\Delta U(298)$:

$$\Delta U(298) = \Delta U(0) + \frac{3}{2}RT. \tag{2.53}$$

This internal energy may be converted into an enthalpy, ΔH, the energy factor at constant pressure, if we consider the change in the number of molar volumes in that reaction. Thus, for equation 2.48, there are two mononuclear, gas phase ions that are formed from one and thus (Eq. 2.54):

$$\Delta H(298) = \Delta U(298) + RT. \tag{2.54}$$

From this, the value of ΔG can be calculated if the entropy is also established (Eq. 2.55):

$$\Delta G = \Delta H - T\Delta S. \tag{2.55}$$

Tabulated values of the energies, enthalpies, and entropies for the solid and gas phase ions in Eq. 2.51 are known, and help the changes in the reaction to be estimated, as for the example of entropy under standard conditions (Eq. 2.56):

$$\Delta_r S^o = S^o_{M^+_{(g)}} + S^o_{X^-_{(g)}} - S^o_{MX_{(s)}}. \tag{2.56}$$

However, the conventional thermodynamic values of aquation that are tabulated are relative to the aquation of the proton as the zero point. Thus the conventional value, such as for the enthalpy, differs from the absolute value (Eq. 2.57):

$$\Delta_{hyd}H^\circ(M^+,g) = \Delta_{hyd}H^\circ(M^+,g)^{conv} - \Delta_{hyd}H^\circ(H^+,g). \qquad (2.57)$$

Some conventional values of the ΔH of aquation appear to look endothermic, but really it is generally just because their values may be less exothermic than those of the aquation of the solvated proton.

$$\Delta_{hyd}G^\circ = -\frac{N_A z^2 e^2}{8\pi\epsilon_o r_i}\left(1 - \frac{1}{\epsilon}\right) \qquad (2.58)$$

$$\Delta_{hyd}G^\circ = -6.86.10^4\frac{z^2}{r_i}\ kJ\ mol^{-1}. \qquad (2.59)$$

From Eq. 2.59, predictions of $\Delta_{hyd}H^\circ$ (Eq. 2.60) and $\Delta_{hyd}S^\circ$ (Eq. 2.61) can be made (as for 298.15 K).

$$\Delta_{hyd}H^\circ = -6.98.10^4\frac{z^2}{r_i}\ kJ\ mol^{-1} \qquad (2.60)$$

$$\Delta_{hyd}S^\circ = -\frac{4069z^2}{r_i}\ J\ K^{-1}mol^{-1}. \qquad (2.61)$$

However, the more reliable quantitative estimations require expansion of this equation, which includes a term for the dissolution of a neutral atom ($\Delta_{hyd}G_{neut}$). The ionic terms encompass two regions: the intra-solvation shell region based on the radius of the ion itself, r, and also that of the extra-solvation sphere (Fig. 2.18), of radius Δr (Eq. 2.62):[22]

$$\Delta_{hyd}G^\circ = \Delta_{hyd}G_{neut} + \Delta_{hyd}G_{intrashell} + \Delta_{hyd}G_{extrashell}. \qquad (2.62)$$

The solvation sphere model indicates that the Born assumption of an unperturbed solvent structure needs to be refined. Around the ion is a coordination shell (the primary sphere) in which there may be some charge dissipation through a degree of covalency. This complex ion has its own secondary solvation shell. A simple way to allow for this using Eq. 2.60 is either to vary the ionic radius with an additional factor, Δr_i, or to reduce the full ionic charge with a refineable effective charge, z_{eff}. The results of this approach (Table 2.13) show that a radius correction of c.70 pm is required to adapt Eq. 2.60 for cations, but is very small for hard anions. Alternatively, if the full correction is carried out via the adjustment of the ionic charge, it is apparent that for cations the deviation from ionic behaviour decreases with increasing ionic radius. It is also clear that the largest deviations are for ions with high z/r ratios, implying a greater covalency in the ion–water interaction.

Table 2.12 *Absolute standard entropies of hydration, $\Delta_{hyd}S^\circ$, of a selection of aqueous ions ($J\ mol^{-1}\ K^{-1}$).*

Cation	S°
H^+	−129.8
Li^+	−140.6
Na^+	−109.8
K^+	−72.5
Rb^+	−63.8
Cs^+	−57.9
Be^{2+}	−307.9
Mg^{2+}	−329.1
Ca^{2+}	−249.6
Sr^{2+}	−236.3
Ba^{2+}	−202.5
F^-	−138.5
Cl^-	−76.3

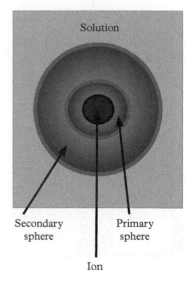

Figure 2.18 *Solvation spheres.*

Table 2.13 *Hydration of s-element ions: Ionic radii (r_i), radius correction (Δr_i), and effective ion charge (z_{eff}), to match the experimental $\Delta_{hyd}H^o$ values.*

Ion	r_i/pm	Δr_i/pm	z_{eff}	$\Delta_{hyd}H^o_{exp}$/kJ mol^{-1}
H$^+$	30	31	0.69	$-1,110$
Li$^+$	69	61	0.75	-538
Na$^+$	102	63	0.79	-424
K$^+$	138	67	0.82	-340
Rb$^+$	149	72	0.83	-315
Cs$^+$	170	73	0.84	-291
Be^{2+}	40	71	1.20	$-2,524$
Mg^{2+}	72	70	1.42	$-1,963$
Ca^{2+}	100	73	1.52	$-1,616$
Sr^{2+}	113	75	1.61	$-1,483$
Ba^{2+}	137	70	1.63	$-1,346$
F$^-$	181	5	-0.98	-504
Cl$^-$	181	14	-0.96	-359

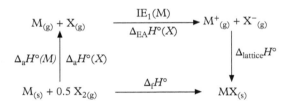

Figure 2.19 *Born–Haber cycle for the formation of MX.*

Lattice energies are not directly measured, but can be estimated using Born–Haber cycles (Fig. 2.19) and Hess's law. The enthalpy of formation of the solid ($\Delta_f H^o$) is experimentally determinable (Table 2.14). In the Born–Haber cycle, the formal steps are carried out with known enthalpy changes. In Fig 2.19, the metal and halogen are first atomized, and then independently ionized. The lattice enthalpy can then be estimated by difference (Eq. 2.63):

$$\Delta_{lattice}H^o = \Delta_f H^o - \left(\Delta_a H^o(M) + \Delta_a H^o(X) + IE_1(M) + \Delta_{EA}H^o(X) \right). \quad (2.63)$$

For NaCl, this equates to Eq. 2.64:

$$\Delta_{lattice}H^o = -411 - (+108 + 121 + 502 - 354) = -788 \text{ } kJ \text{ } mol^{-1}. \quad (2.64)$$

Table 2.14 $\Delta_f G o$, $\Delta_f H^o$, $T\Delta_f S^o$, and lattice energy of alkali metal chlorides.

Salt	$\Delta_f G^o$/kJ mol^{-1}	$\Delta_f H^o$/kJ mol^{-1}	$T\Delta_f S^o$/kJ mol^{-1}	U_{Lexp}/kJ mol^{-1}
LiCl	−384	−409	−25	853
NaCl	−384	−411	−27	786
KCl	−408	−436	−28	715
RbCl	−405	−430	−26	689
CsCl	−404	−433	−29	659

It is worth noting that lattice energies (U) can be tabulated as positive or negative energy values. If we consider the lattice-forming step in Fig. 2.19, whatever the sign of the tabulated value, the energy change will be favourable, that is, negative. This step reduces the number of gaseous particles and for each monoatomic ion we expect the enthalpy to be less favourable on crystallization by RT. For the same reason, the entropy of lattice forming will also be negative.

The values in Table 2.14 show that there is a strong decrease in the lattice energy with increasing ionic radius, but the variation in $\Delta_f G^o$ is relatively small, within 24 kJ mol^{-1}, showing how the lattice energy is largely counterbalanced by the other terms in Eq. 2.63. Indeed it is KCl, the central metal chloride in group 1, which which has the highest $\Delta_f G^o$, this value being dominated by the enthalpy term. It can be a mistake just to fix on one parameter, even one as large as the lattice energy, when aiming to predict the relative stability of chemical compounds.

When the ion is within the solid, it is the sum of the radii which is the important factor. Within the ionic model, the Coulombic attraction between a mole of the cation (charge z_+) and anion (charge z_-) separated by an interatomic distance r at 0 K can be expressed in Eq. 2.65. A correction was made by Born for the fact that the ions are not hard, point charges, with a repulsive term proportional to r^{-n}. This provides a reduction in Coulombic force according to Eq. 2.66, the Born–Landé equation. In that formula, A is called the Madelung constant, and is a function of the crystal structure of the compound. The Madelung constant increases with the coordination number of the ions and the number of ions in the formula unit. In going from the six-coordinate NaCl to the eight-coordinate CsCl, A increases from 1.7476 to 1.7627. The larger increase occurs on going to CaF$_2$ (also eight-coordinate at Ca), with A rising to 2.514.[24,25]

$$U_C = -\frac{N_A A |z_+||z_-|e^2}{4\pi\epsilon_o r} \tag{2.65}$$

$$U = -\frac{N_A A |z_+||z_-|e^2}{4\pi\epsilon_o r}\left(1-\frac{1}{n}\right). \tag{2.66}$$

[24] There are different ways of incorporating the charges in the Born-Landé equation, and these affect the values of the Madelung constant for charges other than unity. See note 25.

[25] D. Quane, 'Crystal Lattice Energy and the Madelung Constant', *J. Chem. Educ.*, 1970, **47**, 396–8.

The values for the lattice energy from the Born–Landé equation for the alkali metal halides are generally *c*.3% lower than those obtained from Born–Haber cycles. Further refinements can be added to reduce that difference, including the Born–Mayer correction of ρ/r (Eq. 2.67); this is of the order of 30 kJ mol^{-1} and can have a significant affect on the outcome of a prediction. However, more general application of the ionic model requires adherence to ionic bonding, which is much less the case in group 11 halides like AgI, and also a knowledge of the crystal structure in order to calculate the Madelung constant.

$$U = -\frac{N_A A |z_+||z_-|e^2}{4\pi\epsilon_0 r}\left(1 - \frac{\rho}{r}\right). \tag{2.67}$$

[26] A. F. Kapustinskii, 'Lattice Energy of Ionic Crystals', *Q. Rev.*, 1956, **10**, 283–94.

It was noted by Kapustinskii[26] that the ratio of of the Madelung constant to the number of ions in a formula unit (v), A/v, varied relatively little between structures, and increased with mean coordination number, that is 0.84 for CaF_2 to 0.88 for CsCl.[27] Hence the Madelung constant could be replaced by v, and more generally applicable equations derived, as in Eq. 2.68, from the Born–Mayer equation, where r is in nm and $\rho \approx 0.0345$.[28]

[27] D. A. Johnson, 'Some Thermodynamic Aspects of Inorganic Chemistry', Cambridge University Press, London, 1968.

[28] W. E. Dassent, 'Inorganic Energetics', Penguin Books Ltd, Harmondsworth, Middlesex, 1970.

$$U = -121.4v\frac{|z_+||z_-|}{r_+ + r_-}\left(1 - \frac{\rho}{r_+ + r_-}\right). \tag{2.68}$$

This has allowed the generalization of the estimation of the calculation of lattice energies to complex ions.[29] The effective, **thermochemical**, radius of these anions can be estimated back from $\Delta_{lattice}H^\circ$ values from Born–Haber cycles, allowing for the correction between lattice energy and standard-state enthalpy (Eq. 2.69),[30] where s_i are the number of the ions in the formula unit and c_i is the number of degrees of freedom of the ion (3 for monoatomic ions, 5 for linear ions, and 6 for nonlinear ones). This relationship assumes the vibrational properties of the ion are unchanged between the gas phase and the crystal. It also accounts for there being an acoustic energy of the ionic motion in the solid, as well as for the differences in degrees of freedom in the gas and solid states.

[29] H. D. B. Jenkins, H. K. Roobottom, J. Passmore and L. Glasser, 'Relationships among Ionic Lattice Energies, Molecular (Formula Unit) Volumes, and Thermochemical Radii', *Inorg. Chem.*, 1999, **38**, 3609–20.

[30] H. D. B. Jenkins, 'Thermodynamics of the Relationship between Lattice Energy and Lattice Enthalpy', *J. Chem. Educ.*, 2005, **82**, 950–2.

$$\Delta H_L = U + \sum_i s_i\left(\frac{c_i}{2} - 2\right)RT. \tag{2.69}$$

For example, the following reaction (Eq. 2.70) effectively requires the vapouriztion of Na_2O and Cs_2CO_3 followed by the condensation of Na_2CO_3 and Cs_2O. Using values for the thermochemical radii of the 4 ions (Na^+ 0.116 nm, Cs^+ 0.181 nm, O^{2-} 0.141 nm, and CO_3^{2-} 0.189 nm), the lattice energy can be estimated from Eq. 2.67, and converted to $\Delta_{lattice}H^\circ$ using Eq. 2.69. The result is an unfavourable reaction enthalpy ($\Delta_r H^\circ = 101$ kJ mol^{-1}); this can also be calculated from the enthalpies of formation of these salts to provide an experimental value of +77 kJ mol^{-1}. The error of ≈ 20 kJ mol^{-1} is about 1% and emanates from the difference between calculated values of the lattice energy.

$$Na_2O_{(s)} + Cs_2CO_{3(s)} = Na_2CO_{3(s)} + Cs_2O_{(s)}. \tag{2.70}$$

In some materials it is difficult to ascribe a unit to a localized complex, particularly when there are chain and lattice structures. Hence the idea of an ionic radius is unhelpful. An alternative approach is consider the volume of the formula unit, V_m. When the crystal structure of a material is known, then this can be calculated readily from the cell volume, V_{cell}, and the number of formula units within that cell, Z (Eq. 2.71):[31]

$$V_m = \frac{V_{cell}}{Z}.\tag{2.71}$$

In turn, V_{cell} can be calculated from the cell dimensions, a, b, and c and angles, α, β, and γ, which in the most general case is given by Eq. 2.72:

$$V_{cell} = abc\sqrt{1 - cos^2\alpha - cos^2\beta - cos^2\gamma + 2cos\alpha cos\beta \cos \gamma}.\tag{2.72}$$

In the absence of a known crystal structure then a measurement of the density of the material, ρ, of molar mass M can provide an estimate of V_m (Eq. 2.73):

$$V_m\left(nm^3\right) = \frac{M\left(g \, mol^{-1}\right)}{602.2\rho\left(g \, cm^{-3}\right)}.\tag{2.73}$$

The Kapustinskii equation, Eq. 2.68, can be adapted to be based upon the molar volume, V_m, for any salt of formula M_pX_q (Eq. 2.74).[29] The parameters α and β were refined from experimental values of lattice energies to values of: $\alpha = 138.7$ kJ mol^{-1} nm and $\beta = 27.6$ kJ mol^{-1}.

$$U = v|z_+||z_-|\left(\frac{\alpha}{\sqrt[3]{V_m}} + \beta\right).\tag{2.74}$$

By taking the standard radii for the cations of groups 1 and 2 as volumes of the cations, V_+ (Table 2.15), then V_m can be apportioned between the cations and anions based upon the molecular formula, and provide estimates of the volumes of the anions, V_- (Table 2.16).

The Kapustinskii equation (2.68) has been reformulated in a different way to remove some empirical constants, using the Eqs 2.75–2.77:

$$U = \frac{AI}{V_{norm}^{\frac{1}{3}}}\tag{2.75}$$

$$\text{where } V_{norm} = \frac{V_m}{2I}\tag{2.76}$$

$$\text{and } I = \frac{1}{2}\sum n_i z_i^2.\tag{2.77}$$

The term, A is the same constant as in 2.68, that is 121.39 kJ mol^{-1} nm. The new term, I, replaces the quantity $|vz_+z_-|$. It is similar in form to the definition

[31] L. Glasser, 'Thermodynamics of Condensed Phases: Formula Unit Volume, V_m, and the Determination of the Number of Formula Units, Z, in a Crystallographic Cell', *J. Chem. Educ.*, 2011, **88**, 581–5.

Table 2.15 *Volumes, V_+, of groups 1 and 2.*

Cation	V_+/nm^3
Li$^+$	0.00199
Na$^+$	0.00394
K$^+$	0.00986
Rb$^+$	0.01386
Cs$^+$	0.01882
Mg^{2+}	0.00199
Ca^{2+}	0.00499
Sr^{2+}	0.00858
Ba^{2+}	0.01225

Table 2.16 *Volumes, V_-, of a selection of anions.*

Cation	V_-/nm^3
F$^-$	0.025
OH$^-$	0.032
Cl$^-$	0.047
O^{2-}	0.043
O$_2^{2-}$	0.052
CO$_3^{2-}$	0.061
HCO$_3^-$	0.064
PO$_4^{3-}$	0.090
SO$_4^{2-}$	0.091

[32] L. Glasser and H. D. B. Jenkins, 'Lattice Energies and Unit Cell Volumes of Complex Ionic Solids', *J. Am. Chem. Soc.*, 2000, **122**, 632–8.

of the ionic strength in solutions, where m_i is the ionic molality (mol kg^{-1}) (Eq. 2.78):[32]

$$and\ I\ /mol\ kg^{-1} = \frac{1}{2}\sum m_i z_i^2. \tag{2.78}$$

A similar approach has also been adopted to relate the standard entropy of materials to their density and formula volume (Eq. 2.75). The regression analysis provides estimates of the constants k and c which varied somewhat between materials (Table 2.17). Nevertheless, taken with the methodology for deriving the lattice energy and enthalpy, then the $\Delta_{\text{lattice}}G^\circ$ values for a material can be predicted.[33]

[33] H. D. B. Jenkins and L. Glasser, 'Standard Absolute Entropy, S°$_{298}$, Values from Volume or Density. 1. Inorganic Materials', *Inorg. Chem.*, 2003, **42**, 8702–8.

$$S^\circ_{298} = kV_m + c. \tag{2.79}$$

Returning to the example reaction in Eq. 2.70, the exchange of ions between two ionic solids, we can now estimate $\Delta_r G^\circ$ (Table 2.18). For this solid-state reaction, unsurprisingly, the entropy change is calculated to be essentially zero. In this case it is the $\Delta_{\text{lattice}}H^\circ$ which determines the direction of reaction. This mode of calculation affords $\Delta_r G^\circ$ value of 90.6 kJ mol^{-1}. The drive in the reaction is to maintain the solid with the smallest V_m, which is the combination of the smallest cation and anion. The loss in lattice enthalpy of going from Na_2O to Cs_2O (320 kJ mol^{-1}) cannot be fully recovered by switching from Cs_2CO_3 to Na_2CO_3 (229 kJ mol^{-1}).

Returning back to Eq. 2.50, the dissolution of a metal halide into water, we can compare the free energy change of lattice formation against the hydration of the

Table 2.17 *Values of the empirical constants k and c relating standard entropy to V_m.*

Material	k	c
Inorganic ionic salts	1,360	15
Ionic hydrates	1,579	6
Minerals	1,262	13

Table 2.18 *Prediction of the thermodynamics of the reaction between Na_2O and Cs_2CO_3 based upon the molar volume, V_m.*

Solid	V_m	U	$\Delta_{\text{lattice}}H^\circ$	S°
	nm^3	kJ mol^{-1}	kJ mol^{-1}	J K^{-1} mol^{-1}
Na_2O	0.05088	2,411.4	$-2,413.7$	84.2
Na_2CO_3	0.06888	2,195.8	$-2,194.3$	108.7
Cs_2O	0.08064	2,091.8	$-2,094.1$	124.7
Cs_2CO_3	0.09864	1,966.7	$-1,965.2$	149.2

cation and anion (Table 2.19). It is evident that the differences between these two large terms are relatively small. Indeed they are smaller than the entropy terms, let alone the much larger enthalpies. Thus it is not surprising that our ΔG^o values do not follow a monoatomic trend. All values are very favourable for dissolution, but the most favourable values are for Li and Cs. This mirrors the molar solubilities of the metal chlorides, all of which are very high.

2.7 Oxidation State Stability

In a Born–Haber cycle for the formation of a metal halide from the elements, such as that shown in Fig. 2.19, the thermodynamic stability of that halide will come from a balance between the energy required to create the gaseous ions and the energy gained in forming the lattice from those ions. Hence the ionization energies provide a first guide to the stability of oxidation states. The first eight ionization energies of the group 16 elements in periods 2 and 3 are shown in Fig. 2.20. It is evident that for both elements the values rise to extremely high values ($> 10^4$ kJ mol^{-1}) at the seventh ionization, corresponding to breaking into the core shell of the $1s$ orbitals. This rise could not be offset by chemical bonding, and so the maximum oxidation state is six. The equivalent ionization energies are lower for sulphur than oxygen, and would be expected for the element with the higher atomic radius. The break point in oxidation states will be attained for the energy gained in reactions with more electronegative elements. For sulphur, this would be with oxygen and fluorine. Sulphur can attain the oxidation state of six for both of these elements, as in SF_6, SO_3, and SO_4^{2-}. We can anticipate a lower maximum positive oxidation state for oxygen, occurring for its fluorides. The maximum occurs for O^{II}, as in OF_2.

A particularly useful means of comparing oxidation state stabilities in solutions is from the electrochemical potentials of reduction-oxidation or **redox**,

Table 2.19 *Estimated $\Delta_{dis} G^o$ from lattice and hydration values versus molar solubility of group 1 chlorides.*

Salt	$\Delta_{dis} G^o$ kJ mol^{-1}	Solubility M (303 K)
LiCl	−57	20.3
NaCl	−23	6.2
KCl	−23	5.0
RbCl	−30	8.1
CsCl	−39	11.7

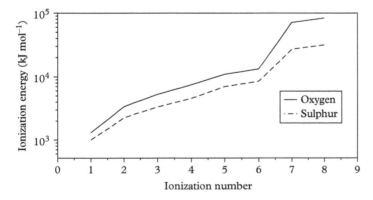

Figure 2.20 *Ionization energies of oxygen and sulphur.*

processes. Conventionally these are the **standard electrode potentials**, E^o, tabulated as **reduction potentials**, as in Eq. 2.80 for the hydrogen reference half-reaction at an activity of the hydrogen ion of 1, and a pressure of 10^2 kPa at 298.15 K.

$$H^+_{(aq)} + e = \frac{1}{2}H_{2(g)}. \tag{2.80}$$

For the electrolysis of water, the partner half-reaction is Eq. 1.81:

$$\frac{1}{4}O_{2(g)} + H^+_{(aq)} + e = \frac{1}{2}H_2O_{(l)}. \tag{2.81}$$

The key link between the standard reduction potential and the change in the standard free energy is as in Eq. 2.82, where n is the number of electrons involved in the reaction, and F is the **Faraday constant** (96,485 C mol^{-1}):

$$\Delta G^\circ = -nFE^\circ. \tag{2.82}$$

Electrode potentials can be used to show thermodynamic stability in voltage equivalent, or Frost, diagram for example for copper as in Fig. 2.21. These plot values of nE° against oxidation number. As can be seen from Eq. 2.82, this is equivalent to plotting $\Delta G^\circ/F$.[34] The origin in Fig. 2.21 is taken by copper in its standard state of the metal in zero oxidate state. The standard reduction potential for Cu^{2+} is 0.34 eV. So for oxidation from Cu(0) to Cu(II), $nE^\circ = 2 \times 0.34$. The other known standard electrode potential is for $Cu^{2+/+}$ (0.15 eV). This is a chemical reduction from Cu^{2+}, and hence the change in oxidation number is -1. Hence the nE° value for Cu^+ is 0.68—0.15 = 0.53 V. This value is above the mean for copper metal and Cu^{2+}. Under standard conditions (pH 0) Cu^+ should disproportionate according to Eq. 2.83:

$$Cu^+_{(aq)} = \frac{1}{2}Cu + \frac{1}{2}Cu^{2+}_{(aq)}. \tag{2.83}$$

However, practical situations occur under non-standard conditions and the ΔG° and thus the E° of reactions which involve hydrogen ions will change with pH. This can be accommodated by a **Nernst equation**, for example Eq. 2.84, where Q is the product of the activities of the reagents divided by those of the reagents:

$$E = E^\circ - \frac{RT}{nF}\ln(Q) = E^\circ - \frac{2.3026RT}{nF}\log_{10}(Q). \tag{2.84}$$

Taking the temperature as standard, this equation simplifies to Eq. 2.85 for the reduction of the hydrogen ion, for which E° is 0, as in Eq. 2.80:[35]

$$E\left(H^+/\frac{1}{2}H_2\right) = 0 - 0.0592\log_{10}\left(\frac{1}{a_{H^+}}\right) = -0.0592\text{pH}. \tag{2.85}$$

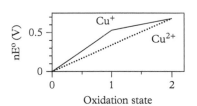

Figure 2.21 *Frost diagram for copper.*

[34] Frost diagrams are drawn as oxidations, so the sign of the reduction potential must be changed.

[35] pH is defined as $-\log_{10}a_{H^+}$.

Similarly, the reduction of O_2 (Eq. 2.81), for which E° is 1.23 V, can be quantified as a function of pH by Eq. 2.86:

$$E\left(\frac{1}{2}O_2/H_2O\right) = 1.23 - 0.0592 log_{10}\left(\frac{1}{a_{H^+}}\right) = 1.23 - 0.0592 pH. \quad (2.86)$$

Equations 2.85 and 2.86 set the theoretical boundaries over varied pH vales of electrode potentials for species to be stable in water (Fig. 2.22). This plot assumes that there is no kinetic barrier preventing the equations 2.80 and 2.81 from occurring. Such barriers do exist and can be represented by an overpotential. This represents an increased barrier in voltage (and hence in energy) to be overcome for water to be oxidized to O_2 or reduced to H_2. Hence the dotted lines might be expected to provide a practical range to the stability of ions in aqueous media. The standard electrode potential of $[Co(OH_2)_6]^{3+}$ (Eq. 2.87) is sufficiently high that, thermodynamically, this ion should oxidize water according to Eq. 2.88. In practice this process is extremely slow due to a substantial kinetic barrier. Such barriers can be extremely useful. For example, as in Eq. 2.89, the potential for the reduction $Al^{3+}_{(aq)}$ is firmly in the range that aluminium should reduce water and dissolve. However, the oxidation passivates the metal surface as it forms an impervious and stable oxidic layer preventing reaction of the bulk metal. A different situation arises under strongly alkaline conditions under which the oxidic layer may dissolve as hydroxides.

$$Co^{3+}_{(aq)} + e = Co^{2+}_{(aq)}, \; E^\circ + 1.92 \, V \quad (2.87)$$

$$Co^{3+}_{(aq)} + \frac{1}{2}H_2O_{(l)} = Co^{2+}_{(aq)} + \frac{1}{4}O_{2(g)} + H^+_{(aq)} \quad (2.88)$$

$$Al^{3+}_{(aq)} + 3e = Al_{(s)}, \; E^\circ = -1.66 \, V. \quad (2.89)$$

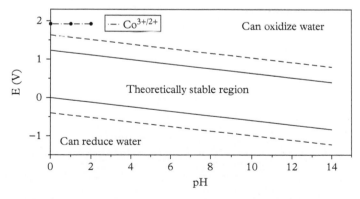

Figure 2.22 *Limits of stability of ions in water. Dashed lines with a 400 meV overpotential.*

The pH range for $[Co(OH_2)_6]^{3+}$ is shown in Fig. 2.22 as terminating at the value of 2. This had been considered a limiting value of the stability of this ion in diagrams showing the stability of species over a map of electrode potential and pH, called **Pourbaix diagrams**, invented by Marcel Pourbaix. In practice these diagrams vary with temperature and metal concentration. The nature of these diagrams is still a matter of study due to the difficulty of speciation mixtures of solids and solution and the thermodynamics of the reactions, as exemplified by studies on cobalt.[36,37]

A partial Pourbaix diagram at 298.15 K can be constructed as in Fig. 2.23, again with a plot of potential, E on the hydrogen scale plotted against pH. The theoretical range for water stability is shown as dotted lines. The updated thermodynamic data indicate that $Co^{3+}_{(aq)}$ is not within the plotted scale. Instead $Co^{IV}O_2$ is the most oxidized species. The most reduced species is cobalt metal. The line separating the region from $Co^{2+}_{(aq)}$ is horizontal. This corresponds to the half-reaction in Eq. 2.90:

$$Co^{2+}_{(aq)} + 2e = Co_{(s)}, \; E^\circ - 0.288 \text{ V}. \tag{2.90}$$

Since the hydrogen ion is not included in this equation, there is no dependence on pH. The dividing line is when the two cobalt species are in equilibrium. Since the state of the cobalt is as the solid metal, its activity is unity. The cobalt concentration is chosen to be 10^{-6} M. Hence, from Eq. 2.84, the potential of the cell for this 2-electron change is given by Eq. 2.91:

$$E\left(Co^{2+}/Co\right) = -0.288 - \frac{0.0592}{2} log_{10}\left(\frac{1}{a_{Co^{2+}}}\right) = -0.47 \text{ V}. \tag{2.91}$$

The division between $Co^{2+}_{(aq)}$ and $Co(OH)_2$ can be seen to be a vertical line, and corresponds to Eq. 2.92:

$$Co^{2+}_{(aq)} + 2H_2O = Co(OH)_{2(s)} + 2H^+_{(aq)}. \tag{2.92}$$

[36] J. Chivot, L. Mendoza, C. Mansour, T. Pauporté and M. Cassir, 'New insight in the behaviour of Co-H$_2$O system at 25–150°C, based on Revised Pourbaix diagrams', *Corr. Sci.*, 2008, **50**, 62–9.

[37] M. Bajdich, M. García-Mota, A. Vojvodic, and J. K. Nørskov, and A. T. Bell, 'Theoretical Investigation of the Activity of Cobalt Oxides for the Electrochemical Oxidation of Water', *J. Am. Chem. Soc.*, 2013, **135**, 13521–30.

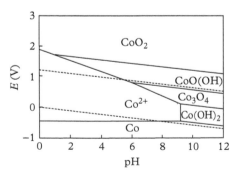

Figure 2.23 *A partial Pourbaix diagram for Co at 25 °C and 10^{-6} M.*

There is no redox process involved and so the change is only dependent upon the pH. The equilibrium for the reaction will be given by Eq. 2.93:

$$K = \frac{a_{Co(OH)_2} a_{H^+}^2}{a_{H_2O}^2 a_{Co^{2+}}} = \frac{a_{H^+}^2}{a_{Co^{2+}}}. \tag{2.93}$$

Since water and solid $Co(OH)_2$ have unity activities, the pH at equilibrium can be calculated form the equilibrium constant and the cobalt ion concentration.

The remaining lines are sloped, with several having the same slope as those for the electrolysis of water. This is called the **Nernst slope** (Eq. 2.94). For example, in the equilibrium between cobalt and Co(II) hydroxide, this ratio will also be unity (Eq. 2.95), but will be twice as large for the boundary in Eq. 2.96. However, it must be added that the ability to fully reach Co(IV) in such a system is open to question.

$$\text{Nernst slope} = -\frac{2.3026RT}{F} \frac{n_{H^+}}{n_e} \tag{2.94}$$

$$Co(OH)_{2(s)} + 2e + 2H^+ = Co + 2H_2O \tag{2.95}$$

$$CoO_{2(s)} + 2e + 4H^+ = Co^{2+} + 2H_2O. \tag{2.96}$$

2.8 Terms and States

When we discussed the lifetime of species in the atmosphere (Section 1.8) we referred to Russell Saunders terms, particularly for the oxygen atom. Being in group 16, the valence shell of atomic oxygen will have a configuration of $2s^2 2p^4$ in its states of lowest energy. For the $2s^2$ subshell there is only one possible configuration in which the two electrons are paired in the single orbital. There are choices to be made for the $2p^4$ subshell. The choices are called **microstates**. In the (p) subshell, with orbitals having the l, orbital, quantum number of 1, we have three orbitals which differ in having magnet quantum number, m_l, taking values of 1, 0, and -1. Each electron can adopt one of two spin numbers, $m_s \pm 1/2$. As can be seen in Fig. 2.24, we can envisage the two holes in the subshell. There are six choices over where to locate the first hole, leaving five for the location of the second one. Since we cannot distinguish between electrons, or between holes, the total number of combinations, or microstates is equal to $\frac{6.5}{2}$, that is, 15.

Two possible microstates for p^4 are depicted in Fig. 2.24. The relative energy of these can be gauged using Hund's rules of maximum multiplicity:

1. The most stable state will be that with the maximum spin quantum number, M_S.
2. Then for states with the same M_S value, the most stable state will be that with the largest M_L value.

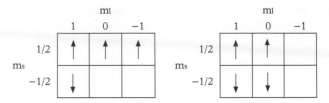

Figure 2.24 *Microstates of a p^4 subshell.*

Table 2.20 *Term symbols for the L values.*

L	Symbol
0	S
1	P
2	D
3	F
4	G
5	H
6	I

The left-hand example in Fig. 2.24 has two unpaired electrons. For this one the sum of the m_s values $= \frac{3}{2} - \frac{1}{2}$. Hence the overall spin quantum number of the atom, M_S equals 1. These overall values are presented with upper-case values, M_S in this case. The maximum overall spin quantum number, S, for this high-spin p^4 state will be 1, and there will be three possible M_S states: $+1, 0$, and -1. States are labelled by their spin multiplicity, given by $2S+1$. In this case the spin multiplicity is 3, and the state is termed a triplet. For this microstate the overall orbital quantum number, $M_L = 2 - 1$, so M_L also equals 1. In parallel, this is the maximum value of M_L that can be achieved, hence the L value for this state overall is also 1, and there are associated M_L values of 1, 0, and -1. The labelling of states parallels that of orbitals and so is based upon that used for orbitals, but in upper case (Table 2.20). As a result, the term symbol here is ^3P. Given the three possible values for M_L and M_S, this state accounts for nine of the fifteen microstates. In the highest spin state with electrons in different orbital space, the inter-electronic distance is relatively high and so the **pairing energy** is minimized. Also, the number of exchangeable electrons (ones with the same m_s) is maximized. Since this is a mechanism for increasing the space available for electrons, this **exchange energy** also makes the state more stable.

The second microstate in Fig. 2.24 is a low-spin state, with a maximum M_S value of 0. It has an M_L value of 2, and hence it has a term symbol of ^1D. That will incorporate five possible M_L values—2, 1, 0, -1, -2—and so account for a further five microstates. Only one microstate remains, and this must be due to a singlet state for both L and S, so must be ^1S. By Hund's second rule, this will be of higher energy than the ^1D.

With this approach, the term symbols for the most stable states of ions like those of cobalt, discussed in Section 2.7, can be assessed. Using Hund's first rule, the maximum S value for a free Co^{3+} ion with a d^6 subshell can be shown to be 2 (Fig. 2.25). The L value for this configuration will be that possible for the odd electron, since those of the first five will cancel out to 0. Hence from an L value of 2, we can ascribe the term symbol for the most stable state to be ^5D. Adding an extra electron to create the d^7 Co^{2+} ion will reduce the S value by $\frac{1}{2}$ to $\frac{3}{2}$. The L value will increase by 1 to 3. Hence the term symbol for the most stable Co^{2+} ion will be ^4F.

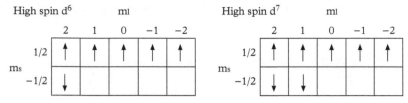

Figure 2.25 *High spin states for $Co^{2+/3+}$.*

2.9 Embodied Energy, Carbon, and Water

The ecological factors associated with converting a mineral to a material are manifold. They will include extraction from the Earth together with the associated infrastructure to allow that and transferral of the mineral to processing. There will be further steps to convert a spectrum of materials into a finished product such as a refrigerator, car, or laptop. Three key factors of global significance are: the energy required to create a unit of a material (generally per kg), the CO_2e evolved, and the water required. These are called **embodied** parameters. The energy is not just that required from the thermodynamics of the equations, but includes material losses, and processing energy. Similarly, the evolution of CO_2e and the use of water include but extend beyond those of the chemical equations. In practice then **embodied energy** (EE), **embodied carbon** (EC), and **embodied** or **virtual water** (VW) are dependent upon a particular installation and will vary from source to source, and plant to plant. The relationship between EE and EC will depend strongly upon the primary sources of energy at that installation. There are databases and collections of these values for a wide range of materials, giving global means of ranges. Different data sources can differ in their valuations, which may differ by several tens of percentage points. Values for virtual water are still very limited. Even with these caveats, these factors provide a basis for comparing alternative materials, and give targets for improving processes.

The embodied factors do not provide the total requirements for construction of a final consumer product and are not **life-cycle analyses** (LCA). But they provide the basis for estimating the environmental burden of a bill of materials. From the known mass (m_i) of the components of the material, the total demands can be estimated, for example for the total embodied energy as in Eq. 2.97:

$$EE_{material} = \sum (m_i EE_i). \tag{2.97}$$

2.10 Questions

1. Show the radioactive decay pathway for the isotope $^{232}_{90}Th$ based upon following order of decay emissions: $\alpha, \beta, \beta, \alpha, \alpha, \alpha, \alpha$.

2. Estimate the Z_{eff} of sodium and caesium using the Slater rules, and predict their first ionization potentials (eV). Compare the results between the two elements.

3. Place the following elements and compounds on a van Arkel–Ketelaar triangle:

   Cs, CsF, F_2, WC, AgBr, CuZn, NbZr, OsO_4.

   Comment on the bonding in the transition metal-containing compounds.

   [*Allen electronegativities,* χ_A: *Cs 0.659; O 3.610; F 4.193; C 2.544; Br 2.685; Cu 1.85; Zn 1.59; Zr 1.32; Nb 1.41; Ag 1.87; W 1.47, Os 1.65.*]

4. (i) Construct a Born–Haber cycle for the dissolution of solid LiF in water.

   (ii) Calculate values for the ΔG° of the dissolution of the ions and for the formation of the solid from the gaseous ions. Hence estimate ΔG° for dissolution of LiF in water.

   (iii) Comment on the results given that the solubility of LiF at 25 °C is reported as 1.3 g/L.

   [*Properties of lithium and fluorine. Gaseous ions:* $\Delta_f H^\circ$ *(kJ mol^{-1})* Li^+ *686,* F^- *−255;* S° *(J mol^{-1} K^{-1}):* Li^+ *133.1,* F^- *145.6; ionic volumes (nm^3):* Li^+ *0.00199,* F^- *0.025; ionic radii (pm):* Li^+ *69,* F^- *181; solvation radius correction (pm):* Li^+ *61,* F^- *5.*]

5. (i) Construct a Frost diagram for two actinide elements from the following E° values in acidic aqueous solution.

   [*Uranium:* $U^{3+/0}$ *−1.642;* $U^{4+/3+}$ *−0.577;* UO_2^{2+}/U^{4+} *0.39;* $UO_2^{2+/+}$ *0.16 V.*

   Americium: $Am^{2+/0}$ *−1.95;* $Am^{3+/0}$ *−2.07;* $Am^{4+/3+}$ *2.62;* AmO_2^+/Am^{4+} *0.82;* $AmO_2^{2+/+}$ *1.59 V.*]

   (ii) Comment on the stability of the following oxidation states of these two elements in acidic solution: (III), (IV), (V).

6. Construct a partial Pourbaix diagram for iron at 25 °C and a concentration of 10^{-6} M, using the data below.

   E° *values:* Fe_2O_3/Fe^{2+} *0.728;* $Fe^{3+/2+}$ *0.771;* Fe^{2+}/Fe_s *−0.44 V. Log K* *($2Fe^{3+}/Fe_2O_3$) 1.43.*

7. (i) What is the electronic configuration of the oxygen atom in the ground state?

   (ii) Derive the Russell–Saunders term symbols for the ground state and an excited state.

   Extension: by considering the microstates available with the electronic configuration, identify a second excited state.

8. A model catalytic converter has been fabricated from a low-alloy steel container (1 kg), an alumina honeycomb support for the catalyst (0.5 kg), and the platinum catalyst (1 g).

(i) Using the data below, estimate the embodied energy and carbon of the materials in the catalyst structure.

(ii) Calculate the changes in these values if (a) the platinum is substituted by palladium, and (b) the low-alloy steel is substituted by stainless steel.

(iii) Comment on these results.

Embodied energy (MJ/kg): low-alloy steel 33; stainless steel 85; alumina 26; platinum 387,000; palladium 304,000. Embodied carbon (kg CO_2e/kg): low-alloy steel 2.0; stainless steel 5; alumina 2.1; platinum 33,000; palladium 25,000.

3 Earth: Minerals to Materials

In this chapter we can examine most of the elements on Earth as they can be most readily found in the Earth's crust. We assess how they can be converted from source to material, and the impact of these processes on the environment. We will conclude with a comparison of the environmental impact of some of the important materials in our civilization.

3.1 *s*-Elements

3.1.1 Lithium and Sodium

The two lightest elements in group 1 differ most markedly by their variation in abundance in the Earth's crust.

Sodium is the seventh most abundant element (2.277%) and does not figure on the Risk List of the BGS. The vast majority of the uses of sodium are in commonly used salts: $NaCl$, Na_2CO_3 (soda ash), $NaHCO_3$ (baking soda), $NaOH$, $NaNO_3$, Na_3PO_4, Na_2HPO_4, $Na_2S_2O_3 \cdot 5H_2O$, and $Na_2B_4O_7 \cdot 10H_2O$ (borax). Uses include glass, soaps, water softeners, and food. Future supply is not an issue. For example, even an annual production of *c.*54 Mt represents the region of 0.1% of the known reserves. The annual production of $NaCl$ is in the region of 280 Mt. By comparison, the annual production of sodium metal is $\approx$ 100 kt. The majority of sodium metal is utilized in chemical synthesis. Products include indigo dyes (blue denim) and sodium methoxide (for biodiesel manufacture).

Lithium on the other hand is present only at 16 ppm and is in an intermediate risk zone according to the BGS. Annual mine production is in the region of $\approx$ 45 kt, with reserves in the region of 16 Mt. The most abundant deposit is as the mineral *spodumene*, $LiAlSi_2O_6$. Extraction from this mineral and a related mineral, *petalite*, $LiAlSi_4O_{10}$, is also reported. The most significant proportion of production is from brines, a mixture of metal chlorides. These occur in arid regions, such as the Atacama and Nevada deserts, as might be anticipated given their solubility in water. The carbonate, Li_2CO_3, is the most common basic reagent. Developments in battery technology (Chapter 5) mean that lithium supply is a priority. However, presently this amounts to about 40% of the use of lithium and its compounds, with ceramics and glasses the destination for *c.*32% and lubrication for *c.*9%.

Elements of a Sustainable World. John Evans, Oxford University Press (2020). © John Evans.
DOI: 10.1093/oso/9780198827832.003.0003

The two metals both have a body-centred cubic structure (*bcc*), shown in part in Fig. 3.1. Each atom has eight nearest neighbours at the vertices of the cube. Above each of the six faces of the cube there is a second coordination sphere 15% longer than the inner one. The packing efficiency of spheres in a simple (primitive) cube is 52%, but the body-centring increases this to 68%, approaching the maximum of the close-packed structures (74%).

The production of the metals from readily available salts is an uphill task, as expected for such electropositive elements. The two reactions in Table 3.1 are alternatives.

The carbothermal reduction of Na_2CO_3 by carbon (the most readily available of reductants) was developed in the nineteenth century, operating at 1,100 °C. The equilibrium constant at constant pressure, K_p, can be calculated for the Gibbs free energy change for the reaction as in Table 3.1 (Eq. 3.1):

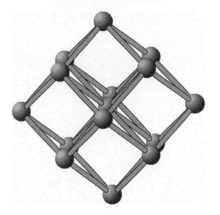

Figure 3.1 *Partial bcc structure.*

$$\Delta_r G_{298} = -RT \ln K_p. \tag{3.1}$$

The standard ΔG values for the elements are defined as 0, and thus for this reaction, K_p can be estimated from Eq. 3.2 and on to Eq. 3.3:

$$RT \ln K_p = -\frac{3\Delta G^{\circ}_{CO}}{2} - \frac{\Delta G^{\circ}_{Na_2CO_3}}{2} \tag{3.2}$$

$$\ln K_p = \frac{-205.8 + 522.8}{8.314.298.1000} = -127.7. \tag{3.3}$$

The result is a vanishingly small value of the equilibrium constant. The reaction has a large positive entropy due to the production of CO, and also of sodium (bp 882.9 °C). Under those conditions the reaction becomes feasible, and the products can be removed from equilibrium as vapours, an example of the Le Chatelier principle. However, this is at the cost of considerable input of heat to achieve the reaction temperature, as well as the evolution of CO.

The present commercial process is based upon the second reaction in Table 3.1, but is carried out with an electrochemical energy input. It is necessary to carry out the reaction in a molten salt to allow the ions to flow between the

Table 3.1 *Prediction of the thermodynamics for the production of Li and Na under standard conditions.*

Reagents	Products	$\Delta_r H^{\circ}$ kJ mol^{-1} Lithium	$\Delta_r H^{\circ}$ kJ mol^{-1} Sodium
$0.5\ M_2CO_{3(s)} + C_{(s)}$	$M_{(s)} + 1.5CO_{(g)}$	402	316
$MCl_{(s)}$	$M_{(l)} + 0.5Cl_{2(g)}$	411	414

electrodes and avoid reaction with water. In a Downs cell, the anode is constructed of carbon to carry out the following reaction (Eq. 3.4):

$$2Cl_{(l)}^- \rightarrow Cl_{2(g)} + 2e \ (E° \ 1.36 \ V).$$ (3.4)

At the iron cathode, the sodium is formed by Eq. 3.5:

$$2Na_{(l)}^+ + 2e \rightarrow 2Na_{(l)} \ (E° \ -2.71 \ V).$$ (3.5)

The temperature of the melt is lowered from the melting point of sodium chloride (801 °C) by mixing it with 58% $CaCl_2$, allowing operation at $\approx$ 600 °C. This saves energy and allows sodium (mp 97.8 °C) to separate and float on the surface of the melt above the cathode. A membrane in the melt keeps the Cl_2 in a separate compartment to be collected from the anode. The voltage can be set at a level which does not reduce the much more electropositive calcium ($E°$ −2.87 V). In fact sodium is collected at *c.*115 °C, so the calcium (mp 842 °C) can be filtered from molten sodium. The current efficiency is $\approx$ 85%, and the electrical energy required is $\approx$ 10 kWh kg^{-1} of sodium (828 kJ mol-Na^{-1}). On the basis of mean electric supply conversion factors for the UK in 2018, this involves 65 g CO_2e per mol of sodium (23 g).

Compared to sodium, lithium has more demanding standard higher electrode potential (−3.05 V) and melting and boiling points (180.5 and 1324 °C, respectively). A similar cell system is used as for the electrochemical manufacture of sodium, but the melt used this time is $\approx$ 40 mole percent of KCl with LiCl. This is a liquid at 400 °C and so the overall cell reaction in Eq. 3.6 can be achieved:

$$LiCl_{(l)} \rightarrow Li_{(l)} + \frac{1}{2}Cl_{2(g)}.$$ (3.6)

[1] G. Hammond, C. Jones, *Inventory of Carbon and Energy, v. 2*, 2011, University of Bath.

The energy consumption required is 35 kWh kg^{-1} of lithium, which is slightly higher than that required for sodium manufacture on a molar basis (874 kJ mol Li^{-1} or 69 g CO_2e mol-Li^{-1}). The embodied energy of lithium is reported as 853 MJ/kg, compared to $\approx$ 20 MJ kg^{-1} for steel).[1]

3.1.2   Magnesium

Magnesium is considered to be present in higher abundance than sodium in the Earth's crust (2.81%). However, it is similarly placed on the BGS Risk List to lithium due to a restricted set of supply streams. Based on MgO (magnesia) equivalents, the production of magnesium salts amounted to 27 Mt, with estimated reserves of 7,800 Mt; hence long-term supply should not be in question. The most utilized sources are the carbonates *magnesite*, $MgCO_3$, and *dolomite*, $CaMg(CO_3)_2$, and also brines, $MgCl_2$. Magnesia itself is one of the most used commodities in refractory materials such as furnace linings. There is also increasing use in waste-water treatment, animal feeds, and fertilizers.

Elemental magnesium does have a close-packed structure—hexagonal close packed (hcp) (Fig. 3.2). This figure shows three close-packed layers. The highlighted atom can be seen to have six nearest neighbours in the plane of the paper. Above that plane there are an additional three nearest neighbours. The angle of view shows that there is another triangle of coordinated atoms below the middle hexagonal coordination. Hence the total nearest-neighbour coordination is 12. Also the stacks of the close-packed layers alternate in location following a pattern *..ABAB.....*

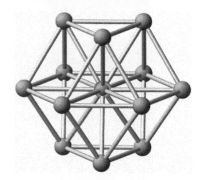

Figure 3.2 *Partial hcp structure.*

Production of magnesium metal is an order of magnitude larger than that of sodium, and can be expected to increase due to its demand in lightweight alloys for structural use. In 2017 world production was ≈ 1.1 Mt. The vast majority in manufactured in China, with the USA and Russia being the other main sources.

As for the group 1 metals, the options for reduction are chemical and electrochemical. Again, taking the values of $\Delta_f H°$ of the reagents and products, all of the processes are endothermic (Table 3.2). The carbothermal method is little used, substantially due to the difficulty in avoiding reoxidation of the metal by CO on cooling.

The least endothermic is the silicothermic (or Pidgeon), process in which the reductant is an iron silicide. That and *dolomite* are the precursors. Calcination is required of the *dolomite* which will generate CO_2 (Eq. 3.7):

$$CaCO_3 \cdot MgCO_3 \rightarrow CaO \cdot MgO + 2CO_2. \tag{3.7}$$

This appears to be the predominant process used in China. To drive the reaction forward, the reaction is performed at $\approx 1,160\,°C$ under reduced pressure ($c.10^{-4}$ atm). Under those conditions, magnesium (mp $650\,°C$, bp $1090\,°C$) can be separated as a vapour from solid iron (mp $1,538\,°C$, bp $2,861\,°C$) and the ceramic, calcium silicate. On that basis, for every kg of magnesium produced, 3.6 kg CO_2, 2.3 kg of iron, and 6.9 kg of Ca_2SiO_4 (also known as slag) will also be generated; a plant has a daily production in the order of 50 kg of magnesium. The 'slag' may be reprocessed for remediation of acid mine drainage, but often

Table 3.2 *Prediction of the thermodynamics for the production of magnesium under standard conditions.*

Reagents	Products	$\Delta_r H°/\text{kJ mol}^{-1}$
$MgO_{(s)} + C_{(s)}$	$Mg_{(s)} + CO_{(g)}$	491
$CaO \cdot MgO_{(s)} + 0.5\,FeSi_{(s)}$	$Mg_{(s)} + 0.5Ca_2SiO_{4(s)} + 0.5\,Fe_{(s)}$	120
$MgCl_{2(s)}$	$Mg_{(l)} + Cl_{2(g)}$	646

ends in landfill sites. Within the overall Pidgeon process, the manufacture of the iron silicide in an arc furnace can consume $\approx 1/3$ of the overall energy required. A newer process, the Mintek Thermal Magnesium Process, operates at higher temperatures (1,600–1,700 °C) and pressures ($c.$0.85 atm). It can operate with a much higher daily productivity (100 t). It also includes the option of using aluminium as the reductant; however, aluminium itself has a high embodied energy (EE = 211 MJ kg-Al^{-1}).

The principal alternative method is electrolysis of the molten salt at 700–800 °C (1 atm); an example of this is the Dow process. This starts from a brine mixture of metal halides, from which calcium oxide is first used to precipitate out magnesium hydroxide (Eq. 3.8):

$$Mg_{aq}^{2+} + CaO \rightarrow Ca_{aq}^{2+} + Mg(OH)_2. \tag{3.8}$$

Second, the hydroxide is converted into the hydrated chloride (Eq. 3.9):

$$Mg(OH)_2 + HCl \rightarrow MgCl_2 \cdot nH_2O. \tag{3.9}$$

The Dow process involves dehydration to $MgCl_2 \cdot H_2O$, which requires less energy than full dehydration. However this degrades the carbon anode more quickly and contaminates the Cl_2. The electrochemical step of these cells consumes 37.8–47.5 MJ kg-Mg^{-1} (920–1,150 kJ mol Mg^{-1}).

Overall these are very energy-intensive processes; the aim of their use is that lower-density structural materials will afford long-term gains in the energy required, particularly for transport.[2] Estimates about the effect on the environment vary, and also suggest that the type of manufacturing process does have an influence on the embodied energy of an ingot of magnesium.[3] Estimates published in 2008 indicated that the Pidgeon process had the highest embodied energy (366 MJ kg-Mg^{-1}), with an electrochemical method using an aluminium-manufacture type of electrochemical cell faring rather better (272 MJ kg-Mg^{-1}). Other chemical reductions related to the Mintek process could reduce this further to ≈ 200 MJ kg-Mg^{-1}, similar to the value reported for aluminium metal (211 MJ/kg-Al^{-1}). In 2004, the GHG emission (embodied carbon) associated with the Pidgeon process overall was estimated at 42 kg CO_2eq kg-Mg^{-1},[4] but now strategies have been developed aiming at a $c.$40% reduction in this impact.[2–4]

3.2   *p*-Elements

The 'earthy' set of seventeen elements are those unshaded in Fig. 2.6. They comprise the stable elements of groups 13–16, with the exception of the main atmospheric gases of nitrogen and oxygen. The electronegativity extremes (Allen χ values) are carbon at the top of group 14 (2.64), and aluminium (1.61) second in group 13. The former is the classic element for which covalent bonding

[2] H. Li, W. Zhang, Q. Li, and B. Chen, 'Updated CO_2 emission from Mg production by Pidgeon process: Implications for automotive application life cycle', *Resour. Conserv. Recycl.,* 2015, **100**, 41–8.

[3] F. Cherubini, M Raugei, and S. Ulgiati, 'LCA of magnesium production: Technological overview and worldwide estimation of environmental burdens', *Resour. Conserv. Recycl.,* 2008, **52**, 1093–100.

[4] S. Ramakrishnan and P. Koltun, 'Global warming impact of the magnesium produced in China using the Pidgeon process', *Resour. Conserv. Recycl.,* 2004, **42**, 49–64.

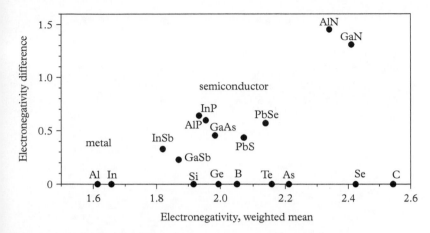

Figure 3.3 *Electronegativity triangle of some p-block materials.*

Table 3.3 *Band gaps (300 K) of p-block materials.*

Material	Band gap /eV
AlN	6.0
diamond	5.47
GaN	3.36
AlP	2.45
Se	1.85
GaAs	1.42
Si	1.14
Ge	0.74
GaSb	0.68
Te	0.34
PbSe	0.27
InSb	0.17

Table 3.4 *Volumes, V_-, of a selection of p-block anions.*

Ion	V_-/nm^3
N^{3-}	0.062
P^{3-}	0.083
S^{2-}	0.067
Se^{2-}	0.072
SeO_4^{2-}	0.103
TeO_4^{2-}	0.110
PO_3^-	0.067
AsO_4^{3-}	0.088
SbO_4^{3-}	0.071
$AlCl_4^-$	0.156
$SnCl_6^{2-}$	0.234

dominates, whilst the latter is quite electropositive and will exhibit considerable ionic character in its bonding. As can be seen in Fig. 2.12, on descending the groups the variation in electronegativity across the groups reduces as the atomic radius increases and the bonding becomes softer. The majority of these elements have χ values of 2.0 ± 0.25. This means that many materials containing these p-block elements will span across arc between metallic and covalent in the electronegativity triangle (Fig. 2.14). Thus a plot similar in type to Fig. 2.15 but concentrating on the p block shows a concentration of materials in this intermediate behaviour region (Fig. 3.3). This can be compared with the band gaps of a selection of these materials (Table 3.3). Three of the materials— diamond, aluminium nitride, and gallium nitride—have the largest band gap and are to the right-hand side of the electronegativity triangle. A large proportion of the compounds in this list have band gaps near and below 1 eV, and thus are likely to have useful semiconductor properties. At the left-hand end of the x-axis fall aluminium and indium. Aluminium has the fourth-largest electrical conductivity of all metallic elements, after silver, copper, and gold. That of indium is about 1/3 of that of aluminium, but the value lies between that of lithium and potassium, and so indium too shows metallic behaviour.

The thermodynamic volumes of some anions, in addition to those in Table 2.16 and including p-block elements, are shown in Table 3.4. These too may be incorporated into Eq. 2.74 to provide an estimate of the lattice energy of salts of complex ions.

3.2.1 Carbon

Carbon has the fourth-largest abundance in the Universe (5,000 ppm), but ranks considerably lower in its estimated crustal abundance (200 ppm). It occurs in a vast variety of forms and so recurs throughout this book. The increase in the

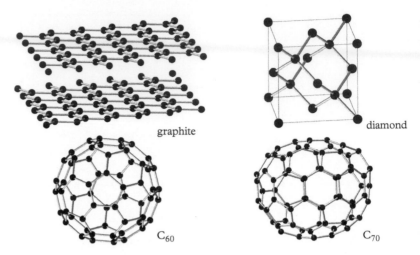

Figure 3.4 *Some of the allotropes of carbon.*

quantities of carbon dioxide and methane in the troposphere is described in Chapter 1 (Planet Earth), and further topics will be covered in Chapters 4 (Air) and 5 (Fire). It has a wide array of allotropes (Fig. 3.4), with graphite, diamond, and amorphous carbon the most abundant. But various classes of additional structures have been identified in recent years including fullerenes (e.g. C_{60}), carbon nanotubes (CNT), and graphene. Graphite has (only just) the lowest free energy of these structures (by 2.9 kJ mol^{-1} from diamond). The diamond structure is related to the *fcc* structure of aluminium (Fig. 3.8) in which half of the tetrahedral holes in the structure are filled with carbon. The result is a three-dimensional lattice structure of C–C single bonds (1.544 Å). The graphite structure is radically different, comprising single-atom-thick layers of hexagons which are polyaromatic (1.424 Å). The interlayer spacing is 3.645 Å, indicating a much weaker interaction called π–π stacking. A detached single layer is called graphene. The classic fullerene C_{60} incorporates twelve five-membered rings, and these pentagons afford icosahedral symmetry equivalencing every atom in the structure. With additional hexagonal rings this structure can be stretched, as in C_{70}. Continuing this process to a very high degree will create closed carbon nanotubes, and chemical etching can open the relatively strained tube ends.

Much of the available supplies of carbon are as fuels: coal, oil, and natural gas. Coal is substantially carbon, but with complex mixtures of termination groups and impurities. Hydrocarbons predominate in oil and natural gas, with their biological roots affording compounds of the essential elements, notably as organo-sulphur and -nitrogen. In Table 3.5, the supply landscape is simplified and summarized. In addition to being the sources of fuels, the fossil legacy also provides the source of petrochemical materials. These include synthetic diamond (99% of industrial diamond is made that way), CNTs, fullerenes, and graphene.

Table 3.5 *Carbon resources types, production (2017) and reserves.*

Resource	Production	Reserves
coal	3,768.6 Mt	1,035,012 Mt
oil	4,387.1 Mt	239,300 Mt
gas	2,454 Mt	129,000 Mt
natural graphite	1.2 Mt	270 Mt
natural industrial diamond	12.4 t	240 t
activated carbon	1.73 Mt	–
carbon nanotubes, fullerenes	10,000 t	–
graphene	≈ 1,000 t	–

The quantity of graphene produced is also expressed as area of supported films rather than mass, as this is the viable form for devices. Activated carbon is a major use of amorphous carbon with high surface area and active sites which have strong adsorbent properties. A variety of sources is used including fossil fuels and biomaterials (wood, nutshells). Generally the supply of oil is the dominant influence on petrochemicals, with ≈ 87% being used as fuel and *c*.4% for each of plastics and chemicals.

The environmental impact of the synthesis of an organic thin film for a solar cell has been developed for C_{60} and C_{70} derivatives.[5] Two broad classes of fullerene production were compared: pyrolysis and plasma decomposition. The major energy input for the plasma methods is in electricity, and resulting embodied energy of separated C_{60} was calculated to be in the region of 100 GJ kg^{-1}. This is nearly an order of magnitude higher than that derived for pyrolysis methods. For that method the embodied energy varied with precursor. Tetralin, $C_{10}H_{12}$, provided a 16% reduction in embodied energy over toluene, C_7H_8.

The tetralin pyrolysis method affords a soot with an embodied energy estimated to be 1.6 GJ kg^{-1}, most of which is that of the original tetralin (Fig. 3.5). Extraction of the fullerene mixture requires additional chemical and solvent input, increasing the embodied energy by a factor of 4, largely due to the unwanted material in the soot. Chemical separation and subsequent stages of purification to afford a purity of 99.9% require additional solvent and chemicals and result in the mass loss due to impurities and inefficiencies. As a result the embodied energy increases by a further factor of 4. Finally, use of the fullerene in an organic solar cell necessitates a functionalization for attachment of a thin layer, and the additional chemicals and solvent raise the embodied energy by a further 250%. Overall the embodied energy of this specialized material is ≈ 65,000 MJ kg^{-1}, compared to 853 MJ kg^{-1} for lithium. However, its application would be as a monolayer film, and thus can be applicable.

[5] A. Anctil, C. W. Babbitt, R. P. Raffaelle and B. J. Landi, 'Material and Energy Intensity of Fullerene Production', *Environ. Sci. Technol.*, 2011, **45**, 2353–9.

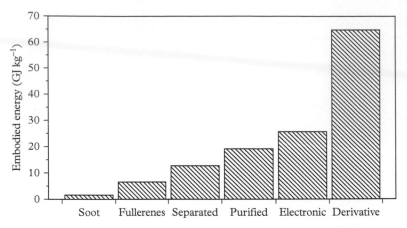

Figure 3.5 *Embodied energy during synthesis of a C_{60} derivative for organic solar cells.*

3.2.2 Aluminium

Although aluminium has only 1/10 of the abundance of carbon in the Universe (50 ppm), but its estimated crustal abundance (84,149 ppm) is exceeded only by oxygen and silicon. The principal mineral for extraction is *bauxite*, which is a mixture of hydrous aluminium oxides, aluminium hydroxide, clays, and some insoluble materials. These can include more specific minerals such as *gibbsite*, $Al(OH)_3$, and the oxyhydroxides, $AlO(OH)$, *böhemite* and *diaspore*. An example of these structures in *gibbsite* (Fig 3.6). The aluminium ions are octahedrally coordinated to the hydroxyl groups which themselves bridge two aluminiums. Some protons of the hydroxyls are pointed away from this sandwich and hydrogen bond to oxygen atoms of the next layer (O..H 2.120 Å). This is a basic structure type for many layered clays.

The bauxite is converted in large scale to alumina, Al_2O_3, by the Bayer process. In 2017, the production of alumina was 126 Mt, which would have required ≈ 300 Mt of bauxite. In addition to this, worldwide production of cement was 4,100 Mt. Two components of Portland Type 1 cement are $Ca_3Al_2O_6$ (tricalcium aluminate, termed C3A) and Ca_2AlFeO_5 (*brownmillerite*), and this cement will include 114 Mt of aluminium.

The Bayer process involves the digestion of the bauxite in a superheated aqueous caustic soda solution to form $[Al(OH)_4]^-$. For gibbsite, this involves a temperature of $c.140\,°C$ (Eq. 3.10). However, for the more oxidic böhemite and diaspore, higher temperatures in the region of $240\,°C$ are necessary (Eq. 3.11).

$$Al(OH)_3 + Na^+ + OH^- \rightarrow Na^+ + [Al(OH)_4]^- \tag{3.10}$$

$$AlO(OH) + Na^+ + OH^- + H_2O \rightarrow Na^+ + [Al(OH)_4]^-. \tag{3.11}$$

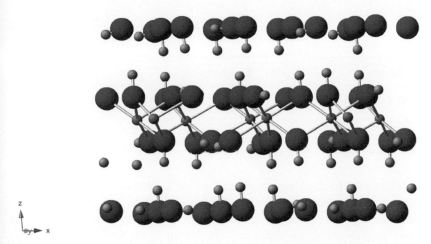

Figure 3.6 *Structure of α-Al(OH)₃, gibbsite; blue = Al, red = O, pink = H.*

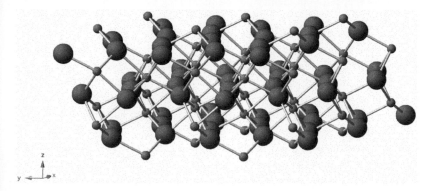

Figure 3.7 *Structure of α-Al₂O₃, corundum; large = O, small = Al.*

Sedimentation is required to remove the insoluble impurities, and then Al(OH)₃ is precipitated from the liquor, essentially by the reverse of Eq. 3.10. In the final, calcination, stage, the hydroxide is roasted at 1,100 °C (Eq. 3.12)

$$2Al(OH)_3 \rightarrow Al_2O_3 + 3H_2O. \qquad (3.12)$$

Aluminium oxide itself has considerable application as a hard, refractory material. *Corundum*, α-alumina, can be viewed as comprising a hexagonal close-packed array of oxides with aluminium ions located in 2/3 of the octahedral holes (Fig. 3.7). The oxides are coordinated to four aluminium ions, providing a very strong lattice.

Aluminium oxide and aluminium sulphate are manufactured on a Mt scale, the latter principally for water treatment. But the vast majority (≈ 90%) is converted

Table 3.6 *Prediction of the thermodynamics for the production of aluminium under standard conditions.*

Reagents	Products	$\Delta_r H^\circ$ /kJ mol^{-1}
$0.5Al_2O_{3(s)} + 1.5C_{(s)}$	$Al_{(s)} + 1.5CO_{(g)}$	672
$0.5Al_2O_{3(s)} + 0.75C_{(s)}$	$Al_{(s)} + 0.75CO_{2(g)}$	543
$0.5Al_2O_{3(s)}$	$Al_{(s)} + O_{2(g)}$	838

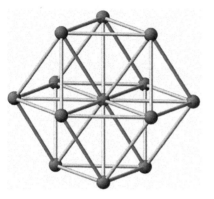

Figure 3.8 *Partial ccp (fcc) structure.*

into aluminium metal. Aluminium metal adopts the second of the close-packed structure, termed cubic close-packed (ccp). In crystal terms it is a face-centred cubic (fcc) structure. The orientation in Fig. 3.8 shows the central atom has six nearest neighbours in one plane, and a further six in the forms of two triangles above and below that plane. In this case these triangles are mutually staggered rather than eclipsed as in the hcp structure, making the alternative layer structure *..ABCABC...*

As for *s*-block metals, a carbothermic method to manufacture the metal could be envisaged (Table 3.6), but in this case it is even more endothermic. Direct conversion of alumina into the metal has an even higher energy demand. Instead, an electrochemical version of the carbon reduction is utilized. In 2017 63.4 Mt of aluminium metal was produced in this way, the most of any non-ferrous metal.

The electrochemical production process is called the Hall–Héroult. Alumina is dissolved in the mineral *cryolite*, Na_3AlF_6 (mp 1,012 °C). The mining deposits of cryolite, in Greenland, were depleted in 1987. Hence a synthetic version is now utilized, which is formed either from fluoridation of alumina (Eq. 3.13), or using hexafluorsilicic acid, H_2SiF_6. That acid itself is manufactured by reaction of silica with HF (Eq. 3.14).

$$6NaOH + Al_2O_3 + 12HF \rightarrow 2Na_3AlF_6 + 9H_2O \tag{3.13}$$

$$SiO_2 + 6HF \rightarrow 2H_3O^+ + SiF_6^{2-}. \tag{3.14}$$

To reduce the melting point, CaF_2, *fluorite*, is added, allowing the operating temperature to be lowered to ≈ 960 °C. At this temperature, aluminium in liquid (density *c*.2.3 g cm^{-3}, mp 660 °C), and forms a lower layer below the electrolyte mixture. At first sight this seems odd, since the density of cryolite is 2.95 g cm^{-3}. However, at this temperature the density of the melt is lower (≈ 2.1 g cm^{-3}), possibly due to the partial dissociation of the anion (Eq. 3.15):

$$AlF_6^{3-} \rightarrow AlF_4^- + 2F^-. \tag{3.15}$$

The cell itself is outlined in Fig. 3.9. The cell outer chamber contains a ceramic liner within a steel casing, within which the molten cryolite mixture sits. The

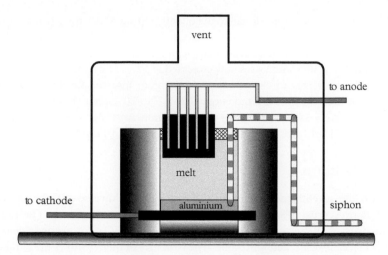

Figure 3.9 *Schematic of the Hall–Héroult cell for aluminium production.*

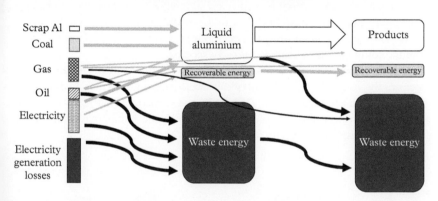

Figure 3.10 *Schematic of the exergy flow for aluminium production.*

cathode is graphite, but once liquid aluminium is formed it becomes the electrode surface. The anode is also carbon on a support which allows it to be removed and replaced, as it is sacrificed during the anodic reaction forming CO_2. Liquid aluminium can be siphoned off, and the gaseous mixture that escapes through the crust formed on the surface is vented off to a gas treatment plant.

The large scale of aluminium production means that it has a significant impact on global energy consumption. For example, it is estimated that 5% of the electricity use in the USA is devoted to it. Analysing the flows through its production can identify opportunities for improvement. A simple schematic for flows of **exergy**[6] for aluminium production worldwide is illustrated in Fig. 3.10. This is akin to a Sankey diagram.[7] In those the flows are depicted through stages showing productive use as straight lines, and losses typically curling down. The

[6] Exergy is defined as the useful energy. While energy is conserved, useful energy may be lost.

[7] J. M. Allwood, J. M. Cullen, *Sustainable Materials with both eyes open*, UIT Cambridge, Cambridge, 2012.

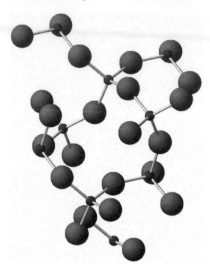

Figure 3.11 *Partial quartz structure; Large = O, small = Si.*

width of the shapes is proportional to the quantity. In Fig. 3.10, the left-hand side shows the exergy inputs to the initial reagents. Overall, $\approx 61\%$ of the input is in electrical energy. By the stage of forming liquid aluminium metal, the exergy of that product is $\approx 36\%$ with losses now increased to 57%. The additional treatment and fabrication stages to convert liquid aluminium into the finished products allows some increase in recovered energy (from 7 to 10%), but also increases the energy losses to $\approx 69\%$. Overall, this renders aluminium a high embodied energy of 170–210 MJ kg^{-1}, with each kg being responsible for 10 kg of CO_2.

3.2.3   Silicon

Silicon is the seventh most abundant element in the Universe and the second in the Earth's crust. Element supply is therefore not in question. The most common set of chemical forms are based upon silica, SiO_2, for example *quartz* (Fig. 3.11), *cristobalite*, and sandstone. The structure of quartz shows the tetrahedral coordination at silicon with each silicon bridged by a two-coordinate oxygen. It is a relatively open structure and many alternatives are possible. This low-temperature phase of quartz includes twelve-membered rings of alternating Si and O atoms. In counting ring size in microporous solids, the oxygens are omitted from the count and this would be termed a six-ring.

There are many complex salts—silicates—derived from silica. The simplest of these are the orthosilicates, which typically have isolated SiO_4 tetrahedra, They include calcium silicate, Ca_2SiO_4 (Fig 3.12), and *garnets*, $[M_3M'_2(SiO_4)_3]$. The silicon atoms are coordinated to oxygens for which the remaining coordination sites are all occupied by calcium ions.

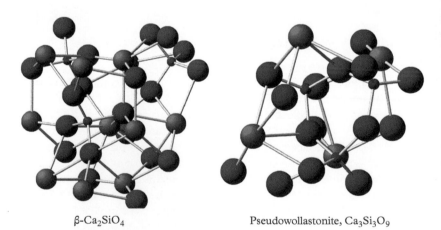

β-Ca_2SiO_4 Pseudowollastonite, $Ca_3Si_3O_9$

Figure 3.12 *Examples of silicates with Si:O ratios of 1:4 and 1:3; red = O, large blue = Ca, small blue = Si.*

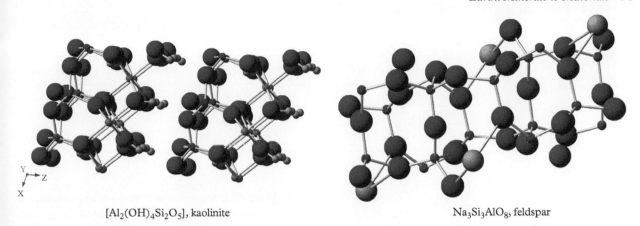

[Al$_2$(OH)$_4$Si$_2$O$_5$], kaolinite   Na$_3$Si$_3$AlO$_8$, feldspar

Figure 3.13 *Examples of layered and 3-D condensed aluminosilicates; red O, yellow Na, pink H; left: dark blue Si, light blue Al; right: dark blue Si/Al.*

Chain-linked silicates, of formula [SiO$_3$]$^{2-}$, are based on corner-sharing tetra-hedra. These structures can include ring structures such as the six-membered Si$_3$O$_3$ ring (a three-ring) in the polymorph of CaSiO$_3$ in Fig. 3.12. Each silicon is linked to two bridging oxygens and the remaining two, terminal, oxygens are otherwise coordinated to calcium.

Cross-linking of chains or planar condensation leads to layer silicates and are the basis of clay minerals and micas such as [Al$_2$(OH)$_4$Si$_2$O$_5$], *kaolinite*. The clay has an aluminosilicate bilayer, with octahedral aluminium sites and tetrahedral silicon sites (Fig. 3.13). The aluminiums are coordinated to the hydroxyl groups to achieve a local charge balance, and the protons link to the silicate side of the bilayers via hydrogen bonding. There are matching Si and Al six-rings at these junctions.

Where all vertices of the tetrahedra are shared, three-dimensional lattices are formed. Rather than revert to a quartz or cristobalite SiO$_2$ structure, a vast array of aluminosilicate materials are known in which up to 50% of the silicon is replaced by aluminium. Each replacement of Si(IV) by Al(III) creates a negative charge on the lattice, which requires a balancing counter ion. *Feldspars*, the most abundant mineral in the crust, are of this type, with formulae like NaAlSi$_3$O$_8$ (Fig. 3.13). In this material the silicon and aluminium sites could not be distinguished, and disorder of these atoms is common. This requires the aluminium also adopt a tetrahedral site in this instance. Even in this relatively condensed structure there are four-rings. Three-dimensional lattice aluminosilicates also include *zeolites*, the adsorbent and catalytic properties of which feature in the following chapters. As is described in Section 3.5 (Materials), very large-scale uses of silicates are in the fields of the structural materials glass and cement.

Elemental silicon, which also adopts the diamond structure in Fig. 3.4, is also a key element due to its semiconductor properties (Fig. 3.3, Table 3.3), which

Table 3.7 *Prediction of the thermodynamics for the production of silicon under standard conditions.*

Reagents	Products	$\Delta_r H^\circ$ /kJ mol^{-1}
$SiO_{2(s)} + 2C_{(s)}$	$Si_{(s)} + 2CO_{(g)}$	689
$SiO_{2(s)} + 3C_{(s)}$	$SiC_{(s)} + 2CO_{(g)}$	623
$SiO_{2(s)} + \frac{4}{3}Al_{(s)}$	$Si_{(s)} + \frac{2}{3}Al_2O_{3(s)}$	−208

are highlighted in Chapter 5 (Fire). The sources used are SiO_2 in the forms of quartzite or sand. Reduction to silicon and ferrosilicon amounted to $\approx$ 7,400 kT in 2017, with the majority of this silicon (c.61%) as ferrosilicon in various Fe:Si ratios). We noted above that ferrosilicon was utilized in the manufacture of magnesium, but the majority is used in the iron and steel industries. The majority of elemental silicon is utilized in the production of aluminium alloys and in the chemical industry. It is estimated that the market in Europe, North America, and Japan for silicon compounds is c.1,140 kt, of which 25% is directed towards precursors for solar-grade silicon and 19% to silicone sealants.

As shown in Table 3.7, an exothermic reduction of SiO_2 can be achieved by ignition with aluminium powder (the thermite process). However, we have also seen that aluminium is very energy intensive, so carbon is a more attractive alternative. Reduction is very endothermic at ambient temperatures. The evolution of gaseous CO gives the process a large positive entropy, and thus will become more favourable at high temperature. Generally the silica is in excess to avoid the alternative product of silicon carbide (SiC). A schematic of the furnace has some similarities to that of aluminium production (Fig. 3.9). In this chemical reduction of SiO_2 by carbon, the reagents are solids and are heated to c.2,000 °C by an electric arc which passes from carbon electrodes submerged into the solid charge to the carbon lining of the base of the furnace. The arc is to provide heat and thus is generally an alternating current, unlike the direct current required for the electrochemical reduction of aluminium. The silicon (mp 1,414 °C) also forms a pool at the base of the furnace, and is drawn off. The gaseous products, CO and some O_2, are vented and collected.

The embodied energy of this metallurgical silicon has been estimated as c.72 MJ kg^{-1},[8] about half that of aluminium. However, the purity of this material, 98–99%, is insufficient for electronic purposes. It contains about 14 impurity elements at a level of over 10 ppm, with the most prevalent being aluminium and iron (1,000s of ppm). One route to the production of electronic-grade silicon is through further refinement of metallurgical silicon.

To achieve this purification, the silicon is reacted with dry HCl gas at 300 °C (Eq. 3.16):

$$Si + 3HCl \rightarrow SiHCl_3 + H_2. \tag{3.16}$$

[8] D. Holst, *Embodied energy and solar cells*, The Green Initiative Blog, 2017.

The impurities generally form anhydrous halides, such as Al_2Cl_6 (bp 180 °C, sub.) and $FeCl_3$ (bp 316 °C, dec), of very different volatility to $SiHCl_3$ (bp 31.8 °C), reducing the atomic level of the impurities to under 1 ppb. Eq. 3.16 is then reversed at 1,100 °C for 200–300 h to form high-purity silicon rods. The embodied energy of this electronic-grade silicon is increased to $\approx$ 430 MJ kg^{-1}. Crystallization requires breaking up these rods, and annealing at 1,400 °C, which increases the embodied energy of the solar-grade monocrystalline silicon to $c.$1,660 MJ kg^{-1}.

An alternative route to ultra-pure silicon involves the catalytic disproportionation of trichlorosilane (Eq. 3.17). The catalyst is often a Lewis acid, such as Al_2Cl_6, but it can also be a Lewis base. The components can be separated on the basis of their volatility. Silane, SiH_4, is extracted and the equilibrium re-established to increase conversion.

$$SiHCl_3 \rightarrow SiH_4 + SiH_3Cl + SiH_2Cl_2 + SiCl_4. \qquad (3.17)$$

Silane can then be thermally decomposed at 650–800 °C to afford ultra-pure silicon (Eq. 3.18), a process called chemical vapour deposition (CVD):[9]

$$SiH_4 \rightarrow Si + 2H_2. \qquad (3.18)$$

3.2.4   Germanium

In structure and properties, germanium is very similar to silicon. It adopts the diamond structure (Fig 3.4) and is also a semiconductor (Fig. 3.3). It has a larger radius and higher electronegativity than silicon and its band gap is lower at 0.74 eV (Table 3.3). However, it contrasts strongly with silicon in its occurrence, with an estimated crustal abundance of 1.3 ppm. As a result, germanium is rather high in the supply risk list (Table 2.1). The production of germanium worldwide was in the region of 134 t in 2017. The predominant materials and applications are GeO_2, for glasses including optical fibres ($\approx$ 4 wt%) and wide-angle lenses, $GeCl_4$ (bp 86.5 °C) for chemical synthesis and elemental germanium.

Germanium is extracted generally from sulphidic ores, consistent with the softer characteristics emanating from its larger atomic radius. Generally these ores are zinc-containing, such as the mineral *sphalerite*, (Zn,Fe)S. The starting points for the thermodynamics of the processes are presented in Table 3.8. First, the sulphides are roasted, at temperatures between 700–1,000 °C to form the oxides and liberating the acid gas SO_2. This is often collected to manufacture sulphuric acid, H_2SO_4; for germanium this step is highly exothermic. The GeO_2 is highly impure and can then be converted into $GeCl_4$ by treatment with HCl (or by Cl_2). Pure $GeCl_4$ can then be isolated from the mixture by fractional distillation. This in turn can afford a source of pure GeO_2 after hydrolysis with deionized water. Elemental germanium can be manufactured by reduction by carbon or hydrogen. The carbon method tends to be employed for metallurgical germanium,

[9] W. O. Filtvedt, A. Holt, P. A. Ramachandran and M. C. Melaaen, 'Chemical vapour deposition of silicon from silane: Review of growth mechanisms and modelling/scaleup of fluidized bed reactors', *Sol. Energy Mater. Sol. Cells*, 2012, **107**, 188–200.

Table 3.8 *Prediction of the thermodynamics for the production of germanium under standard conditions.*

Reagents	Products	$\Delta_r H^o$ /kJ mol^{-1}
$GeS_{2(s)} + 3O_{2(s)}$	$Ge_{(s)} + 2SO_{2(g)}$	-955
$GeO_{2(s)} + 2Cl_{2(g)}$	$GeCl_{4(l)} + O_{2(g)}$	19
$GeO_{2(s)} + 4\,HCl_{(aq)}$	$GeCl_{4(l)} + 2H_2O_{(l)}$	116
$GeO_{2(s)} + C_{(s)}$	$Ge_{(s)} + CO_{2(g)}$	158
$GeO_{2(s)} + 2H_{2(g)}$	$Ge_{(s)} + 2H_2O_{(l)}$	-21

and reduction by pure hydrogen utilized for electronic materials. Much of the germanium is lost in the fly-ash during the roasting process, and it is estimated that only 3% of the germanium in the zinc ores is recovered. That said, *c.*30% of germanium products are from recycled material, with over 60% of germanium metal in optical devices being recycled. The additional steps due to extraction of a low-concentration element contribute to increasing the embodied energy to $\approx$ 900 MJ kg^{-1}.

3.2.5   Indium

Indium has a similar relationship to aluminium as germanium has to silicon. It is very much less abundant in the Earth's crust ($\approx$ 50 ppb) even compared to germanium. Like germanium, it is extracted as a side-product from sulphidic ores, typically sphalerite which mainly affords zinc. World production in 2017 was *c.*720 t. Unsurprisingly, it also has a high estimated embodied energy ($\approx$ 3,000 MJ kg^{-1}).

Indium has a slightly higher electronegativity than aluminium (Fig. 3.3) and a significantly higher atomic radius. It adopts a related crystallographic form to aluminium, being based on a cubic close-packed structure (Fig 3.8). It displays a tetragonal distortion, giving four closest neighbours at 3.252 Å and eight at a slightly longer distance (3.377 Å). Its larger atomic radius and softer characteristics contribute to it being used for components in some semiconductors with lower band gaps than aluminium and gallium, such as InP (1.34 eV) and InSb (0.17 eV) (Table 3.3).

The major indium product (*c.*50% of all indium used) is indium tin oxide (ITO), in which the predominant oxide is In_2O_3. In_2O_3 might be anticipated to adopt a structure like α-alumina (corundum) (Fig. 3.7) in which the cation is located at a regular octahedral site. However, with a larger cation, there is a tendency to increase the coordination number. This can be achieved in the fluorite structure (CaF_2) (Fig. 3.14), there the cations can be seen in an fcc structure. Within the unit cell there are four cations.[10] There are eight tetrahedral holes in the unit cell which are occupied by fluorides. If we pick the cation on the top

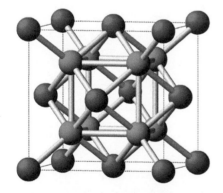

Figure 3.14 Fluorite *structure unit cell; paler grey = F, darker = Ca.*

[10] Each of the eight corner sites is shared by eight unit cells, so each contains one atom. There are six atoms at the centre of faces, each shared between two cells. Hence the cell contains a total of four cations.

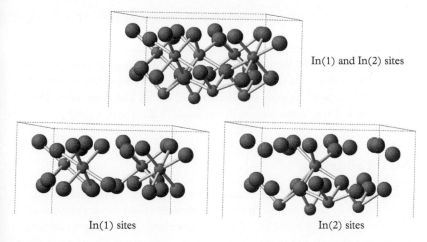

In(1) and In(2) sites

In(1) sites   In(2) sites

Figure 3.15 *Indium oxide sites in the* bixbyite *structure; large = O, small = In.*

Table 3.9 *Band gaps (300 K) of tin and group 13 oxides.*

Material	Band gap /eV
SnO	2.5
SnO_2	3.6
α-Al_2O_3	8.8
α-Ga_2O_3	4.9
In_2O_3	2.9
ITO	≈ 4

face in Fig. 3.14, we can see that it has four anions as nearest neighbours. The next set of atoms in the cell above will be identical to those at the bottom of this cell. Adding the two coordinations together will generate eight-coordinate cubic coordination for the cation. The hexagonal, corundum structure is known for In_2O_3, but the more stable structure is a cubic one named *bixbyite* (Fig. 3.15), which is related to fluorite. There are two indium sites in the structure, which are more easily visualized when plotted separately. Both are six-coordinate, with the oxygen sites being four-coordinate. Given that the anions are four-coordinate, there are insufficient anions to achieve a coordination of eight so there are $\frac{1}{4}$ missing. In effect the coordination at each indium is $\frac{6}{8}$ of a cube. For site In(1) ($\frac{1}{4}$ of the indium), the vacancies are at opposite corners of the cube resulting in a flattened octahedron. The distortion on the predominant site In(2) is more evident, and is caused by the anion vacancies being on the diagonal of one face of the cube.[11] The preference is for tin to adopt a higher than random preference for the less distorted site.[12]

These oxides are relatively wide-band-gap semiconductors (Table 3.9), although less so than the lower Z elements in Group 13. The mixed metal ITO materials are reported to have a band gap of around 4 eV, which is wide enough to be optically transparent. If the tin atom adopts an indium site, with an anticipated tripositive charge, then its extra electron can contribute to it being an electron carrier (*n*-type semiconductor). Substituting In^{III} (ionic radius 0.94 Å) by Sn^{IV} (ionic radius 0.83 Å) would be expected to reduce the size of the unit cell.[13] In fact the opposite happens. The unit cell size *increases* with tin substitution up to the solubility limit of $\approx 6\%$ Sn. In the soluble region, all the tin is present as Sn^{IV}, releasing one electron per tin atom. A proportion is trapped in regions of the structure, but the remainder is in the conduction band. That proportion is higher for thin films than for powdered or single-crystal

[11] N. Nadaud, N. Lequeux, M. Nanot, J. Jové and T. Roisnel, 'Structural Studies of Tin-Doped Indium Oxide (ITO) and $In_4Sn_3O_{12}$', *J. Solid State Chem.*, 1998, **135**, 140–8.

[12] With a high level of tin, as in $In_4Sn_3O_{12}$, tin(IV) entirely adopts an octahedral site. The second site is now seven-coordinate and that is shared by tin and indium.

[13] Vegard's law implies that for a disordered structure there should be a linear change in the unit cell size with the degree of substitution with the slope related to the difference of the ionic sizes.

samples, suggesting surface effects enhance the conductivity. The result is that thin films of ITO have a combination of highly advantageous properties: transparent in the visible region (and absorbing in the UV and reflecting in the IR), a high electrical conductivity, good adhesion to glass and some polymers, and chemical resistance. The applications are legion, including touchscreen displays antireflection coatings, windscreen defrosting, and solar cells. Indium is in relatively low abundance but its prime use is as films with a thickness in the region of 0.1 μm.

3.2.6   Tellurium

Tellurium is estimated to have a much lower abundance than indium in the Earth's crust, only being at $\approx$ 1 ppb. The estimated world production for 2017 is in the region of 420 t (*c.*12% of that of selenium), with reserves estimated at 31,000 t. Although it does occur in minerals with group 11 metals, such as $AuTe_2$, Ag_3AuTe_2, and $AgAuTe_4$, the dominant method of production is as a side-product to production of other elements. This includes the electrochemical refining of copper (in anodic deposits) and blast furnace refining of lead (in waste dusts). In the anodic deposit route, the tellurides of group 11 elements, M_2Te, will be mixed with selenides. These can be roasted with oxygen at *c.*500 °C according, to Eq. 3.19:

$$M_2\,Te/Se + O_2 + Na_2CO_3 \rightarrow Na_2\,Te/SeO_3 + 2M + CO_2. \qquad (3.19)$$

The tellurate/selenate(IV) mixture can be extracted with water and treated with sulphuric acid. That precipitates TeO_2 but selenium remains in solution (Eq. 3.20):

$$HTeO_3^- + OH^- + H_2SO_4 \rightarrow TeO_2 + SO_4^{2-} + 2H_2O. \qquad (3.20)$$

Elemental tellurium can be formed by electrochemical reduction of TeO_2, or by its reduction using SO_2, reforming sulphuric acid (Eq. 3.21):

$$TeO_2 + 2SO_2 + H_2O \rightarrow Te + 2SO_4^{2-} + 4H^+. \qquad (3.21)$$

Tellurium has a very intermediate electronegativity (2.16, see Fig. 3.3). This and its rather high atomic radius (Fig. 2.12) contribute to it being an intrinsic semiconductor of low band gap (Table 3.3). As may be expected from an element in group 16, its primary coordination is two, forming helical chains in the solid (Te–Te 2.835 Å) (Fig. 3.16). The interchain interactions mean that there is a set of four non-nearest Te–Te interactions of 3.491 Å, considerably shorter than the next-nearest neighbour along the chain (4.441 Å).

The largest-volume application of tellurium is in semiconductor and electronics uses, with cadmium telluride (CdTe) of prime importance due to its high

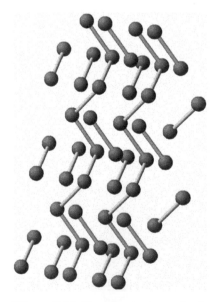

Figure 3.16 *Partial structure of tellurium.*

efficiency in solar cells. The embodied energy of tellurium is estimated to be similar to that of aluminium, *c.*200 MJ kg^{-1}.[14] This is a case where the availability of an element could restrict its application in a key field. This will be addressed in more detail in Chapter 5 (Fire).

3.3 *d*-Elements

The *d*-elements, taken as groups 3 to 12 here (Fig. 2.7), all have low electronegativities and are metallic (Fig. 3.17). The electronegativities do increase significantly across the rows as the filling of the *d* subshell has quite a low degree of self-screening allowing Z_{eff} to increase. Accordingly, the atomic radii decrease across the rows. Unsurprisingly, the radii of the elements in period 5 (*4d*-elements) are larger than those in period 4. However, this trend is very muted between periods 5 and 6. This is ascribed as being due to the lanthanide contraction. The *5d*-elements immediately after the lanthanide row are affected by the filling of the *4f* subshell for which self-screening is very low and thus the radius shrinks.

The most striking feature of the *d*-block is the wide range of oxidation states. Something of that is captured in Fig. 3.17. There are compounds of intermediate oxidation states which fall within the intermediate regions and would be expected to have semiconductor or other electronic properties. With small, charged (hard) ligands like oxide and fluoride higher oxidation states are prevalent, and these materials are likely to have a mixture of ionic and covalent bonding characteristics. These ligands have both σ and π donor properties. The figure includes $W^{VI}O_3$ and $Os^{VIII}O_4$, both in the highest oxidation states of their respective groups. WO_3 adopts three-dimensional lattice structures and has a high melting point ($1,473\,°C$), whilst OsO_4 has a monomeric, tetrahedral structure and a low boiling point ($129.7\,°C$). Indeed for some elements in groups 4 to 8, oxidation states can

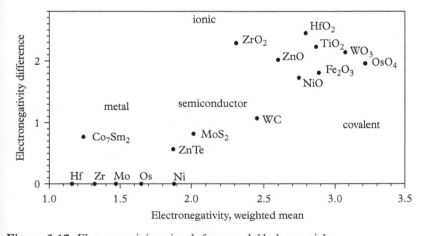

Figure 3.17 *Electronegativity triangle for some* d-*block materials.*

span 11 values representing all options of d^{0-10} configurations. For the lowest oxidation states, ligands with σ donor and π acceptor properties are required.

The variations across the d-block can be rationalized from both covalent and ionic viewpoints. From a covalent viewpoint, the strength of the bonds to a metal can be related to the ability of the valence orbitals to overlap. Thus in period 4, the $3d$ orbitals are thought to extend relatively little beyond the atomic cores, so limiting the degree of bonding overlap. But on descending through periods 5 and 6, the d orbitals extend more relative to the core so increasing the bonding overlap. Hence, it would be expected that bond energies increase on descending a group, in contrast to the norm in the s- and p-blocks. Evidence for this can be seen from the variations in the $\Delta_{at}H^\circ$ values across the d-block (Fig. 3.18). The enthalpies of atomization tend to peak with partially filled d shells in groups 5–7 of periods 5 and 6; the most refractory of metals are Ta, W, Re, and Os. At the end of this block, in group 12 with the outermost electronic configuration of $d^{10}s^2$, the trend switches to that of the p-block and the atomization enthalpies decrease with period number.

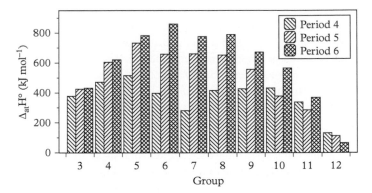

Figure 3.18 $\Delta_{at}H^\circ$ *of the* d-*block elements.*

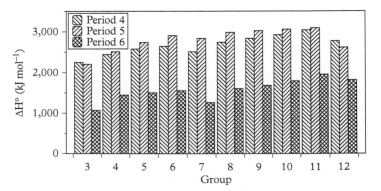

Figure 3.19 $\Delta_f H^\circ$ *of* $M^{2+}{}_{(g)}$ *of the* d-*block elements.*

The majority of gaseous atoms in the d-block have the electronic configuration $md^n(m+1)s^2$. On attaining oxidation state of 2+, the valence electrons retained by the ion will be md^n. For example, double ionization of titanium atoms (Ti is in group 4) will leave an electronic configuration of $3d^2$. The enthalpy of formation of this ion from the metal will be given by Eq. 3.22:

$$\Delta_f H^\circ(M^{2+}) = \Delta_{at}H^\circ + IE_1 + IE_2. \qquad (3.22)$$

The trends in these $\Delta_f H^\circ$ values across the d-block are shown in Fig. 3.19. They are quite distinct from those in Fig. 3.18 as the ionization energies are much larger than the heats of atomization. It is evident that formation of the M^{2+} ions requires much less energy for period 6 as compared to periods 4 and 5, which tend to be similar. On traversing the periods, the increase in Z_{eff} is evidenced by the upwards trends in the $\Delta_f H^\circ(M^{2+})$ values.

The enthalpy trends to attaining gaseous 4+ ions are shown in Fig. 3.20. Not all of these elements can attain this oxidation state. This is most evidently so for the group 3 elements Sc, Y, and La. Removing the fourth electron will break into the more core-like subshell, for example affording an outermost configuration of $3p^5$ for Sc^{4+}. This energy cannot be recovered by bond formation. Hence there is a considerable reduction in the enthalpy of formation moving to group 4. Generally similar qualitative trends are evident to those in Fig. 3.19. The oxidation state is more attainable for period 6 over period 5 over period 4. The increase in Z_{eff} on crossing the periods affords an increase in ionization energies. The increases are more relentless for the higher oxidation state, and for groups 10, 11, and 12 the values of $\Delta_f H^\circ(M^{4+})$ become similar to those of group 3. Not surprisingly then there are no reliable reports of isolatable samples of M^{4+} of groups 11 and 12. There are M(IV) compounds of group 10, such as $K_2[NiF_6]$, PdF_4, and $PtCl_4$.

There is a very wide range of coordination numbers (CNs) which have been identified for compounds of d-block elements: from 1 to 9. But the most

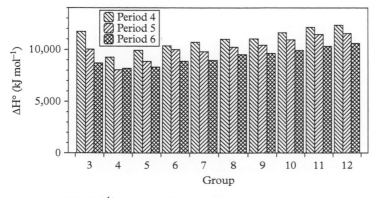

Figure 3.20 $\Delta_f H^\circ$ of $M^{4+}{}_{(g)}$ of the d-*block elements.*

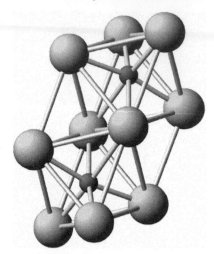

Figure 3.21 *Structure of MoCl₅; large = Cl, small = Mo.*

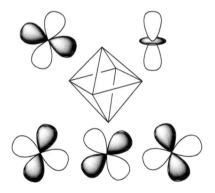

Figure 3.23 *Orientation of the d-orbitals in an octahedral site.*

important geometry is octahedral (CN = 6); four-coordinate structures tend to adopt tetrahedral geometries, with some special cases favouring the square planar option. An example of this is the structure of $MoCl_5$ (Fig. 3.21). Rather than being five-coordinate, there are two adjacent coordination octahedra (edge-sharing), giving a molecular formula of $[Mo_2Cl_{10}]$, linked by two bridging chlorides in *cis* positions.

The discussion so far would suggest that structural and thermochemical factors would change smoothly across the *d*-block as Z_{eff} increases. However, there are clear divergences from this, as can be seen from the trends in ionic radii of the dications in period 4 (Fig. 3.22). Scandium is absent from this plot as the cation is unstable towards disproportionation to Sc(0) and Sc(III). There is a linear trend from Ca^{2+} ($3d^0$), through the higher manganese value ($3d^5$) to Zn^{2+} ($3d^{10}$). Hence these three elements follow an expected trend of a reduced ionic radius due to the increase in Z_{eff} across the period. From the reduced atomic volume, V_+, the lattice energy will increase (Eq. 2.74).

All the other cations, which do not have empty, half-filled, or full *d*-subshells, show ionic radii with values below this line. This can be ascribed to the filling of the *d*-orbitals within the ligand field of the octahedral coordination geometry. Two of the five *d*-orbitals are oriented towards the ligands, namely the $d(x^2-y^2)$ and $d(z^2)$ (Fig. 3.23). These are shown at an upper level as they have metal–ligand $\sigma^\star$ character. They are doubly degenerate and are often called by their symmetry label of e_g under the O_h point group. The remaining three orbitals of the subshell are the $d(xy)$, $d(xz)$, and $d(yz)$, which are oriented between the ligands towards the edge of the octahedron. This triply degenerate set has the symmetry label t_{2g}. They are σ non-bonding and thus have a lower energy.

An overall energy level diagram is shown in Fig. 3.24 for an ion in an octahedral site. There are filled orbitals at low energy which are predominantly ligand in character. Some will be σ in character and some π. Then the next levels are those primarily of metal *d* character. The initial electrons are added into the t_{2g} set and they will have parallel spins in different orbitals to afford the lowest energy state

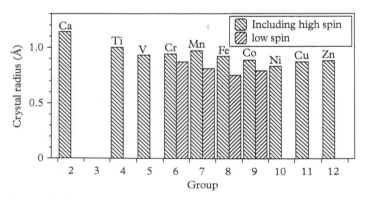

Figure 3.22 *Ionic radii of M^{2+} in octahedral sites in crystals.*

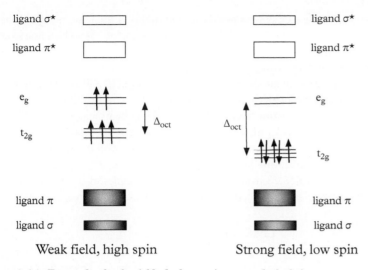

Figure 3.24 *Energy levels of a* d *block element in an octahedral site.*

(in accord with Hund's rules of maximum multiplicity). That will apply to V^{2+}, $3d^3$. These electrons have been added to orbitals pointing between the ligands as in each stage the atomic charge increases. As a result there is a larger reduction in ion size as compared to having the electrons smeared over a sphere (Fig. 3.22).

Adding the fourth electron opens up two options. The first of these is called **high spin**, in which all four electrons are parallel. This is displayed in Fig. 3.24 for a d^5 example. To create a high spin state, an electron must be added to a higher-level e_g orbital. This will only happen when the energy required to pair electrons in the same orbital space is more than that needed to move the electron into the e_g level. It will occur if the ligand field splitting, Δ_{oct}, is small: if the ligands are weak σ donors or they are also π donors. This π donor interaction will stabilize the orbitals labelled as ligand π in Fig. 3.24, and the metal t_{2g} orbitals will acquire $\pi^\star$ character, raising their energy. The addition of these extra electrons into the e_g orbitals which point directly towards the ligands has the effect of increasing the metal–ligand repulsion more than a spherical average, overcoming the increase in nuclear charge and increasing the ionic radius. Adding the fifth electron in the e_g levels restores the spherical average.

The other option occurs when the energy required to pair up the electrons is less than that required to add the extra electron into the e_g orbital. The spin pairing reduces the value of the spin quantum number, S, of the ion, and hence the term **low spin**. This will occur in a number of circumstances. First, that would be when the metal–ligand bonding makes Δ_{oct} large. This can be when the ligand is a strong σ donor, and also when it behaves as a π acceptor. In that case it is the interaction with the empty ligand $\pi^\star$ orbitals that stabilizes the t_{2g} set. The second circumstance is when the pairing energy is reduced because of expansion of the

d-orbitals. For the elements in periods 5 and 6, this is the case. So the electron–electron repulsion is reduced and also the enhanced metal-ligand bonding strength means that Δ_{oct} is also larger. As a result, it is the low-spin option that occurs in the d-block for these two periods. The maximum lowering of the ionic radius will be when the configuration reaches t_{2g}^{6}, as in Fe(II). This also implies that there is an enhanced thermodynamic stability for d^6 sites. This is called the **ligand field stabilization energy** (LFSE). Taking the two examples in Fig. 3.24, we can consider a zero-LFSE line at a weighted mean of the t_{2g} and e_g levels. Hence the former would be $\frac{2}{5}\Delta_{oct}$ below the zero-LFSE line, and the latter $\frac{3}{5}\Delta_{oct}$ above it. That way the LFSE will be zero for high-spin d^5 and d^{10}. So the LFSE can be calculated using Eq. 3.23 using the electron occupancies of the two levels. Thus for high- and low-spin Mn^{2+}, the LFSE will be 0 and $2\Delta_{oct}$, respectively.

$$\text{LFSE} = \left(\frac{2}{5}n(t_{2g}) - \frac{3}{5}n(e_g) \right)\Delta_{oct}. \tag{3.23}$$

The values of Δ_{oct} can be derived from the electronic absorption spectra of transition metal compounds. These are known for their colours, due to the fact that the electronic absorptions span from the near infrared through the visible into the ultraviolet regions. Generally the most intense absorptions are charge transfer bands. In Fig. 3.24, promotion of an electron from the filled ligand σ or π orbitals to an empty space in the t_{2g} or e_g orbitals would largely move the electron from the ligands to the metal, and this is termed a *ligand-to-metal charge transfer* (LMCT) band. Alternatively, for example in the strong field case, promotion of an electron from the filled t_{2g} set into the ligand $\pi^\star$ set would cause the electron transfer to proceed in the opposite sense: a *metal-to-ligand charge transfer* (MLCT) transition. There can also be intra-ligand transitions from filled to empty orbitals. These would be of high energy in this figure.

Transitions within the d-orbitals tend to be somewhat weaker in intensity. In principle they are electric dipole forbidden since $\Delta l = 0$, rather than ± 1, and this reduces the probability of absorption. For a centro-symmetric site, such as one with O_h symmetry, then g to g transitions are Laporte forbidden (this criterion is not relevant to tetrahedral complexes as they do not possess a centre of symmetry and so their d–d bands tend to be more intense). It generally requires the mixing of the electronic transition with an antisymmetric vibration to remove the centre of symmetry to provide a mechanism for absorption. This is called *vibronic coupling*. A further restriction on the intensity of bands is that they should be spin-allowed, that is $\Delta S = 0$. We might envisage a transition between the two states shown in Fig. 3.24, but from the left- to the right-hand side $\Delta S = -5/2$. Indeed, if we look at the weak field case there is no spin-allowed transition possible within the d-orbitals. Hence high-spin d^5 octahedral complexes, such as $Mn^{2+}{}_{aq}$, are weakly coloured.

The energies of the electronic transitions are dependent upon the two factors which are the basis of the options between high- and low-spin complexes, namely the repulsive energy between d-electrons and the ligand field splitting, Δ_{oct} for an octahedral complex. As a result the values of Δ_{oct} can be extracted from these spectra. The values vary with metal, oxidation state, and ligand type. As a rule, the ratio of Δ_{oct} for periods 4, 5, and 6 are 1:1.5:2, consistent with the order of bonding strengths. Generally, for a σ donor ligand, changing from an oxidation state of 2 to 3 will also increase Δ_{oct} by $\approx 50\%$. Δ_{oct} also tends to decrease with the presence of e_g electrons—hence it will be smaller for Ni(II) than V(II).

The values for ligands follow a general trend which is substantially independent of the metal, and is called the **spectrochemical series**. As discussed above, the σ donor ability is the major factor for creating Δ_{oct} by a ligand. Thus increasing the electronegativity of a donor atom will *decrease* Δ_{oct}. The other major factor is the π bonding type. Ligands like F^- and O^{2-} are π donors, and this will reduce the magnitude of Δ_{oct}. As a result Δ_{oct} is not a measure of bond strength for such ligands: the stronger the multiple bond character, the lower Δ_{oct} will be. On the other hand, ligands with low-lying acceptor orbitals, typically those with $\pi^\star$ orbitals like CN^- and CO, will have larger values of Δ_{oct} as the trends caused by metal–ligand σ and π bonding are in the same direction. A partial spectrochemical series can then be set out, as in Eq. 3.24:

$$O^{2-}, F^- < OH_2 < NH_3 < CH_3^- < CN^-, CO. \tag{3.24}$$

The alternative predominant coordination geometry is the tetrahedron. There is not such a clear-cut difference between the two types of d-orbital in this case. The doubly degenerate set, labelled e in T_d symmetry, are oriented towards the faces of the tetrahedron, and it is the triply degenerate set, labelled t_2 in T_d symmetry, which orients more closely to the ligands (towards the edges of the tetrahedron). As a result, the energy level diagram is inverted, as shown in Fig. 3.25 for a high-spin d^5 ion. Hence the LFSE for a tetrahedral site is given by Eq. 3.25:

$$\text{LFSE} = \left(\frac{3}{5}n(e) - \frac{2}{5}n(t_2)\right)\Delta_{tet}. \tag{3.25}$$

Intrinsically, due to their being four rather than six ligands in the tetrahedral site, if all other structural aspects are equal, then Δ_{tet} will have a value of $\frac{4}{9}\Delta_{oct}$. Using this relationship, the LFSE could be re-expressed in terms of Δ_{oct} (Eq. 3.26). This lower value of Δ_{tet} means that low-spin tetrahedral complexes are rare, and are predominantly of very strong field, π acceptor, ligands like CO.

$$\text{LFSE} = \left(\frac{3}{5}n(e) - \frac{2}{5}n(t_2)\right)\Delta_{tet} = \left(\frac{4}{15}n(e) - \frac{8}{45}n(t_2)\right)\Delta_{oct}. \tag{3.26}$$

In looking at the binding of a ligand to a metal, there is good reason to think that this would follow a pattern like that predicted from the ionic radii for metal dications in Fig. 3.22. The metal–ligand bond strength would be expected to increase as the ionic radius and hence V_+ is reduced. In the case of a ligand binding to aqueous metal ions, ligands to the right of water in the spectrochemical series

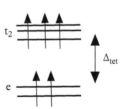

Figure 3.25 *d-orbital splitting in a tetrahedral site.*

[15] H. Irving and R. J. P. Williams, 'The Stability of Transition Metal Complexes', *J. Chem. Soc.*, 1953, 3192–210.

would benefit from increased LFSE due to the higher Δ values. This is largely followed experimentally in the Irving–Williams series ordered stabilities for the dications (Eq. 3.27):[15]

$$Mn^{2+} \; < \; Fe^{2+} \; < \; Co^{2+} \; < \; Ni^{2+} \; < \; Cu^{2+} \; > \; Zn^{2+}. \qquad (3.27)$$

The structures of Cu(II) sites tend to show distortions. For example, $CuCl_2$ is highly stretched along one axis (Fig. 3.26). There is a chain of square planes with Cu–Cl distances of 2.262 Å. Each chain is linked to other chains by two longer Cu–Cl distances of 2.963 Å. In a closely related structure of $(NH_4)_2[CuCl_4]$ the four shorter distances average at 2.315 Å and the longer two are 2.790 Å. However, the hexa-aquo complex in $[Cu(OH_2)_6](ClO_4)_2 \cdot 2H_2O$ has three *trans* pairs of distances at shorter (1.969 Å), intermediate (2.084 Å) and longer (2.202 Å), distances.

Copper(II) has a d^9 electronic configuration, and in an octahedral site there are two options for the location of the unpaired electron—in the $d(z^2)$ or in the $d(x^2-y^2)$; a similar option occurs for high-spin d^4 (Cr^{II}). The Jahn–Teller effect arises from the prediction that such a situation is unstable for a non-linear site.[16] A distortion is predicted to occur, but its nature can vary. The effect of an elongation along the z-axis is shown in Fig. 3.27. The repulsion between the ligands and the $d(z^2)$ electrons will be reduced and this is offset by

[16] H. A. Jahn and E. Teller, 'Stability of polyatomic molecules in degenerate electronic states-1-Orbital degeneracy', *Proc. Roy. Soc. A*, 1937, **161**, A905, 220–35.

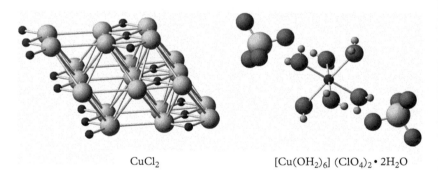

CuCl₂ $[Cu(OH_2)_6](ClO_4)_2 \cdot 2H_2O$

Figure 3.26 *Structures of distorted octahedral sites of Cu(II).*

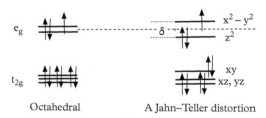

Octahedral   A Jahn–Teller distortion

Figure 3.27 *Effect on energy levels of a Jahn–Teller elongation of a Cu(II) site.*

an increase in the energy of the $d(x^2-y^2)$ electron. If the splitting between these two levels is termed δ then the system will overall gain an increased stability of the factor for the extra electron in the lower-energy orbital. The distortion will also split the t_{2g} orbitals, albeit to a lesser extent due to their non-bonding character. A compression along the z-axis would result in splitting the e_g orbitals in the opposite sense. This effect can account for the enhanced stability of Cu(II) complexes, which may be an opportunity for the purification of copper.

3.3.1 Titanium

Titanium has a relatively high crustal abundance as the ninth most abundant element. World mining of titanium was $c.$7,100 kt in 2017, of which 6,200 kt was of the mineral *ilmenite*, $FeTiO_3$, Fig. 3.28. The figure shows that the metals share faces of distorted O_6 octahedra (Fe on the left). *Rutile*, one of the polymorphs of TiO_2 (Fig. 3.29) is also mined to a major extent ($\approx$ 900 kt in 2017). The structure of rutile shows a regular octahedral coordination at titanium with trigonal coordination at oxygen. These minerals are in widespread supply, with reserves estimated as 2 Gt. Hence titanium rests within the "green" band of the risk register. Over 90% of the titanium is used as its dioxide in pigments.

The metal itself, which has the hcp structure, has utility due to being less dense than steel and having a higher tensile strength than aluminium. Considering the thermodynamics of its formation (Table 3.10), it is evident that attempted reduction by carbon is very endothermic. In preference, the metal is formed by reduction of titanium tetrachloride, a highly reactive liquid (b.p. 136.4 °C). $TiCl_4$ is first formed by chlorination, which also requires carbon to make the process exothermic. It can be purified by fractional distillation. The metal is then formed by the Kroll Process in which the vapour of $TiCl_4$ is reduced by liquid magnesium at 800–850 °C under argon as an inert atmosphere. Further purification is required by melting in an arc furnace, similar to the process for purification of silicon. This all contributes to the embodied energy of titanium which has been variously estimated in the range 360–750 MJ kg^{-1}, in the region of three times that of aluminium.

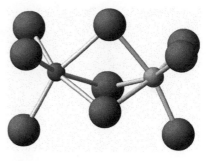

Figure 3.28 *Partial structure of FeTiO$_3$; large = O, small dark = Fe, small light = Ti.*

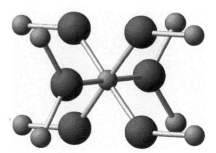

Figure 3.29 *Partial structure of rutile, TiO$_2$; large = O, small = Ti.*

Table 3.10 *Prediction of the thermodynamics for the production of titanium under standard conditions.*

Reagents	Products	$\Delta_r H°$ /kJ mol^{-1}
$TiO_{2(s)} + 2\ C_{(s)}$	$Ti_{(s)} + 2\ CO_{(g)}$	723
$TiO_{2(s)} + 2\ Cl_{2(g)}$	$TiCl_{4(l)} + O_{2(g)}$	140
$TiO_{2(s)} + 2\ Cl_{2(g)} + C_{(s)}$	$TiCl_{4(l)} + CO_{2(g)}$	−254
$TiCl_{4(g)} + 2\ Mg_{(l)}$	$Ti_{(s)} + 2\ MgCl_{2(l)}$	−449

3.3.2 Chromium

Although of higher abundance than titanium in the Universe (15 ppm), the crustal abundance of chromium is about 30 times lower (135 ppm). Mine production in 2017 was c.31,000 kt, based upon the formula of *chromite*, $FeCr_2O_4$. There is considerable recycling, estimated to be $\approx 31\%$ in the USA. Reserves are substantial, considered to be in the region of 510 Mt. Any supply concerns may then be attributed to the localization of the resources, with 95% being in Kazakhstan and southern Africa. The major use of chromium is in the production of steel, chromium plating, and in the high-performance-temperature alloys called *superalloys*. Reduction of the Cr(III) oxides is via carbon in an electric arc furnace, and these types of processes will be considered in more detail in Section 3.3.3. As a result there is considerable energetic benefit in recycling chromium alloys. For example, the embodied energy of NiCr alloys has been estimated to be ≈ 180 $MJ\ kg^{-1}$, but the energy intrinsic to the recycling process is $\approx 33\ MJ\ kg^{-1}$. The embodied energy of chromium itself is estimated as 300 $MJ\ kg^{-1}$.

Other prominent exploitations of the properties of chromium are in gemstones and pigments. Ruby contains Cr(III) ions dispersed in the corundum structure (Fig. 3.7), where the transition metal ions replace some of the Al^{3+} in the octahedral sites. Both the electronic absorption spectra and fluorescence effects from the three *d*-electrons in the *d*-orbitals combine to provide the deep red colouration. In other environments, such as glasses, the ligand field differs and affords a green colouration. The well-known yellow pigments chromates are a result of tetrahedral Cr(VI) centres. These are d^0 sites, and as a result the yellow colouration must be due to ligand-to-metal (LMCT) charge transfer transitions. Although chromium is known to be an essential element to humans, in the wrong biochemical location it can be toxic. The chromate anion $[CrO_4]^{2-}$ can be mistaken for sulphate or phosphate in some sites. However, the Cr(VI) can effect unwanted oxidation at the binding sites.

3.3.3 Iron

Iron is the fifth most abundant element in the universe, and the fourth most abundant in the crust. Supply should not be an issue so iron appears very low down in the risk index (Table 2.1). In 2017, mine production of usable ore was 2,400 Mt with an iron content of 1,500 Mt. Current reserves are estimated at 170,000 Mt (83,000 Mt of Fe). By far the largest use of the ore is conversion into pig iron and steel. In 2017 their production was estimated at 1,200 and 1,700 Mt, respectively. Iron ores generally contain Fe(II) or Fe(III) in oxidic environments, including oxides, hydroxides, carbonates, and silicates. The oxides *magnetite*, Fe_3O_4, and *haematite*, Fe_2O_3, are the main precursors extracted from mining sites.

Haematite adopts the same crystal structure as corundum, Al_2O_3 (Fig. 3.7). In magnetite, the irons adopt both Fe(II) and Fe(III) oxidation states. The ambient temperature structure of magnetite (and also that of *chromite*) is related to that of the mineral *spinel*, $MgAl_2O_4$. That structure is based upon a cubic close-packed array of oxide ions. In that array, there will be one octahedral and two tetrahedral holes (inclusion sites) per oxide anion. The element of lower charge, Mg^{2+}, adopts the lower coordination site and so will occupy $\frac{1}{8}$ of the tetrahedral sites, leaving the higher-charged Al^{3+} to occupy $\frac{1}{2}$ of the octahedral sites. The unit cell contains eight formula units, and thus will include eight magnesium, sixteen aluminium, and thirty-two oxygen sites (Fig. 3.30). The site preference is dominated here by the higher electrostatic attraction of having the cation with the higher charge close to the higher number of oxide anions. The second factor is the size of the cavity. Normally cavity sizes increase with coordination number, and so the preference might be for the larger ion to adopt the octahedral site. In this case, the radius of four-coordinate Mg^{2+} (58 pm) is similar to that of six-coordinate Al^{3+} (53 pm) and thus provides similar M–O distances through the lattice. Switching the sites around would give a substantial size disparity (The radii of six-coordinate Mg^{2+} and four-coordinate Al^{3+} are 72 and 39 pm, respectively.).

In the case of magnetite, the ions in question will be Fe(II) and Fe(III). On the basis of charge, then, the higher-charged ion would be anticipated to adopt the octahedral site, as in spinel itself. Since the ligands are of a weak

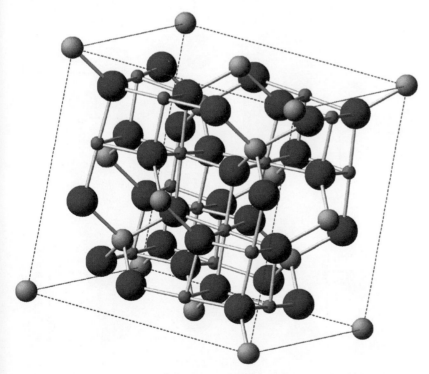

Figure 3.30 *The unit cell of $MgAl_2O_4$; red = O, yellow = Mg, blue = Al.*

Table 3.11 *Crystal ionic radii of high-spin iron (II) and (III).*

Site	Radius /pm
Fe(II), tet.	63
Fe(II), oct.	78
Fe(III), tet.	49
Fe(III), oct.	65

field type (O^{2-} has both σ and π donor properties), both the oxidation states of this $3d$ metal are expected to be in a high-spin state. The ionic radii for Fe(II) and Fe(III) sites (Table 3.11) portray a similar trend to those of magnesium and aluminium. However, in this example there is a further site-related energy preference. Returning to the ligand field energy level diagram in Figs 3.23 to 3.25, it is evident that there is no ligand–field stabilization energy for a high-spin d^5 ion in either octahedral or tetrahedral sites. However, the extra electron in the d^6 will afford LFSE in both sites: $\frac{2}{5}\Delta_{oct}$ and $\frac{3}{5}\Delta_{tet}$ for six- and four-coordinate sites respectively. Using the relationship between the two values of Δ (Eq. 3.26), the LFSE for tetrahedral site can be re-expressed as $\frac{4}{15}\Delta_{oct}$. This provides a site preference of $\frac{2}{15}\Delta_{oct}$ for Fe(II) in the octahedral site. The result of this is that in magnetite all of the Fe^{2+} ions adopt octahedral sites and half of the Fe^{3+} ions move to tetrahedral sites, in what is called an *inverse spinel* structure (Fig. 3.31).

The presence of these three different sites for ions with unpaired electrons gives rise to one of the main characteristics of magnetite. All materials have a type of magnetic behaviour called *diamagnetism*. In that the electronic clouds around

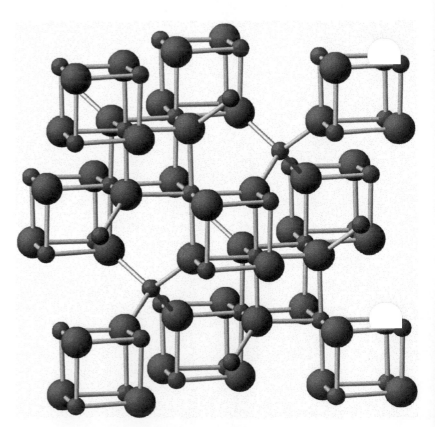

Figure 3.31 *The structure of Fe_3O_4; large = O, small = Fe.*

a nucleus distort slightly to oppose a magnetic field applied to the sample. An isolated site with unpaired electrons gives rise to *paramagnetism* (Fig. 3.32). The magnetic moment created by the spin-free electrons will align parallel with an applied magnetic field. However, once that field is removed the orientation of the moments rapidly randomizes, as depicted in Fig. 3.32. However, if ions interact with each other then there is an energy difference between having neighbouring moments aligned parallel (*ferromagnetism*), as in metallic iron, or antiparallel (*antiferromagnetism*), as in nickel oxide, NiO. As the temperature is raised, thermal motions tend to oppose this ordering, and eventually at a transition temperature the spins will become randomized as the sample changes into a paramagnetic material. These are called the *Curie* (T_C) and *Néel* (T_N) temperatures for ferro- and antiferro-magnetic materials, respectively. For metallic iron, which is ferromagnetic, T_C is 1,043 K.

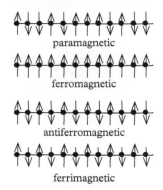

Figure 3.32 *Classes of magnetism.*

The simplest means of estimating the magnetic moment of an ion is using the *spin only* formula (Eq. 3.28), where S is the spin quantum number of the unpaired electrons on the ion. This is a reasonable approximation for most *3d* metal ions, although those with high symmetry sites also display an additional (orbital) contribution.

$$\mu_{eff} = 2\sqrt{[S(S+1)]}. \tag{3.28}$$

Using Eq. 3.28, the magnet moments for high-spin Fe(II) ($S = \frac{4}{2}$) and Fe(III) ($S = \frac{5}{2}$) can be estimated to be 4.9 and 5.9 μ_B, respectively. In Fe_3O_4, the moments in the octahedral sites are opposed (antiferromagnetically coupled) to those in the tetrahedral sites. The tetrahedral sites are occupied one Fe(III) site per formula unit, whilst the octahedral ones contain one each of Fe(II) and Fe(III). Hence there is a net permanent moment for the excess Fe(II) site of ≈ 4.9 μ_B. This is an example of *ferrimagnetism*. Magnetism will be considered further in Chapter 4 (Fire) in the context of turbine materials.

The production of steel is one of the major processes on Planet Earth, in large part extending the structures on its surface and providing transport. In terms of structural materials, it is only exceeded by cement in terms of annual production. An outline of the steps involved is presented in Fig. 3.33. The generally oxidic ores are first treated in a blast furnace with coke as the primary reductant. Hence evolution of CO_2 is intrinsic to the chemical reduction to iron. The iron formed in this process has a high carbon content ($\approx 4\%$). In the steel-making step the carbon content is reduced to a much lower level ($\approx 0.4\%$), and thus a further, smaller production of CO_2 occurs. The recycling of ferrous goods provides scrap iron and steel. This does not need to be incorporated into the blast furnace, but instead undergoes a purification procedure in an electric arc furnace.

The blast furnace (Fig. 3.34) incorporates chemical steps listed in Table 3.12. The overall reduction of ores like haematite is carried out by CO, hence CO_2 is a by-product. The CO is formed from carbon as coke introduced through the top of the furnace, and from hot air injected towards the bottom. The oxidation of carbon is highly exothermic, and the region near the air injection may be

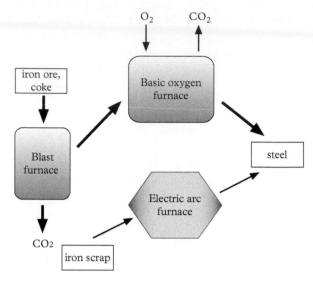

Figure 3.33 *The outline of the processes of the manufacture of steel.*

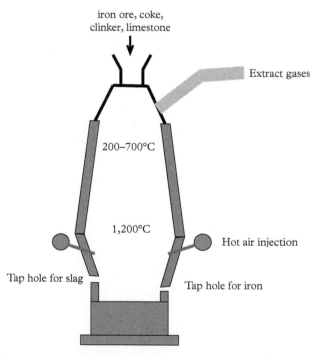

Figure 3.34 *Schematic of a blast furnace for the manufacture of pig iron.*

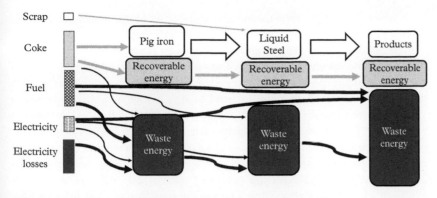

Scrap

Coke

Fuel

Electricity

Electricity losses

Figure 3.35 *Schematic of the exergy flow for steel production.*

Table 3.12 *Prediction of the thermodynamics per Fe atom for the production of iron under standard conditions.*

Reagents	Products	$\Delta_r H^{\circ}$ /kJ mol^{-1}
$\frac{1}{2} Fe_2O_{3(s)} + \frac{3}{2} CO_{(g)}$	$Fe_{(s)} + \frac{3}{2} CO_{2(g)}$	-3.4
$\frac{1}{2} Fe_2O_{3(s)} + \frac{1}{6} CO_{(g)}$	$\frac{1}{3} Fe_3O_{4(s)} + \frac{1}{6} CO_{2(g)}$	1.1
$\frac{1}{3} Fe_3O_{4(s)} + \frac{1}{3} CO_{(g)}$	$1\, FeO_{(s)} + \frac{1}{3} CO_{2(g)}$	6.6
$FeO_{(s)} + CO_{(g)}$	$Fe_{(s)} + CO_{2(g)}$	-11.1
$C_{(s)} + \frac{1}{2} O_{2(g)}$	$CO_{(g)}$	-55.3
$\frac{1}{2} C_{(s)} + \frac{1}{2} CO_{2(g)}$	$CO_{(g)}$	86.3

near 2,000 °C. Towards the top of the furnace, the Fe(III) is partially reduced to Fe_3O_4, which itself is reduced to FeO_{1-x} at temperatures near 850 °C. The final reduction step occurs at $\approx$ 1,200 °C. The CO is in equilibrium with its favoured disproportionation products C and CO_2 in the *Boudouard reaction*. The iron forms a melt at the base of the furnace and is tapped off from there. The other additives, a clinker based upon Portland cement (Section 3.5.2) and limestone, are used to bind with the impurities in the ores. These form a slag which floats on the surface of the denser iron, and is skimmed off periodically.

Overall the chemical process is slightly exothermic, but the steps from the ores through to the formation of pig iron, the decarbonisation to form liquid steel, and then its processing into final products are all energy consuming, as shown in a simplified Sankey diagram (Fig 3.35). The exergy of these finished products amounts to *c*.18% of the total in the process worldwide. The electricity generation

Table 3.13 *Embodied energy (EE) and carbon (EC).*

Material	EE	EC
	MJ/kg	kgCO$_2$e/kg
iron	25	2.0
steels:		
average	20.1	1.46
virgin	35.4	2.89
recycled	9.46	0.47

Table 3.14 *Mine production in 2017 and reserves (kt).*

Element	2017	Reserves
cobalt	110	7,100
nickel	2,100	74,000
copper	19,700	790,000

losses amount to *c.*25% of the total exergy at the onset of the processing, and this rises to *c.*73 % overall; about 10% of the exergy can be recovered.

The embodied energy of iron and steels follow an expected trend (Table 3.13). There is an increase in the EE going from iron to virgin steel (from 25 to 35.4 MJ kg^{-1}). Also the incorporation of recycled steel, which avoids the blast furnace process, reduces the mean value for steel to 20.1 MJ kg^{-1}. This is very much lower than the value for the average mixture of virgin and recycled aluminium (155 MJ kg^{-1}). Intrinsically, the stoichiometry of Eq. 3.29 implies that each kg of iron will result in the evolution of 1.18 kg of CO$_2$. This is more than doubled by the processes in manufacturing virgin steel to 2.71 kgCO$_2$, with additional contributions by other greenhouse gases to a value of 2.89 kgCO$_2$e. Again, this can be much reduced by recycling. Overall, the emissions of CO$_2$ in 2013 were estimated to have been 3.50 Gt (2.82 Gt from the direct process and 0.69 Gt from associated power requirements). Because of this scale, optimization of steel manufacture can effect significant reductions in global emissions.

$$\frac{1}{2}Fe_2O_3 + \frac{3}{2}CO = Fe + \frac{3}{2}CO_2. \tag{3.29}$$

3.3.4   Cobalt, Nickel, and Copper

The crustal abundances of all these elements are very similar, $\approx$ 27 ppm. The supply risk is envisaged as being quite different, substantially due to the location of the major supplies; cobalt supply is seen to be the most fragile (Table 2.1). The production of all of these elements is substantial (Table 3.14) with significant reserves, but the values differ by about an order of magnitude from cobalt to nickel and again to copper, hence the sensitivity about cobalt supplies. In some locations, copper is present as the native metal, but the sulphides are the most commonly used ores. The commercial ores *chalcopyrite* (CuFeS$_2$) and *bornite* (Cu$_5$FeS$_4$) are are iron-containing, but the copper sulphides, CuS (*cloverite*) and Cu$_2$S (*chalcocite*) are also available. Nickel ores also tend to be associated with iron, with two classes being hydrous oxides, for example *laterite* (Fe,Ni)O(OH), and sulphides, principally *pentlandite* (Ni,Fe)S$_8$. Cobalt is also present in many copper and nickel minerals, but is also associated with arsenic as in *cobaltite* (CoAsS), *erythrite*, (Co$_3$(AsO$_4$)$_2$·8H$_2$O), *glaucodot* (Co,Fe)AsS, and *skutterudite* (CoAs$_3$).

The consumption of refined copper in 2017 is estimated to have been 24 Mt, in excess of production in that year. The vast majority of this was to manufacture the metal and its alloys, notably bronze and brass (5% of total use). Copper's high electrical and thermal conductivity are key to some applications, as is its relative inertness to oxidation. The main uses are in electrical wire (60%), buildings (20%), and machinery (15%). Coinage is also a significant application of copper and nickel, as alloys. Most applications of nickel are as alloys, either ferrous (48%) or non-ferrous (40%). The latter include superalloys with refractory metals like tungsten and rhenium, which are for very high-temperature applications such as in aircraft turbines. An increasing use has been in rechargeable batteries using nickel metal hydride (NiMH) and lithium nickelates. Cobalt too has been most used in high-performance alloys like superalloys, but its applications have been transformed by its being a component of rechargeable lithium ion batteries as lithium cobaltates, for example $LiCoO_2$. It is estimated that 80% of the consumption of cobalt by China is for this application.

Extraction of the metals from the ores again requires energy input, especially as their ores may be dilute. There are also problematic byproducts like SO_2 and As_2O_5. In the main extraction is via a *pyrometallurgical* approach. The ore is concentrated to $\approx$ 30% and then smelted with air or oxygen. With a chalcopyrite mineral, $CuFeS_2$, the smelting process converts the iron to oxides, mainly as FeO which is converted into a slag with silica. At high temperatures the CuS will be converted to impure copper and SO_2. SO_2 is collected for conversion into sulphuric acid. The impure copper is converted back to Cu(II) in a separate process and deposited as highly pure copper via electrochemical reduction (electrowinning). Pyrometallurgical methods are also employed in the extraction of cobalt and nickel. Often all three metals coexist and require separation.

An alternative is the *hydrometallurgical* extraction method. As noted above (Eq. 3.27), metals show different binding strengths to ligands according to the Irving–Williams series, and Cu^{2+} often has the strongest binding of *3d* dications. This process operates on oxidic precursors, such as the ores *azurite* ($2Cu(CO_3)\cdot Cu(OH)_3$), *brochantite* ($CuSO_4$), *chrysocolla* ($CuSiO_3\cdot 2H_2O$), and *cuprite* (Cu_2O), and also on oxidized materials from sulphidic mining sites. Oxime ligands of the type shown in Fig. 3.36 will establish an equilibrium Eq. 3.30: a competition between the metal ion and the proton for the phenolic site. The complex with the highest binding constant will compete most strongly and thus form at the lowest pH. Hence by changing the pH, differing metals will bind to the chelating ligand and be selectively extracted. Selectivity between copper and iron is a critical factor. Hydroxy-aldoximines (R' = H) and -ketoximines (R' = Me or Ph) are employed as mixtures, all having long-chain, for example nonyl or dodecyl, substituents *para* to the phenol group. There is a wide range of solvent/oxime mixes employed due to the variation in local requirements.[17]

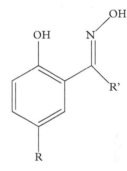

Figure 3.36 *Copper extractant agents.*

$$M^{2+} + 2HL = ML_2 + 2H^+. \tag{3.30}$$

Hydrometallurgical methods have also been developed for extraction of cobalt and nickel. In particular these have proven to be effective for the laterite type of ore, which has been difficult to concentrate (Table 3.15).[18] It is evident from comparison with Table 3.13 that the environmental impact per unit mass is an order of magnitude larger for these elements. In large measure this is due to the increased dilution of the ores and the added complexity of element separation. For nickel the pyrometallurgical approach from ferronickel sulphide ore delivers 95% recovery of the nickel and provides baseline EE and EC estimates.

[17] G. A. Kordosky, 'Copper recovery using leach/solvent extraction/electro-winning technology: Forty years of innovation, 2.2 million tonnes of copper annually', *J. South African Inst. Mining Metall.*, 2002 (Nov./Dec.), 445–50.

[18] T. Norgate and S. Jahanshahi, 'Assessing the energy and greenhouse gas footprints of nickel laterite processing', *Mineral. Eng.*, 2011, **24**, 698–707.

Table 3.15 *Embodied energy (EE) (MJ kg^{-1}), embodied carbon (EC), and metal recovery (%) for nickel laterite processing.*

Type	Process	EE	EC	Ni recovery
Pyrometallurgical	Ferronickel	236	22	95
Hydrometallurgical	HPAL	272	27	92
Pyro/hydrometallurgical	Caron process	565	45	80

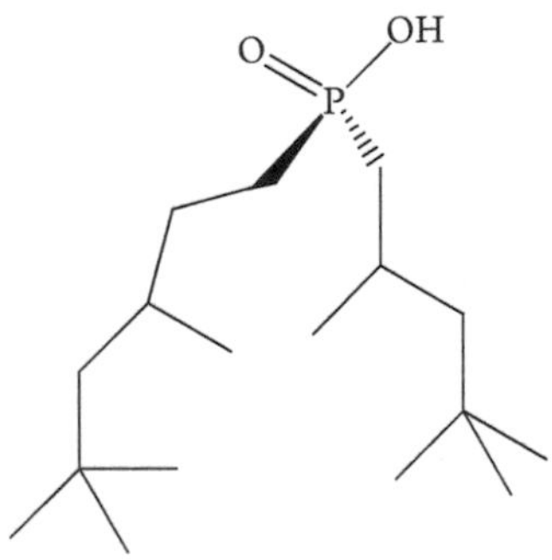

Figure 3.37 *Cobalt/nickel separation agent.*

Table 3.16 *Embodied energy (EE) and carbon (EC).*

Material	EE	EC
	MJ/kg	kgCO$_2$e/kg
cobalt	280	
nickel	164	13
copper:		
average	42	2.71
virgin	57	3.81
recycled	16.5	0.84

[19] Gadolinium is the other element which is ferromagnetic near room temperature (T_C 292 K).

The high-pressure acid leach (HPAL) hydrothermal process affords a similar extraction efficiency, increases the range of viable minerals, but, at present, does not effect an improvement in the EE and EC values. One combined process (Caron process), which incorporates high- and low-temperature steps and includes conversion to the metal twice, was shown to have much higher energy and carbon requirements. Optimizing these processes can have environmental significance.

Related solvent extraction systems have been developed around phosphinic acids, including the commercial extractant Cyanex 272 (Fig. 3.37). Again, once deprotonated, the extractant can form a monoanionic chelating ligand. This has a high affinity for cobalt(II), and can effect high selectivity against calcium and nickel at relatively low pH (3–4). The difficulty of separating metals from their complex ores emphasizes the gain in recycling. From Table 3.13 it was evident that the embodied energy and carbon in scrap steel help reduce the average requirements of steel manufacture. In the cases of these elements, which have much higher EE and EC values, this should be increasingly rewarding. As can be seen in Table 3.16, the mean value for copper is also strongly reduced by the recycling rate ($c.\frac{1}{3}$). Equivalent figures are not available for nickel and cobalt, but this should become evident in future. In 2017, the recycling rate for nickel was estimated to be 68%, with that figure also reported for recycling of cobalt in the EU.

As metals, copper and nickel adopt an fcc structure (Fig. 3.8). For cobalt this structure occurs at $\approx$ 700 K, below which point Co favours an hcp structure (Fig. 3.12). Like iron, both cobalt and nickel display ferromagnetic properties, with T_C values of 1,388 and 627 K, respectively.[19]

3.3.5 Platinum Group Metals (PGMs)

The Platinum Group Metals (PGMs) comprise six elements from periods 5 and 6, in groups 8 to 10: ruthenium, rhodium, palladium, osmium, iridium, and platinum. All of these elements are considered to occur in the ppb range in the Universe. Within the Earth's crust they are also in low abundance. Estimates fluctuate with time, but currently palladium and platinum are considered to be

Table 3.17 *Production and use of PGMs in 2017 (t).*

Element	Catalysts	Jewellery	Total demand	New supply	Recycled
Platinum	102	69	244	186	62
Palladium	262	6	316	205	91
Rhodium	27	–	33	23	10
Iridium			8	1.6*	
Ruthenium			37	13*	
Osmium				0.09*	

*Imported into USA.

in highest abundance (0.0015 ppm) and rhodium, osmium, and iridium the least abundance ($\approx$ 0.00004 ppm). Four of the metals (Rh, Pd, Ir, and Pt) adopt an fcc structure (Fig. 3.8); the other two, ruthenium and osmium favour the hcp structure (Fig. 3.2).

PGMs are placed at the same level of supply risk as lithium (Table 2.1). The major supplies are from RSA, Russia, and Zimbabwe. Mine productions of platinum and palladium were similar in 2017, amounting to $\approx$ 200 t. The overall resources of the PGMs are estimated to be in excess of 100,000 t. By comparison, the mine production of gold in 2017 has been estimated to be 3,150 t, with reserves totalling 54,000 t. More detail of the variation in PGM demand and supply in 2017 is given in Table 3.17.[20] The total demand for platinum was essentially met by a combination of new and recycled supplies. The major demand was for catalysts, primarily for automotive exhaust remediation, with fuel cells also of increasing importance. For platinum there is a considerable demand for the metal in jewellery. For palladium and rhodium catalytic applications dominate the demand. Recycling provided about 160 t of platinum, palladium, and rhodium in 2017, largely from automobile catalytic converters. Examples of these uses will be developed in Chapters 4 (Air) and 5 (Fire). The other three elements are considered to be the minor PGMs. Demand for ruthenium though is similar to that for rhodium, mainly in electronic applications either as a thin metallic film to harden contacts of palladium or platinum, as RuO_2, or as ruthenates of lead and bismuth in thick-film chip resistors. Applications for iridium span electrical, electrochemical, and other chemical sectors. As can be seen the demand for osmium is very low. To a significant extent this can be correlated with the ready formation of its volatile and hazardous oxide, OsO_4.

PGMs are generally associated with sulphide ores containing nickel and copper as the prevalent metals. Gold is often in these deposits, along with chromium, cobalt, silver, selenium, and tellurium. Typically the ores may contain 1–2% of nickel and copper but only 2–10 ppm of PGMs. Ore grade is classified on the concentration of four major precious elements: Au, Pt, Pd, and Rh (4E in g

[20] *PGM Market Report 2018*, Johnson Matthey PLC, February 2018.

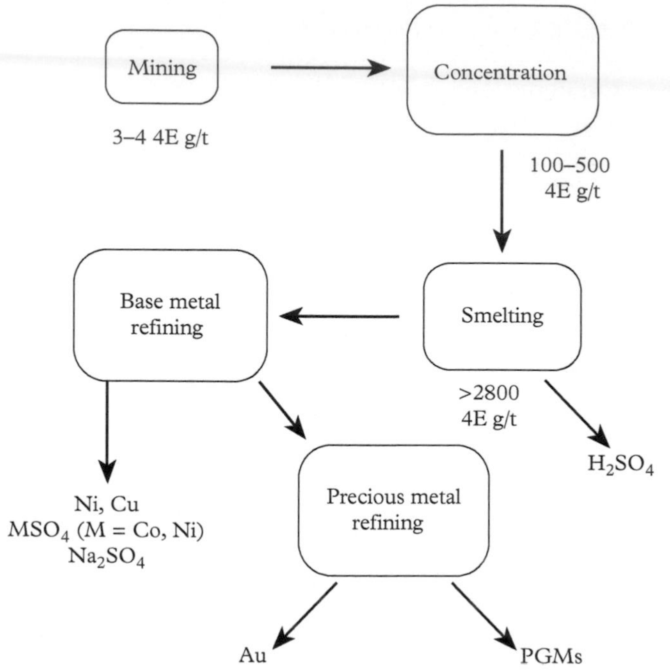

Figure 3.38 *Schematic of the processes in refining PGM ores.*

t^{-1}). An outline of the refining process is shown in Fig. 3.38. Ores typically with a PGM concentration of 3–4 ppm are crushed, milled, and concentrated in a froth flotation unit. The concentrate powder is dried and smelted in an electric furnace with air or O_2 blown through converting the sulphide to SO_2, from which sulphuric acid can be manufactured. Refining processes as described in Section 3.3.4 allow separation of base elements as metals or sulphates, leaving an extra stage for the separation of the PGMs and gold.

The classical way of separating the elements was through a sequence of precipitation steps in high-chloride-containing media (Fig. 3.39).[21] Treatment with concentrated HCl/HNO_3 dissolves out Ag, Au, Pd, and Pt in different oxidation states. Silver chloride is then precipitated with further addition of HCl, and subsequently gold(I) is reduced and precipitated as the metal. The remaining group 10 elements are then separated in turn, with $(NH_4)_2[PtCl_6]$ precipitated first on addition of NH_4Cl. Palladium is extracted as $[PdCl_2(NH_3)_2]$ by sequential treatment with ammonia to form $[Pd(NH_3)_4]^{2+}$ and addition of HCl. The residue from the initial dissolution is smelted with PbO and carbon, and additional silver can be extracted from the resulting solid mixture with nitric acid. The remainder is combined with the residues from the previous stream and this is oxidized with sodium peroxide/carbonate at high temperature. From this process the group 8 elements form $[M^{VI}O_4]^{2-}$. Once separated, these can be oxidized

[21] F. L. Bernardis, R. A. Grant and D. C. Sherrington, 'A review of methods of separation of the platinum-group metals through their chloro-complexes', *React. Funct. Polym.*, 2005, **65**, 205–17.

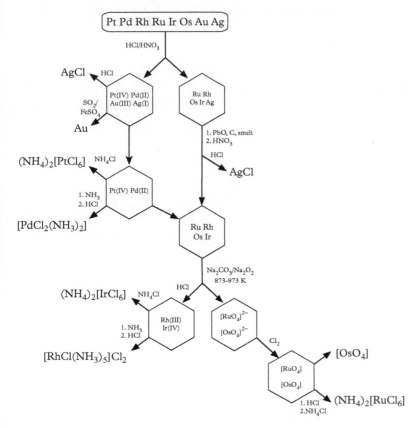

Figure 3.39 *Schematic of one of the classic PGM separation processes.*

to their volatile tetroxides. Osmium can be stored in that form, but RuO_4 is explosive and so is reduced, for example, to the more tractable $(NH_4)_2[Ru^{IV}Cl_6]$. The group 9 elements are extracted after the oxidation step by HCl, and in a similar manner to the extraction of the group 10 pair, sequential precipitation of $(NH_4)_2[Ir^{IV}Cl_6]$ and $[Rh^{III}Cl(NH_3)_5]Cl_2$ can be effected.

These processes require several high temperature-steps and also afford rather poor selectivity in the separation steps. There is a tendency for co-precipitation and also entrainment of the liquid phase with the precipitate. Many of these problems are addressed in modern processes (Fig. 3.40) which can also incorporate solvent extraction and ion-exchange procedures. This scheme shows separation of an ore without osmium, which is not uncommon. The precious metal residues after the base metal refining (Fig. 3.38) are dissolved with a high chloride concentration. Under such conditions the silver is dissolved as $[AgCl_2]^-$. Reducing the chloride concentration allows Eq. 3.31 to move towards the left, and thus silver chloride will precipitate. Gold can be extracted as $[Au^{III}Cl_4]^-$ using methyl*iso*butylketone as solvent. Coordination of Pd(II) using a hydroxyoxime

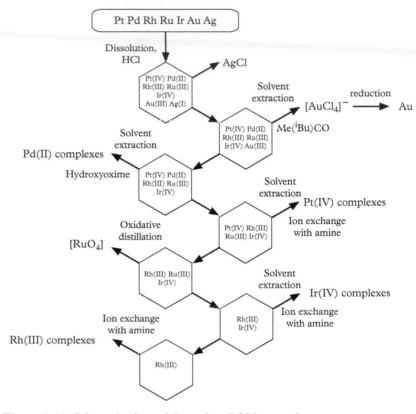

Figure 3.40 *Schematic of one of the modern PGM separation processes.*

ligand (Fig. 3.36) allows it to be removed as a neutral complex by solvent extraction, as in Eq. 3.30. A combination of solvent extraction and ion exchange using an amine-bearing resin can then be used to extract platinum in the +IV oxidation state. As in the classical process, ruthenium (and presumably osmium if present) can be oxidized to its volatile tetroxide and distilled away. This leaves the group 9 elements again, which can be extracted with the appropriate solvent and ion exchange conditions.

$$AgCl + Cl^- = [Ag^I Cl_2]^-. \tag{3.31}$$

The complexity of the processing required to purify compounds from such dilute ores would be anticipated to result in a much higher mass-related environmental profile as compared to the base metals, given that the ores are *c.*10,000 times more dilute in PGMs (Table 3.16). A life cycle analysis (LCA) study of the PGMs has been carried out across the larger part of the global production, with the results collected in Table 3.18.[22] The EE and EC values are in the region of 1,000–2,000 times larger than for cobalt and nickel. It was estimated that $\approx$ 72% of the EE value is due to the power consumption of mining and

Table 3.18 *Embodied energy (EE) and carbon (EC).*

Material	EE	EC
	MJ/kg	kgCO$_2$e/kg
platinum	387,000	33,000
palladium	304,000	25,000
rhodium	3460,00	30,000

[22] T. Bossi and J. Gediga, 'The Environmental Profile of Platinum Group Metals', *Johnson Matthey Technol. Rev.*, 2017, **61**, 111–21.

concentrating. Since the majority of the production is in South Africa the global EC values are very dependent upon the power mix of that country; presently over 90% of electricity there is derived from hard coal. The bulk of the remaining environmental impact (27%) is from smelting and refining processes. Clearly, then, the benefits of recycling are very high.

3.3.6 Molybdenum and Tungsten

Whilst the abundance in the universe of molybdenum (0.005 ppm) is an order of magnitude higher than that of tungsten (0.0005 ppm), their crustal abundances are estimated to be similar (Mo 0.8, W 1.0 ppm); of the *d*-block elements, molybdenum has the third-highest crustal abundance in row 5 (after Zr, 132 ppm, and Nb, 8 ppm) and tungsten is the second most abundant in row 6 (after Hf, 3 ppm). In the BGS Risk List (2015) these two group 6 elements are given the same score as cobalt, and indium. Their production (2017) straddles that of cobalt, with molybdenum higher (290 kt) and tungsten lower (95 kt). Reserves of both are significant: Mo 17 Mt and W 3.2 Mt.

The ores for the two elements are different in type. For molybdenum, the principal source is the Mo(IV) sulphide, *molybdenite* (MoS_2), whilst tungsten is extracted from minerals of the tungstate(VI) anion as in *wolframite* ((Fe,Mn)WO_4), and *scheelite* ($CaWO_4$). In both cases production of the element takes routes through the trioxide, MO_3. WO_3 can be reduced to the metal by hydrogen or carbon. Hydrogen reduction is also employed for molybdenum, but another method utilized is reduction with aluminium in the presence of iron to form the alloys ferromolybdenum.

Many of the uses of these elements relate to their values of $\Delta_{at}H^o$, as evident in Fig. 3.18. This shows the potential for strong binding to these metals. For molybdenum much of the production is used in metallurgical applications, the majority in high-strength and -temperature steels. Other applications are in superalloys and as MoS_2, which has a layered structure and can be employed as a high-temperature lubricant. The lattice consists of a sandwich of hexagonal layers of S–Mo–S, which have weaker S...S interactions to the next sandwich; this provides a mechanism for slippage of the layers appropriate for the lubrication application. For tungsten, which has the highest $\Delta_{at}H^o$, there are applications in very-high-temperature materials, such as rocket nozzles, and also in superalloys. However, the principal application is as its carbide, WC, a material of exceptional hardness. In that structure, carbon forms hexagonal close-packed-type layers, but are eclipsed. These are interspersed by tungsten in trigonal prismatic six-coordinate geometry. This carbide is generally the material of choice for abrasives, cutting tools, and drills. The particular properties of these elements render it difficult to find substitutes. The group 7 elements are not viable due to the low abundance of rhenium and the radioactivity of technetium. Perhaps only the group 4 and 5 elements may be possible substitutes in some applications, albeit with poorer properties.

Table 3.19 *Rare earth elements (REE).*

REE	Abundance
Ce	25th
Nd	3rd REE
Reserves	
China	
USA	
Canada	
Australia	
RSA	
India	
Brazil	
Production	%
China	97
REE	Cost/kg (2017)
Nd	$60

Table 3.20 *Valence electrons of the lanthanides.*

	Ln(g)
La	$5d^1 6s^2$
Ce	$4f^2 6s^2$
Pr	$4f^3 6s^2$
Nd	$4f^4 6s^2$
Pm	$4f^5 6s^2$
Sm	$4f^6 6s^2$
Eu	$4f^7 6s^2$
Gd	$4f^7 5d^1 6s^2$
Tb	$4f^9 6s^2$
Dy	$4f^{10} 6s^2$
Ho	$4f^{11} 6s^2$
Er	$4f^{12} 6s^2$
Tm	$4f^{13} 6s^2$
Yb	$4f^{14} 6s^2$
Lu	$4f^{14} 5d^1 6s^2$

3.4 *f*-Elements

3.4.1 Lanthanides

The lanthanides (Fig. 2.8, Table 3.19) are also called the rare-earth elements (REE). The crustal abundances estimated for the lanthanides vary from 43 ppm (cerium) down to 0.3 (lutetium), and thus all are in much higher abundance than the PGMs. However, their placement at the top of the BGS risk list is due to the restricted range of supply as separated elements.

They occupy their own slice of of the periodic table in period 6 with their own unique set of properties. This arises from the close proximity in energies of the *4f*, *5d*, and *6s* orbitals. They have relatively low electronegativities (Figs 2.10, 2.11) with atomic radii in the region of 180 pm (Figs 2.10, 2.12). The filling of the core-like *4f* orbitals across the series results in an increasing effective atomic number experienced by the valence electrons (Table 3.20). The atomic radii are governed by the radial distribution of the outermost valence orbitals, namely the *6s*, which are poorly screened by the electrons in the spatially oriented *4f* electrons. Hence there is an increase in the ionization potentials and reduction in the covalent radius across the lanthanides. The overall effect in the periodic table is the **lanthanide contraction**, and this results in the radii of elements in the *5d* transition series being virtually the same as those of their congeners in the *4d* series.

The dominant oxidation state is +III with polar donor ligands; there are additional oxidation states near to f^0, f^7, and f^{14}. For the larger ions of the early members of the series, high coordination numbers are common, but these become less prevalent across the series.

The enthalpies of formation of the gaseous M^{3+} ions show a gradual increase across the series (Fig. 3.41), with maxima accompanying breaking the f^7 (Eu, europium) and f^{14} (Yb, ytterbium) subshells. This can be understood in terms of the **exchange energy**. This is the energy released when two or more indistinguishable electrons exchange their positions in a degenerate orbital set. For each element there is a value of the exchange constant, K. The number of K units in the exchange energy, E, depends upon the combinations of pairs possible in the orbital set. For the f block with a degenerate set of 7 orbitals, this effect is high. We can examine this around a half-filled shell using Eq. 3.32. The ionization from f^7 to f^6 would mean a change from exchange energy of 21 K to 15 K. In contrast the ionization from f^8 to f^7 involves no change in exchange energy. The eighth electron is distinguishable from the other seven since it has the opposite spin. Instead there is a loss of pairing energy, $\mathcal{J}$, which is a repulsive term.

$$E = K\frac{n(n-1)}{2}. \tag{3.32}$$

As expected, the fourth ionization energy of lanthanum, La, is very high as it breaks into the the xenon core, but the value for lutetium, at the end of the series and so breaking into the *4f* subshell, approaches that. Lower barriers to M(IV)

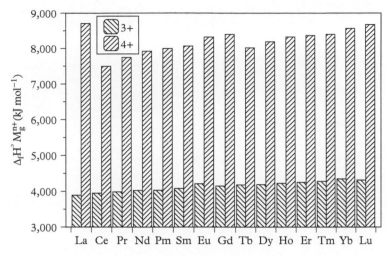

Figure 3.41 *Enthalpies of the formation of the gaseous ions M³⁺ and M⁴⁺ of the lanthanides.*

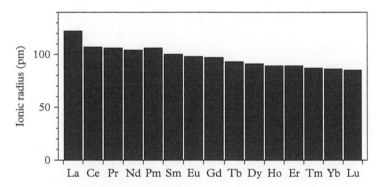

Figure 3.42 *Ionic radii of the Ln³⁺ ions.*

are expected in the early members of the series (cerium, Ce, and praseodymium, Pr) and also at terbium, Tb, which would afford the f^7 configuration.

In the main, the effect on bond strengths by the reduction in radii across the lanthanide series (Fig. 3.42) more than offsets the general increase in ionization energy, as reflected in the heats of formation of the M_2O_3 solids (Fig. 3.43). There are two clear outsiders in the trend at europium and ytterbium, where the increased ionization energies, due to the loss of exchange energies, have an effect. These two elements have well characterized $M^{II}O$ phases. Conversely, the lower values of the fourth ionization energies of cerium, praseodymium, and terbium allow for the stability of $M^{IV}O_2$ crystalline solids. However, the most striking characteristic of the lanthanide series is the broad similarity of the chemistry, and thus the series will be considered together.

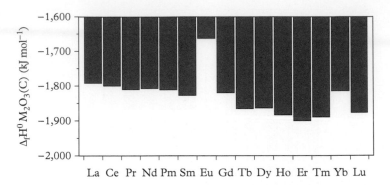

Figure 3.43 *Standard heats of formation for crystalline M_2O_3 of the lanthanide elements.*

[23] Y. Ni, J. M. Hughes, and A. N. Mariano, 'Crystal chemistry of the monazite and xenotime structures', *Amer. Mineral.*, 1995, **80**, 21–6; 'The atomic arrangement of bastnäsite-(Ce), Ce(CO₃)F, and the structural elements of synchysite-(Ce), röntgenite-(Ce), and parisite- (Ce)', *Amer. Mineral.*, 1993, **78**, 415–8.

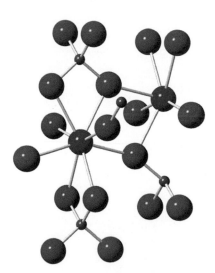

Figure 3.44 *Local structure of xenotime–Tb; red = O, purple = Tb, grey = P.*

The most common ores of the lanthanides (given the generic element symbol, Ln) are the phosphates, $LnPO_4$, and carbonates, for example $Ln(CO_3)F$, *bastnäsite*.[23] The most common $LnPO_4$ minerals are *monazite* and *xenotime*. The coordination site created in xenotime is eight-coordinate, and is preferred for the later members (Tb–Lu) (Fig. 3.44). It is isostructural with *zircon*, $ZrSiO_4$. The eight-coordinate structure can be seen at the purple site to the left of the figure. Overall the coordination geometry is that of a triangulated dodecahedron. That itself consists of two interconnected tetrahedra. The first of these is elongated in the vertical direction, and is formed by two chelating PO_4 ligands. In between them are four other monodentate phosphate ligands which form a flattened tetrahedron. Neighbouring metal ions are linked by four-membered rings. In monazite the phosphate tetrahedra are arranged to produce a coordination site of nine oxygen atoms at the lanthanide, and this preferentially incorporates the larger early members of the series (La–Gd). The lower of the two metals in Fig. 3.45 shows this nine-coordinate geometry. Again there are two *trans* bidentate phosphate ligands which are in orthogonal planes. In this case though there are five monodentate phosphate ligands in the central belt. Again there are four-membered rings linking neighbouring metal ion sites. Bastnäsite also provides a nine-coordination site for the lanthanides (three F and six O donors) (Fig. 3.46). This geometry is shown for the metal site on the left-hand side of the figure. All the carbonate ligands are monodentate to a particular metal. The overall geometry is of a tricapped trigonal prism, with each layer composing two oxygen and one fluorine donor. The prism is made up from the uppermost triangle being eclipsed by the bottom one, and one fluorine is above the other. The two triangles are linked by three square faces, and over each of those is a capping ligand. The fluorine is at the maximum distance from the other two. The neighbouring metal atom is linked by a cage with one fluorine, an oxygen, and an O–C–O link. This illustrates the marked difference to the *d*-elements in terms of coordination preferences.

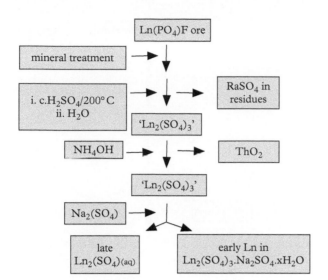

Figure 3.47 *Acid extraction of lanthanides from phosphate ores.*

The bonding for the *f*-elements is much less directional and ligands tend to fit as best they can on steric grounds.

The ores are complex mixtures containing the lanthanide elements, and may also include radionuclides such as thorium. Extraction from both the phosphate and carbonate ores can be achieved by acid digestion, outlined for the phosphates in Fig. 3.47, following pre-treatment of crushing, grinding, and hot froth-flotation. Digestion with hot concentrated sulphuric acid and subsequent treatment with cold water allows the separation of soluble lanthanide sulphates from an insoluble residue which can include the radioactive $RaSO_4$. Neutralization with aqueous ammonia allows the *5f* element thorium to be isolated, as ThO_2. The difference in ionic radii between the trications of the early and late lanthanides (Fig. 3.42) provides a means to separate the late members of the series which co-precipitate with Na_2SO_4. The early members of the series, which will have lower lattice energies, remain in the solution and are subsequently separated from the aqueous host.

An alternative approach is via alkaline digestion (Fig. 3.48). That also isolates thorium and radium from the ores and also allows early extraction of the phosphate as Na_3PO_4.

The large majority of the use of lanthanides ($c.\frac{2}{3}$) is as the elemental mix coming from the ore. A few elements can be extracted by virtue of the redox properties. Cerium(III) can be oxidized to Ce(IV) by chlorine bleach, NaClO, and precipitated as CeO_2; europium(III) may be reduced by zinc amalgam (Zn/Hg) to Eu(II) and precipitated as $EuSO_4$.

However, the major large-scale separation method involves solvent extraction of complexes into organic phases. Extraction into the organic layer can be achieved

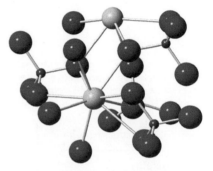

Figure 3.45 *Local structure of* monazite *–Ce; red = O, yellow = Ce, grey = P*

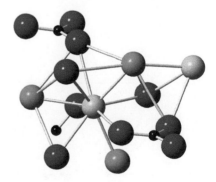

Figure 3.46 *Local structure of* bästnasite *–Ce;   red = O,   yellow = Ce,   green = F, black = C.*

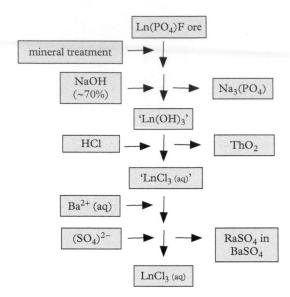

Figure 3.48 *Alkaline extraction of lanthanides from phosphate ores.*

[24] F. Xie, T. A. Zhang, D. B. Dreisinger, and F. Doyle, 'A critical review on solvent extraction of rare earths from aqueous solutions', *Mineral. Eng.,* 2014, **56**, 10–28.

Table 3.21 *Extracting equilibria.*

Cation exchangers:
    anion coordination

$Ln(NO_3)_3 + (NMeR_3)(NO_3)_3 =$

$(NMeR_3)[Ln(NO_3)_4]$

Solvating extractants:
    solvent coordination

$3 L + LnX_3 =$

LnX_3L_3

Anion exchangers:
    anion coordination

$(H_2) = 2 HA$

$3 HA + LnX_3 =$

$LnA_3 + 3H^+ + 3 X^-$

by three types of interaction, used singly or in combination (Table 3.21).[24] Cationic exchangers are typically phosphorus or carboxylic acids which will afford neutral complexes (Fig. 3.49). Solvating extractants are neutral oxygen donors, often phosphine oxides or phosphates, while anion exchangers utilize long-chain ammonium ions to improve the relative affinity with organic solvents.

The method of choice depends upon the pH, auxiliary anions, and elemental mix. Primary extraction, that is of the lanthanides from other elements, and concentration of the elements can be effectively achieved by cation exchangers like the organophosphate D2EHPA. This is due to the fact that the coordination properties of the lanthanide ions are distinct from much of the rest of the periodic table. However, their similarity means that secondary separation of individual lanthanide elements is much harder. Fractionation depends upon small differences in binding constants, largely as a result of the changes in ionic radius across the series (Fig. 3.42). This is similar to the changes in the heats of formation of the oxides (Fig. 3.43) There may be hundreds of separation tanks in which the two phases are mixed, settled, and separated. A large array of fractionation steps are required to achieve good separation of the elements (Fig. 3.50).

Neodymium and Dysprosium

For reasons of applications as magnets within turbines described in Chapter 4, neodymium and dysprosium are employed as an example of the lanthanide elements. The crustal abundances of Nd and Dy are estimated as 20 and 3.6 ppm, respectively. Overall, the mine production of the rare earths as oxides was 130,000 t in 2017, with reserves of 120 Mt, of which neodymium comprises

Cation exchangers

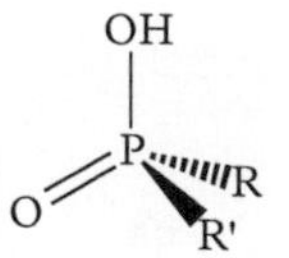

D2EHPA: R = R' = BuCH(Et)CH$_2$O–

HEHEHP: R = BuCH(Et)CH$_2$O–
               R' = BuCH(Et)CH$_2$–

Solvating extractants

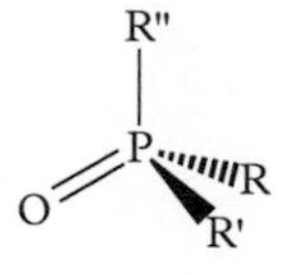

TBP: R = R' = R" = BuO–

Cyanex 921, TOPO: R = R' = R" = CH$_3$(CH$_2$)$_6$CH$_2$–

Anion exchangers

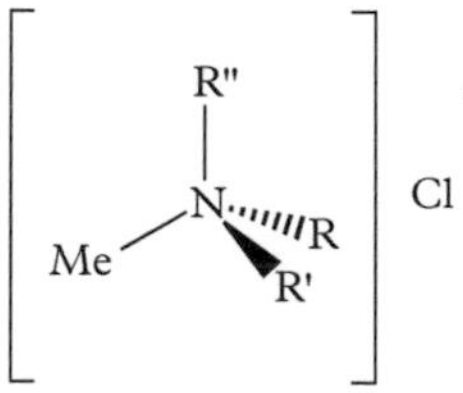

Aliquat 336 R = R' = R' = C$_8$ - C$_{10}$ mixture

Figure 3.49 *Examples of extractants used to separate lanthanide ions from ores.*

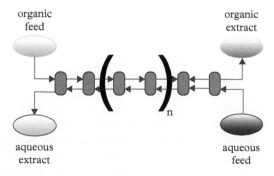

Figure 3.50 *Schematic of counter-flow solvent extraction.*

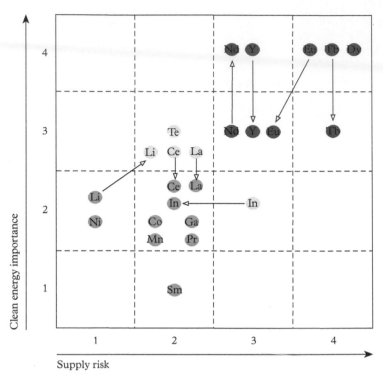

Figure 3.51 *Supply risks of some elements of importance for clean energy for 2015 moving to 2025.*

[25] B. Zhou, Z. Li, and C. Chen, 'Global potential of rare earth resources and rare earth demand from clean technologies', *Minerals*, 2017, **7**, 203.

[26] S. Constantinides, *The important role of dysprosium in modern permanent magnets*, Arnold Magnetic Technologies, 2012 and 2015.

$\approx 16\%$ and dysprosium $\approx 0.5\%$. For clean technologies, neodymium is considered to be in more demand than all of the other lanthanides combined.[25] However, there is considerable performance benefit in adding several wt% of dysprosium to neodymium-iron magnets, and thus the demand for the dysprosium is also high and increasing. Indeed, supply of neodymium and dysprosium, both of which are involved in wind turbines, seems likely to become outstripped by demand. The supply risks that have been estimated for a set of elements envisaged as being important for clean energy are shown in Fig. 3.51.[26] This highlights the importance of neodymium and dysprosium both in terms of applications and supply. It also illustrates the increasing importance of lithium.

The oxides are the most commonly formed compounds of the lanthanides. Production of Nd_2O_3 and Dy_2O_3 in 2015 were estimated to be 33,000 t and 1,700 t, respectively. The two other elements presently in the critical boxes in Fig. 3.51 are europium and terbium; their production was even lower: 550 t (Eu) and 370 t (Tb). In principle, the oxides might be used to afford metal. However, the elements are very electropositive. As a result, reduction of the oxide to the metal by carbon is not viable, for example (Eq. 3.33):

$$\frac{1}{2}Nd_2O_3 + \frac{3}{4}4C = Nd + \frac{3}{4}CO_2. \qquad \Delta G^\circ = 565 \; kJ \; mol^{-1} \qquad (3.33)$$

Instead, reduction of the fluoride by metallic calcium (Eq. 3.34)[27] or by lithium (Eq. 3.35) affords an exoergic process:

$$NdF_3 + \frac{3}{2}Ca = Nd + \frac{3}{2}CaF_2. \qquad \Delta G^\circ = -160 \; kJ \; mol^{-1} \qquad (3.34)$$

$$DyF_3 + 3Li = Dy + 3LiF. \qquad \Delta G^\circ = -460 \; kJ \; mol^{-1}. \qquad (3.35)$$

The majority of Nd metal is manufactured by the electrolysis of Nd_2O_3 in a melt of NdF_3/LiF at $1,000–1,100 \,^\circ C$,[28] reminiscent of the manufacture of aluminium. The two electrode reactions are considered to be:

$$Cathode(tungsten): [NdOF_5]^{4-} - 6e = \frac{3}{2}O_2 + 3Nd^{3+} + 15F^- \qquad (3.36)$$

$$Anode(carbon): 2[NdF_6]^{3-} + 6e = Nd + 12F^-. \qquad (3.37)$$

Each kg of metal requires 12 kWh or 6,200 kJ mol^{-1},[29] as compared to 860 kJ mol^{-1} required on the basis of ΔG°. Some energy loss can be associated with electrochemical degradation of the anode to form CO, CO_2, and some CF_4 (See Section 1.8). However, this is an effective single-stage reaction for converting the oxide to the metal.

Overall, the EE values for the lanthanides fall within the bands anticipated for their concentrations (Table 3.22): the values for cerium are similar to those of cobalt and, although much higher for the rarer lanthanides like europium, the values do not approach those of the PGMs.

3.4.2 Actinides

The differences in chemistry between the lanthanide and actinide elements can be correlated with the trends in the energetics of the atomization of the metals,[30] and their first ionization potentials (Fig. 3.52). For the actinides acquisition of this data is difficult. The heats of atomization remain incomplete, and even the first ionization of a later actinide is a recent addition.[31] The $\Delta_{at}H^\circ$ values for the lanthanides decrease across the first half of the series to europium ($4f^7 6s^2$). This is consistent with the increased electronegativity of the elements shrinking the size of the valence orbitals and so reducing the orbital overlap between atoms. At gadolinium there is an addition of a $5d$ electron (Table 3.20), which presumably can overlap more strongly with neighbouring atoms, and the $\Delta_{at}H^\circ$ value increases. The pattern for the first half of the series is repeated for the later members, and a further increase in $\Delta_{at}H^\circ$ occurs between ytterbium and lutetium ($4f^{14} 5d^1 6s^2$).

It is evident that the $\Delta_{at}H^\circ$ values for the early actinides are higher than those of the lanthanides. From a very similar value at the start of the series there is an

Table 3.22 *Embodied energy (EE) and carbon (EC) of some lanthanides.*

Material	EE	EC
	MJ/kg	kgCO$_2$e/kg
lanthanum	290	19
cerium	510	33
europium	20,000	1,300
neodymium		
dysprosium	1,400	90

[30] R. J. M. Konings and O. Beneš, 'The Thermodynamic Properties of the f-Elements and their Compounds. 1. The Lanthanide and Actinide Metals', *J. Phys. Chem. Ref. Data*, 2010, **39**, 043102 (1–47).

[31] T. K. Sato, *et al.*, 'Measurement of the first ionisation potential of lawrencium, element 103', *Nature*, 2015, **520**, 209–11.

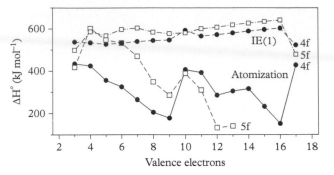

Figure 3.52 *Enthalpies of atomization and the first ionization of the 4f and 5f elements.*

Table 3.23 *Valence electrons of the actinides.*

	An(g)
Ac	$6d^1 7s^2$
Th	$6d^2 7s^2$
Pa	$5f^2 6d^1 7s^2$
U	$5f^3 6d^1 7s^2$
Np	$5f^4 6d^1 7s^2$
Pu	$5f^6 7s^2$
Am	$5f^7 7s^2$
Cm	$5f^7 6d^1 7s^2$
Bk	$5f^9 7s^2$
Cf	$5f^{10} 7s^2$
Es	$5f^{11} 7s^2$
Fm	$5f^{12} 7s^2$
Md	$5f^{13} 7s^2$
No	$5f^{14} 7s^2$
Lr	$5f^{14} 7s^2 7p^1$

[32] It is thought that this stabilization emanates from a relativistic effects. Orbitals which have a significant probability of being close to the nuclei afford a very high kinetic energy to the electrons in elements with Z.

increase from actinium to thorium. As can be seen in Table 3.23, the extra electron has entered the *(n+1)d* subshell rather than the *nf* set that occurs for cerium. This higher occupancy of the *d* orbitals appears to further enhance the bonding strength between the atoms. Progressing through the early actinides, there is again a reduction in the $\Delta_{at}H^\circ$ values as the occupancy of the *5f* orbitals increases. As for the *4f* elements, their shielding is poor and the electronegativity increases. There is also an increase between nine and ten valence electrons which occurs for f^7 configurations and the tenth electron enters the *(n+1)d* subshell. The pattern for the part of the later members of the actinides appears to be following that of the lanthanides, albeit with steeper reductions. That would imply reduced bonding overlap for the later actinides and, indeed, their tendency is for their oxidation state ranges to be lower (II and III). The first ionization energies of periods 6 and 7 follow a similar pattern, with the values for the actinides being generally the higher. There is a sharp change at the end of the series as the ionization energy of lawrencium drops below that of lutetium. This can be ascribed to their differing electronic configuration. For lawrencium, the *(n+2)p* energy level drops below that of the *(n+1)d* and so it is the *7p* electron which is ionized.[32]

Thorium, Uranium, and Plutonium

Thorium has a high crustal abundance for a heavy metal, 5.6 ppm, and the global resources are estimated to be over 6 Mt. Imports and exports into the USA in 2017 were in the region of 100 t. Most thorium is derived as a side-product from production of lanthanides. Global demand is presently much lower than production. In terms of generation of nuclear energy, its weak radioactivity has meant that uranium has been the preferred option. ^{232}Th, the naturally occurring isotope, requires a driver isotope such as ^{233}U, ^{235}U, or ^{239}Pu. With such a driver, however, ^{232}Th can act as a fuel and breed ^{233}U.

Uranium is also relatively abundant in the Earth's crust for a heavy element, at 1.3 ppm. World production of uranium in 2017 is reported as 70,200 t (as U_3O_8). Recoverable resources of uranium are estimated as 5.7 Mt. At the other end of the scale, world generation of reactor-grade plutonium is estimated as 70 t per year. The historical production in spent nuclear fuel is some 1,300 t, from which 400 t has been extracted. Over the last 30 years, 130 t of plutonium has been employed as mixed oxide (MOX) fuel, with a use of 8–10 t per year.[33]

Chemically the simplest of uranium ores is *uranite* or *pitchblende*, UO_2. This has the *fluorite* structure (Fig. 3.14). Other ores include silicates, phosphates, titanates, and ferrates with uranium in oxygen environments in oxidation states 4 or 6. Most often, the ores undergo leaching with sulphuric acid, with the uranium following the following reactions (Eqs 3.38–3.40):

$$UO_2 + 2Fe^{3+} \rightarrow [UO_2]^{2+} + Fe^{2+} \tag{3.38}$$

$$and\ UO_3 + 2H^+ \rightarrow [UO_2]^{2+} + H_2O \tag{3.39}$$

$$then\ [UO_2]^{2+} + 3SO_4^{2-} \rightarrow [UO_2(SO_4)_3]^{4-}. \tag{3.40}$$

[33] World Nuclear Association, London, http://www.world-nuclear.org.

Separation from other elements is carried out-by solvent exchange and/or ion exchange methods similar to those utilized for PGMs and REEs. For solvent exchange, amine or organophosphates like D2EHPA (Fig. 3.49) are utilized to extract into a kerosene-based solvent. Stripping out of the uranium from that phase is carried out with Na_2CO_3 and the uranium deposited as U_3O_8 known as *yellow cake*; the mean oxidation state of uranium is $\frac{16}{3}$. There are two uranium sites in this structure (Fig. 3.53). Each has chains of *trans* uranyl (UO_2) chains. The local geometries are close to a regular seven-coordinate pentagonal bipyramid, but with the site labelled U1 (grey) having one long U–O distance of 245 pm, and this being 271 pm to the U2 equivalent (dark blue). The oxidation states do not appear to be localized.

^{238}U has a natural abundance of 99.27% but it is ^{235}U (0.72%) which is the fuel for applicable nuclear energy. Its own decay pathway is predominantly via α emission to form $^{231}_{90}Th$. However, ^{235}U will undergo capture of a fast (high-energy) neutron to form $^{236}_{92}U$. Activated in this way this isotope rapidly engages in γ emission (15%) and spontaneous fission (85%) and thus forms the basis of energy release (Eq. 3.41). The initiator is a neutron source such as a Be/Po mixture or Ra, and the ^{235}U is the fuel. Eq. 3.41 also affords a supply of fast neutrons and so causes a chain reaction to occur. There are many fission pathways providing different energy releases. The average value per atom is 3.2×10^{-11} J (82 TJ kg^{-1}), which may be compared to that released by burning an atom of carbon as a fuel (6.5×10^{-19} J).

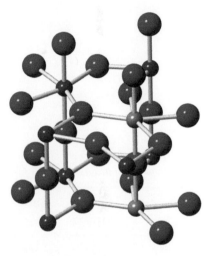

Figure 3.53 *Local structure of U_3O_8.*

$$^{235}_{92}U + ^1_0n \rightarrow ^{236}_{92}U \rightarrow fission\ (e.g.\ ^{139}_{56}Ba + ^{94}_{36}Kr + 3^1_0n) + energy. \tag{3.41}$$

An additional step beyond isolation of the elements is required for this fuel to be viable, namely isotopic enrichment. This is achieved using UF_6. Fluorine has two advantages in this regard. First, it is mono-isotopic (^{19}F) and so the mass differences between molecules are purely due to the isotopes of uranium. Second,

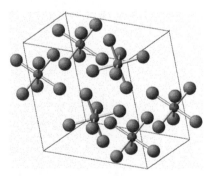

Figure 3.54 *Partial structure of crystalline UF_6; larger = F, smaller = U.*

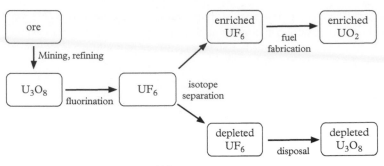

Figure 3.55 *Processes from ore to ^{235}U fuel.*

this is a molecular solid (Fig. 3.54) rather than an extended lattice. The triple point for UF_6 is at 64 °C and 151 kPa, which is slightly above atmospheric pressure. Hence at atmospheric pressure UF_6 sublimes (56.5 °C). Two methods are employed to separate the molecules on the basis of their mass: gaseous diffusion or gas centrifuge. Low enriched uranium (LEU) is defined as having < 20% ^{235}U, and this is what is required for for reactors in nuclear power stations. Light water reactors (LWR) utilize enrichment levels to 3–5%. The required additional steps are outlined in Fig. 3.55. U_3O_8 is converted into UF_6 by the reaction sequence in Eq. 3.42. Initally the oxide is digested in concentrated nitric acid, and is oxidized to uranyl nitrate. Treatment with ammonia affords ammonium diuranate (ADU), $(NH_4)_2U_2O_7$, which is then reduced to U^{IV} oxide. After fluorination with hydrofluoric acid, UF_6 is derived by reaction with F_2. Following isotopic enrichment, the hexafluoride is generally converted back to UO_2 through a series of steps via ammonium diuranate (the ADU process). An alternative process via ammonium uranyl carbonate (AUC), $(NH_4)_4[UO_2(CO_3)_3]$, has also been adopted as this carbonate more readily yields UO_2 on reaction with steam and H_2 (500–600 °C). Much of the depleted uranium is stored as the hexafluoride in steel cylinders. Exposure to moisture will cause hydrolysis to the corrosive mixture of uranyl fluoride, UO_2F_2, and HF, and so methods have been developed to convert this to U_3O_8.

$$U_3O_8 \xrightarrow{HNO_3} UO_2(NO_3)_2 \xrightarrow{NH_3} (NH_4)_2U_2O_7 \xrightarrow{H_2} UO_2 \xrightarrow{HF} UF_4 \xrightarrow{F_2} UF_6. \quad (3.42)$$

These extra steps make contributions to the environmental impact of the manufacture of uranium nuclear fuel from ores (Table 3.24).[34] This study did not include the EE factors for the chemicals and materials in the processes, but provides a baseline estimate. It was evident that the choice of the isotopic separation method was a major factor in the overall figure. The value given for enrichment in Table 3.24 (128 MJ kg^{-1}) was for centrifuge separation, but the estimate for the diffusion process was very much higher (8,760 MJ kg^{-1}). Overall,

Table 3.24 *Embodied energy (EE) and water use per kg of uranium nuclear fuel.*

Process	Energy MJ/kg U	Water L/kg
extraction	332	6,300
conversion	181.5	1,100
enrichment	128	31
fuel (UOX)	33.9	16.8
depleted U	13.2	11
transportation	7.52	
TOTAL	696	7,460

[34] E. Schneider, B. Carlsen, E. Tavrides, C. van der Hoeven, and U. Phathanapirom, 'Measures of the environmental footprint of the front end of the nuclear fuel cycle', *Energy Econ.*, 2013, **40**, 898–910.

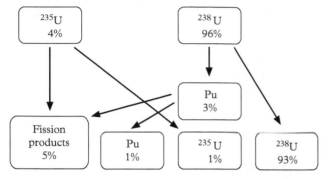

Figure 3.56 *Typical composition changes from burning ^{235}U fuel.*

the EE values for uranium oxide (UOX) fuel are comparable to those for some lanthanide elements.

The changes in nuclear fuel content at a refuelling point are typically as presented in Fig. 3.56.[35] About 75% of the ^{235}U (3%) is converted into fission products. A similar quantity of ^{238}U is converted into a mixture of plutonium isotopes, two of which, ^{239}Pu and ^{241}Pu, are fissile and add to the supply of fission products; there is a net conversion from ^{238}U to plutonium of $\approx 1\%$. The processing of this used fuel provides plutonium, which can be converted into PuO_2. When added with UO_2 this affords MOX (mixed oxide) fuel. After the first recycle this increases the energy output over UOX by $\approx 12\%$. Compared with the figures in Table 3.24, the EE value for UOX is estimated to increase from the figure of 33.9 to 141.5 MJ kg^{-1} with an additional load of 50 L kg^{-1} of water.[34]

The processing of spent fuel is carried out by processes such as PUREX (plutonium uranium redox extraction). Fuel is generally clad with aluminium, zirconium, or stainless steel to prevent reactions between it and the coolants in the reactor. Aluminium can be dissolved away with NaOH, and zirconium- or steel-clad rods are mechanically cut into small pieces. The fuel itself is first treated with concentrated nitric acid, with uranium being oxidized to the soluble U^{VI} (Eq. 3.43) and plutonium forming soluble Pu^{IV} nitrate (Eq. 3.44). The nitrates can be complexed with tributylphosphate (TBP) and thus extracted into a non-aqueous solvent, typically kerosene. The residual material (called the raffinate) is a mixture of fission products (*e.g.* group 1 and 2 metals, 4d metals, and Te), steel corrosion elements, transuranic elements, and gadolinium (^{157}Gd is a strong neutron absorber). Plutonium can be reduced by a variety of reagents, including U^{IV}, to Pu^{III}, which can be back-extracted into an aqueous phase. As we have noted in Eq. 3.42, treatment of uranyl nitrate with ammonia provides a route to reform UO_2 for reuse.

$$UO_2 + 4HNO_3 \rightarrow UO_2(NO_3)_2 + 2NO_2 + 2H_2O \qquad (3.43)$$
$$PuO_2 + 4HNO_3 \rightarrow Pu(NO_3)_4 + 2H_2O. \qquad (3.44)$$

[35] http://www.world-nuclear.org.

PuO_2 can be isolated by treatment of the plutonium-containing aqueous phase with oxalic acid (Eq. 3.45). Calcination effects a dehydration, oxidation to the carbonate, and then removal of CO_2 to afford the dioxide (3.46). The multi-element radioactive residue is stored in stainless steel tanks to allow the most active isotopes to decay, before being converted into a glass for longer-term storage. This aspect was not included in the environmental profile in Table 3.24.

$$Pu(NO_3)_4 + 2H_2O + 2C_2O_4H_2 \rightarrow Pu(C_2O_4)_2 \cdot 6H_2O \downarrow + 4HNO_3$$
(3.45)

$$Pu(C_2O_4)_2 \cdot 6H_2O \rightarrow Pu(C_2O_4)_2 \xrightarrow{O_2} Pu(CO_3)_2 \rightarrow PuO_2.$$
(3.46)

3.5　Structural Materials

Two of the largest-volume structural materials are oxidic with s- and p-block counter ions. Both have complex atomic structures involving regions with some order and others which are disordered, but do so on different length scales. Both start from some carbonate raw materials and so their manufacture intrinsically generates CO_2.

3.5.1　Glass

The quartz structure of silica, SiO_2, contains corner-linked SiO_4 tetrahedra in an ordered fashion forming a macroscopic network (Fig. 3.11). It is transparent in the visible and into the UV, but it is difficult to form having a high melting point near 1,700 °C, varying over a range depending upon the crystal form. This can be overcome by incorporating 'soda', Na_2O, into the structure. Na_2O is generated by decarboxylating sodium carbonate, Na_2CO_3 (soda ash). The sodium acts as a network modifier for the glass-forming silica host; aluminium can also replace some silicon is the glass host. The structures of the local tetrahedra are very ordered, but the network is changed to provide coordination sites for the sodium ions. Hence the sodium acts as a **network modifier**. This occurs only with a local order (< 1 nm), and the longer range order is insufficient for crystallographic diffraction. This soda glass comprises a random network of locally ordered units. However, the coordination sites of the singly charged Na^+ ions are not sufficiently strong as to avoid dissolution by water. If a doubly charged ion is added the extra binding energy essentially eliminates the problem, and typically limestone, $CaCO_3$, is an additional source of network-modifying cations in **soda-lime-glass**. Calcium and magnesium can be introduced together if dolomite is utilized. A balance is found between the singly and doubly charged network modifiers to give the required degree of rigidity. A typical mix is: SiO_2 73%, Al_2O_3 1%, Na_2O 17%, MgO 4%, CaO 5%. That provides a glass transition temperature of $\approx$ 570 °C, and a maximum service temperature range of $\approx$ 170–400 °C. Annual world production is in the region of 84 Mt.

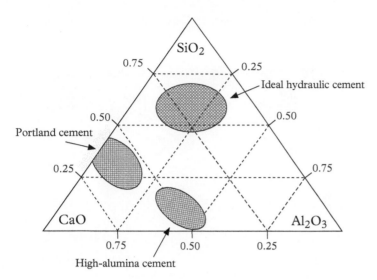

Figure 3.57 *Ternary phase diagram for Ca–Si–Al oxide and regions of some cements.*

Borosilicate glasses have the calcium component replaced by boric acid; a typical composition is SiO_2 74%, Al_2O_3 1%, Na_2O 4%, B_2O_3 15%, PbO 6%. The result of this small, trivalent atom is a tighter structure with much lower thermal expansion coefficient and a higher service temperature range $\approx$ 230–460 °C. About 60% of the use of borates is in these glasses. Accordingly its applications are in kitchen and laboratory ware. A comparison of its environmental footprint with soda-lime glass is given in Table 3.25. Clearly borosilicates have more energy and CO_2 consequences in production by a factor of $\approx$ 2.5, but they are appropriate for their application.

3.5.2 Cement

As for glass, limestone ($CaCO_3$) and silica (SiO_2), as sand, are prime raw materials for the production of cement. The difference is in the use of clay rather than soda. Clays are aluminosilicates (Section 3.2.3), which can also have significant iron content. The three primary metallic elements, Ca, Si, and Al, can be mixed in a variety of proportions, giving different types of cement (Fig. 3.57). For example, calcium silicate cement, often used in dentistry, has a composition close to the 50% point on the Ca–Si and it has active components containing dicalcium silicate, Ca_2SiO_4, and Ca_3SiO_5. A very common form of cement is Portland cement, which is richer in both calcium and aluminium. The Al:Ca ratio is higher in high-alumina cement and the main active constituent is monocalcium aluminate, $CaAl_2O_4$.

The common form of cement, Portland, is made by calcination of the mixture of sand, limestone, and clay at $\approx$ 1,450 °C. Decarboxylation of $CaCO_3$ is

Table 3.25 *Embodied energy (EE) and carbon (EC) of glasses.*

Material	EE MJ/kg	EC kgCO$_2$e/kg
soda-lime:		
production	10.5	0.75
moulding	8.7	0.7
recycling	8.2	0.5
borosilicate:		
production	28.5	1.7
moulding	8.9	0.68
recycling	22	1.3

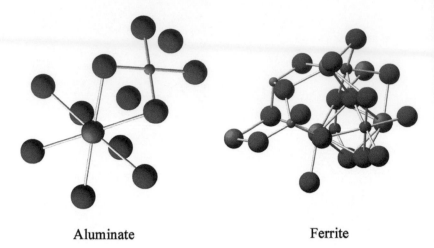

Aluminate   Ferrite

Figure 3.58 *Partial structures of components in Portland cement:* aluminate *and* ferrite; *red = O, large blue = Ca, small blue = Al, brown = Fe.*

endo-thermic and -ergic at ambient temperature, but with the evolution of one CO_2 per calcium it has a high positive entropy and so is viable at such high temperatures. The clinker formed is a mixture of materials which is pelleted. *Gypsum* ($\approx$ 5%) is added and then it is ground to a powder. The powder is a heterogeneous mixture of phases. It is the obverse of a glass in terms of order, having particles with crystallographic order, but it has a heterogeneity of composition on a microscopic scale. The major component (65%) in dry Portland cement is *alite*, Ca_3SiO_5 (abbreviated in this field to C_3S). As expected, the silicon site has a tetrahedral coordination (Si–O 1.59–1.62 Å). The structure is complicated with three different calcium sites: two are six-coordinate and octahedral and the third is seven-coordinate. The second most prevalent phase (14%) is *belite*, Ca_2SiO_4, (C_2S); it also displays isolated SiO_4 tetrahedra (Si–O 1.61–1.65 Å) and a mixture of six- and seven-coordinate calcium sites.

Two other important phases are *ferrite*, $Ca_4Al_2Fe_2O_{10}$, (Ca_4AF, 11.6%) and *aluminate*, $Ca_3Al_2O_6$ (C3A, 3%)(Fig. 3.58); a small proportion of CaO (1%) is also present. In ferrite, the iron(III) can be seen to be in an octahedral site and the aluminium in a tetrahedral one (Al–O 1.77–1.84 Å). Calcium adopts a single, distorted seven-coordinate site (Ca–O 2.3–2.5 Å) with an eighth oxygen atom further away (2.88 Å). The aluminium site in aluminate is different, and is a distorted octahedron with four short (1.906 Å) and two long (2.36 Å) Al–O bonds. The calcium site possesses six oxygen atoms in a plane (Ca–O 2.70 Å), with a further three at a longer distance (2.85 Å). In these oxide structures the main group elements then behave in distinct structural ways: silicon prevails in a tight tetrahedral unit, aluminium can vary between similar tetrahedral units and octahedral structures, and calcium is located in a wide variety of sites with

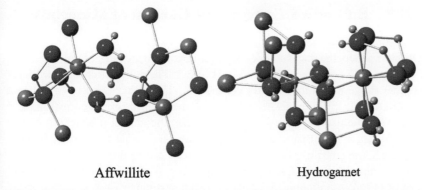

Affwillite   Hydrogarnet

Figure 3.59 *Partial structures of components in hydrated Portland cement:* affwillite *and* hydrogarnet; *red = O, large blue = Ca, small blue = Al, pink = H.*

coordination numbers of six and above and with relatively long Ca–O interactions. Two of the consequences of this are that there are many phases of similar energy and also that there are weak channels in structures that allow hydration.

Following hydration a mixture of hydroxide phases has been identified: $Ca(OH)_2$ (*portlandite*, 29%), $Ca_3Al_2(OH)_{12}$ (*hydrogarnet*, 7%), $[Ca_2Al(OH)_6 \cdot 2H_2O]_2SO_4 \cdot 2H_2O$ (*monosulphate*, 2%), and $[Ca_3Al(OH)_6 \cdot 12H_2O]_2(SO_4)_3 \cdot 2H_2O$ (*ettringite*, 8%). The remainder is calcium-silicate-hydrate (C-S-H) (51%) with some Fe_2O_3 (3.0%).

The calcium-silicate-hydrate (C-S-H) has a composition approximating to $[Ca_3Si_2O_3(OH)_8]$ and is a gel of uncertain structure. A mineral of closely related composition is *afwillite*, $[Ca_3\{Si(OH)O_3\}_2(H_2O)_2]$ (Fig. 3.59). Here the aquation has taken effect in two ways. First, one of the silicate oxygen atoms is protonated and all silicons are as $[Si(OH)O_3]^{3-}$ centres. Second, two of the three seven-coordinate calcium sites have a water ligand. In the hydrogarnet structure, the aluminium sites all comprise of octahedral $[Al(OH)_6]^{3-}$. There is an edge-bridging link via two *cis*-hydroxyl groups to each calcium. The calcium ions are eight-coordinate in essentially a cubic environment.

Overall, the EE value of cement is estimated to be rather lower than that of glass (Table 3.26). Much use of cement is in concrete in which it is mixed with natural materials such as stone and sand, which have much lower EE figures. Hence on these figures brick and baked clay have more embodied energy per unit mass than concrete. Timber has an even higher value, largely due to pre-treatments. However, the sheer scale of cement production means that ways are being found to reduce the EE even further, and to reduce the impact of extraction of virgin raw materials. One such filler is fly-ash from combustion in power stations and incinerators. Since the overall production of cement in 2017 was estimated as $\approx$ 4,000 Mt, over twice that of steel production, small gains in EE and EC values can have large effects. In 2016 cement production was estimated to have been responsible for 1.45 ± 0.2 Gt of CO_2, or 4% of fossil fuel emissions.[36]

Table 3.26 *Embodied energy (EE) and carbon (EC) of cement and building materials.*

Material	EE MJ/kg	EC kgCO$_2$e/kg
cement	4.5	0.74
concrete	0.75	0.1
brick	3.0	0.24
plaster	1.9	0.13
limestone	1.5	0.09
sand	0.08	0.005
baked clay	3.0	0.24
timber	10	0.7

[36] R. M. Andrew, 'Global CO_2 emissions from cement production', *Earth Syst. Sci. Data*, 2018, **10**, 195–217.

3.5.3   Embodied Energy and Carbon of Materials

It is estimated that 64% of the annual emission of CO_2 is due to human activity.[7] Of that fraction, the major sources of the emission are industry (35%), the use of buildings (31%), and the use of transport (22%). The materials considered in this chapter all fall within the industry component. The overall effects of producing materials using the elements need to be considered for the mass-normalized and also production-scaled values.[7,37] For example, the annual production of steel is $\approx 10^7$ larger than that of platinum and this large scale means that $\approx 5.6\%$ of greenhouse emissions can be ascribed to steel production. This is the largest single contribution to industrial emissions (25%), with cement being the second largest (19%). Aluminium (3%), plastic (4%), and paper (4%) also make large contributions.

Both factors have been considered together for production of metallic elements,[38] and a selection of elements highlighted in this chapter feature in Fig. 3.60. The scales on each axis are logarithmic due to the wide differences between the elements. The energy required for extraction and purification (and resulting emissions) is strongly related to the nature and concentration of the ores; this is clearly shown by the location of the PGM elements on the right-hand side of the figure. This has implications for future trends as more difficult ores may have to be used to provide raw materials. As a result, the value of recycling increases.

Overall, taking the contribution of steel production to global emissions as 5%, then the emissions related to the production of each of the elements which may be applicable to the amelioration of emissions, either by clean-up or in alternative energy sources, are in the region of 10^{-4} times smaller. Thus there

[37] M. F. Ashby, *Materials and the Environment*, second edition, Butterworth-Heinemann, Oxford, 2013.

[38] E. van der Voet, R. Salminen, M. Eckelman, G. Mudd, T. Norgate, and R. Hischier, *Environmental Risks and Challenges of Anthropogenic Metals Flows and Cycles: A Report of the Working Group on the Global Metal Flows to the International Resource Panel*, UNEP, 2013.

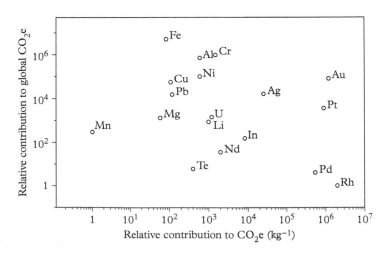

Figure 3.60 *Relative contributions of metal production to CO_2e emissions per kg and overall.*

is possible scope for increasing the use of these materials and having a net benefit on emissions. This will be considered in Chapters 4 (Air) and 5 (Fire).

3.6 Questions

1. The manufacture of strontium metal has been increasing in recent years with the ore strontianite, $SrCO_3$, as raw material. Given the values of $\Delta_f G_{298}$ given below, assess the feasibility of a process involving reduction of the oxide by carbon or by aluminium.

   [$\Delta_f G_{298}$: $SrCO_3$ $-1,144$, SrO -643, CO_2 -394, Al_2O_3 $-1,582$ kJ mol^{-1}]

   What alternative method might you suggest for manufacture of the metal?

2. Cobalt minerals include phosphates and arsenate. A cobalt arsenate of interest for its magnetic properties is $Co_2(OH)(AsO_4)$.
   (i) Assign oxidation states for arsenic and cobalt and the d-electron count for cobalt.
   (ii) Draw an energy level diagram for cobalt with this d^n count in its preferred spin state.
   (iii) The cobalt sites in this compound do not adopt regular octahedral sites. One of the two sites has four short Co–O distances (2.08 Å) and two longer trans interactions (2.23 Å). Account for there being a distortion in this site.
   (iv) The second cobalt site is five-coordinate and adopts a trigonal bipyra-midal structure. Draw an energy-level diagram for this structure and consider whether the site might distort further.
   (v) Calculate the magnetic moments of the cobalt ions using the spin-only formula.

3. A hydroxy oxime ligand of the type in Fig. 3.36 can be used for solvent extraction of $3d$ metal ions. The pH values at which there is 50% extraction from aqueous solution are: Cu(II) 0.1, Ni(II) 2.8, Co(II) 4.5, Zn(II) 6.5.
   (i) What stoichiometry would you propose for the extracted complexes? Draw a structure for such a complex.
   (ii) What does this pH order imply for the relative stability of these complexes?
   (iii) Account for this stability order.
   (iv) How might this be exploited for separation of the elements?
   (v) For Fe(III) 50% extraction occurs at pH 1.5. What effect can this have on the application of this extractant?

4. Derive the Russell-Saunders ground states for the following ions: Ti^{3+}, Fe^{2+}, Ni^{2+}, Sm^{2+}, and Dy^{3+}.

5. The mixed-valence cobalt oxide Co_3O_4 adopts a spinel-type structure, with Co(II) and Co(III) adopting high and low spin states, respectively.

(i)   Predict whether the structure will be a normal or inverse spinel.

(ii) Predict the magnetic moment per formula unit.

(iii) Below 40 K, the magnetic moment drops to a very low value. Account for this observation.

6. (i)   Construct a Frost diagram for chromium in acidic solution (pH 0) using the data below:

[$Cr^{3+/0}$ −0.74; $Cr^{3+/2+}$ −0.424; $Cr^{VI/3+}$ 1.38; $Cr^{VI/V}$ 0.55; $Cr^{V/IV}$ 1.34 V.]

(ii) Comment on the stabilities of the oxidation states.

Air

4.1 The Volatile Elements

The most volatile elements are in an atomic and diatomic form with low Z (Table 4.1). As the atomic number increases the enhanced polarizability of the atoms raises the heat of vaporization and thus the boiling point of the element. Two relatively volatile elements have allotropes which are molecular solids: P_4 and S_8. However, they are insufficiently volatile to have a major effect on the atmosphere and thus are included in Chapter 3 (Earth) (shown shaded in Fig. 4.1).

As an example of the trends with atomic number, whilst F_2 is gaseous under ambient conditions, there are winter temperatures towards the poles at which Cl_2 would liquify. Bromine is liquid under more normal ambient conditions, and I_2 a volatile solid. The four highest-Z elements in groups 17 and 18, At, Rn, Ts, and Og, are all short-lived, and the radioactive properties of their most stable, confirmed isotope are given in Table 4.2.

Astatine (At, $Z = 85$) is considered to be the rarest element in the Earth's crust. Less than 1 g is estimated to be present at any one time, split between four isotopes: astatine-215, −217, −218, and −219 with $t_{1/2}$ values, respectively of 100 μs, 32 ms, 1.5 s and 56 s. ^{217}At emanates from ^{237}Np, ^{218}At is on the decay train from ^{238}U, and both ^{215}At and ^{219}At have ^{235}U as their source. The longest-lived of the these isotopes, ^{219}At, comes via the following sequence (Eq. 4.1):

$$^{235}_{92}U \rightarrow \alpha + ^{231}_{90}Th \rightarrow \beta^- + ^{231}_{91}Pa \rightarrow \alpha + ^{227}_{89}Ac \rightarrow \alpha + ^{223}_{87}Fr \rightarrow ^{219}_{85}At + \alpha. \quad (4.1)$$

Its primary decay pathway is by α loss to form ^{215}Bi, with the minor process being β^- loss affording ^{219}Rn.

Synthesis of the least transient isotopes of astatine is generally carried out by irradiating a bismuth target with high-energy α particles on a scale of $c.10^{-7}$ g. Three isotopes result from the processes in Eqs. 4.2–4.4:

$$^{209}_{83}Bi \xrightarrow{\alpha} ^{211}_{85}At + 2n \quad (4.2)$$

$$^{209}_{83}Bi \xrightarrow{\alpha} ^{210}_{85}At + 3n \quad (4.3)$$

$$^{209}_{83}Bi \xrightarrow{\alpha} ^{209}_{85}At + 4n. \quad (4.4)$$

Elements of a Sustainable World. John Evans, Oxford University Press (2020). © John Evans.
DOI: 10.1093/oso/9780198827832.003.0004

Table 4.1 *Boiling points.*

Element	Z	b.p. (°C)
Hydrogen	1	−252.9
Helium	2	−268.9
Nitrogen	7	−195.8
Oxygen	8	−183.0
Fluorine	9	−188.1
Neon	10	−246.0
Phosphorus	15	280.5
Sulphur	16	444.6
Chlorine	17	−34.0
Argon	18	−185.8
Bromine	35	58.8
Krypton	36	−153.4
Iodine	53	184.4
Xenon	54	−108.1
Astatine	85	336.8
Radon	86	−61.7
Tennessine	117	solid
Organesson	118	solid

The prevalent decay process of the lighter two synthetic isotopes is β^+ loss (e.g. Eq. 4.5), with α emission the minor route (e.g. Eq. 4.6).

$$^{210}_{85}At \rightarrow {}^{210}_{84}Po + \beta^+ \tag{4.5}$$

$$^{211}_{85}At \rightarrow {}^{207}_{83}Bi + \alpha. \tag{4.6}$$

The primary decay mode for ^{211}At also forms polonium, but via the EC mechanism (Eq. 4.7):

$$^{211}_{85}At \xrightarrow{EC} {}^{211}_{84}Po. \tag{4.7}$$

There is a paucity of experimental evidence about the properties of astatine (At) having a lifetime of hours. Even for radon (Rn, $Z = 86$), with a half-life of 3.8 d, it is estimated that there are only a few tens of grams in the atmosphere, but the effect is significant. It is continually generated, principally through the decay chain of ^{238}U, which leads to the longest-lived isotope, ^{222}Rn (Eq. 4.8):

$$^{238}_{92}U \rightarrow \rightarrow \rightarrow {}^{234}_{92}U \rightarrow \alpha + {}^{230}_{90}Th \rightarrow \alpha + {}^{226}_{88}Ra \rightarrow {}^{222}_{86}Rn + \alpha. \tag{4.8}$$

Radon itself, an inert gas, is also an α emitter, affording ^{218}Po on the route to the stable ^{206}Pb (Fig. 4.2). This affords six chemically active elements which are α and/or β emitters.

The nuclides in the tracks to ^{210}Pb ($t_{1/2}$ 22.2 y) will equilibrate fairly quickly, but the approach to equilibrium will take tens of years to achieve. The chemically active elements will tend to adsorb onto particles and be removed from the atmosphere. Empirical evidence indicates that the daughter elements will tend

Group

15	16	17	18
		H	He
N	O	F	Ne
P	S	Cl	Ar
		Br	Kr
		I	Xe
		At	Rn
		Ts	Og

Figure 4.1 *The volatile elements.*

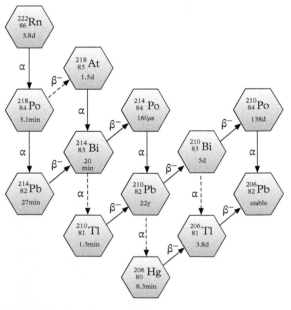

Figure 4.2 *Radioactive decay processes from* ^{222}Rn.

to 40% of the theoretical level. The activity of radon along ocean shores is $c.1$ Bq m^{-3},[1] coming from $c.150,000$ atoms m^{-3}. Concentrations build up in caves, mines, and buildings. The WHO has recommended a reference level of 100 Bq m^{-3} for buildings. The reference level adopted varies between countries, with Public Health England[2] taking 200 Bq m^{-3} as the national reference level. A radon-affected area is defined as a part of the country with a 1% probability of a home having a higher radon level than this. This probability is strongly dependent upon local geology. For large areas of lowland England with sedimentary rocks this probability is not exceeded. In areas like Dartmoor, Cornwall, and the Cairngorms (Fig. 4.3), the probability of exceeding this level is 30% or more. Any building with an activity level 1,000 Bq m^{-3} is recommended to be closed

Table 4.2 *Radioactivity properties.*

Isotope	Half-life	Decay	Product
^{210}At	8.1 h	β^+	^{210}Po
		α	^{206}Bi
^{222}Rn	3.8 d	α	^{218}Po
^{294}Ts	51 ms	α	^{290}Mc
^{294}Og	0.69 ms	α	^{290}Lv

[1] 1 becquerel (Bq) represents a decay rate of 1 nucleus/second.

[2] J. C. H. Miles, J. D. Appleton, D. M. Rees, B. M. R. Green, K. A. M. Adlam, and A. H. Meyers, *Indicative Atlas of Radon in England and Wales*, HPA-RPD-033, 2007, Health Protection Agency, Chilton, UK, http://www.ukradon.org.

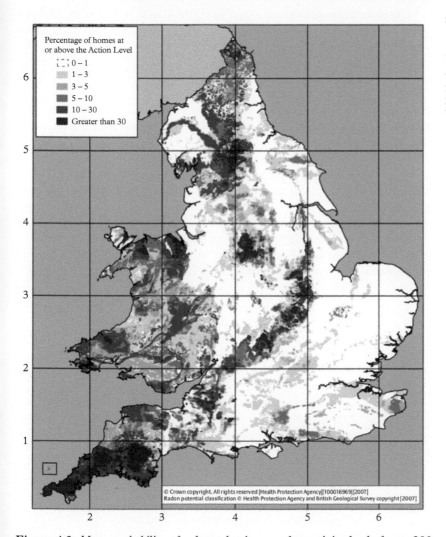

Percentage of homes at or above the Action Level
- 0 – 1
- 1 – 3
- 3 – 5
- 5 – 10
- 10 – 30
- Greater than 30

Figure 4.3 *Mean probability of a home having a radon activity level of over 200 Bq m^{-3}.*

3 http://www.radioactivity.eu.com.

Table 4.3 *Dry air composition.*

Allotrope	Composition (% by volume)
Nitrogen	78.08
Oxygen	20.95
Argon	0.93
Neon	0.0018
Helium	0.00052
Krypton	0.00011
Hydrogen	0.00005
Xenon	0.0000087
Ozone	0.000001

until remedial action is taken.[3] The health risks are associated with lung cancer, albeit at a greatly lower level than that caused by tobacco smoke.

The half-lives of the two highest Z confirmed synthetic elements, tennessine (Ts) and oganesson (Og), are much less than 1 s. Synthesis has involved collisions of ^{48}Ca ions with heavy elements such as ^{249}Bk. This afforded a transient isotope of tennessine, and loss of neutrons generated a few atoms of two isotopes of Ts (Eq. 4.9):

$$^{249}_{97}Bk + ^{48}_{20}Ca \rightarrow ^{297}_{117}Ts^* \rightarrow ^{294}_{117}Ts + ^{293}_{117}Ts. \qquad (4.9)$$

As a result of the low abundance of the volatile elements in row 6 of groups 17 and 18, and the synthetic nature of those in row 7, the elements considered further in this chapter have been reduced to those unshaded in Fig. 4.1.

The mass of the atmosphere is $\approx 5.15 \times 10^6$ Gt. As is evident in Table 4.3, 99% of dry air is made up of N_2 and O_2, with the remainder dominated by argon. The next most abundant gas is CO_2, currently ≈ 410 ppm.

4.1.1   Group 18 Elements

Apart from argon, all of the group 18 gases of periods 1–5 are present in trace amounts in the atmosphere. Neon is the predominant trace gas (18 ppm), with helium (5.2 ppm), krypton (1.1 ppm), and xenon (0.087 ppm) present in increasingly small concentrations. The atmospheric gases are generally separated by virtue of their differing boiling points (cryogenic fractional distillation). The gases are also occluded in the Earth's crust, within natural gas and dissolved in seawater.

Argon is generally used as an inert gas in reactive and high-energy environments such as arc welding, phosphorescent light tubes, and chemical synthesis. It is also employed as the insulating gas in multiply glazed windows. Annual production is ≈ 700 kt, which may be scaled against an atmospheric content of $\approx 6 \times 10^4$ Gt (Table 4.4). Neon is a somewhat scarcer resource with much lower

Table 4.4 *Production (2012) and atmospheric content of Group 18 gases.*

Element	Production (t)	Atmospheric content (Gt)
helium*	30,000	3.7
neon	520	650
argon	700,000	66,000
krypton	380	16
xenon	60	2

*Figures for He are for 2017 production, excluding China.

production of $\approx$ 520 t in 2012. This trend is continued through krypton (380 t) and xenon (60 t). These trace gases are generally employed for specific purposes that cannot be met by argon, such as lasers (Ne), fluorescent lighting (Kr), plasma screens (Ne, Xe), and halogen lights (Xe). The ratio of annual production to atmospheric composition for most of these gases in of the order of 10^7, rising to 10^9 for neon.

For helium this ratio is $\approx 10^5$, but the dilution renders it extremely resource intensive to extract from the atmosphere. Usage is increasing, particularly in medical diagnosis, and the length of the supply of viable sources (2%) is probably about 200 years. The low atmospheric content for helium is due to its low mass, and so an increased tendency to diffuse away from the atmosphere (the residence time is a few weeks). Almost all helium is extracted from fossil fuel reserves, mainly natural gas. In this sense, helium is a mineral as it is not the current atmospheric composition that forms the source. About 30% of the world supply presently comes from stored helium/natural gas in the US Federal Helium Reserve. That facility is due to close in 2021, perhaps to transfer to a different type of enterprise. Helium's very low melting point and low density provide the basis for the two largest spheres of application. The first attribute is required is for maintaining the superconducting nature of the coils in magnetic resonance imaging (MRI) devices (and in NMR spectrometers), and an interruption of supply would compromise this medical diagnostic technique. The second is as the lifting gas in airships and balloons, in their various sizes and applications. Together, these account for nearly 50% of helium use.

The production of crude argon requires a similar amount of energy as for the separation of the two major atmospheric gases (Table 4.5). As for the extraction of minerals from ores, the energy costs of effective cryogenic fractional distillation increase rapidly with dilution. The three highest Z of these elements have all been considered for use in multiply glazed windows. In terms of their contributions to the overall embodied energy of a model double-glazed window system, the filler gas component is very minor for argon, is significant ($c.33\%$) for krypton, and is estimated to be dominant for xenon ($c.81\%$).[4] Thus only dry air and argon are really viable options from this point of view.

The group 18 elements have very similar electronegativities and covalent radii to their neighbours in group 17 (Fig. 2.12), and decrease significantly down the group. The ionization enthalpies accordingly decrease with period (Fig. 4.4). Indeed, the values for xenon, particularly at the higher ionizations, become very similar to those of the heavier metals in group 8. Both ruthenium and osmium exhibit their group number as their maximum oxidation state, for example in their tetroxides. Indeed, compounds of the 'inert' gas xenon with electronegative elements such as fluorine and oxygen are well known. For example, direct fluorination of xenon affords XeF_2 (Eq. 4.10). Xenon difluoride is a molecular solid with linear units (Fig. 4.5).

$$Xe + F_2 \rightarrow XeF_2. \tag{4.10}$$

Table 4.5 *Embodied energy of atmospheric elements.*

Element	EE (MJ/kg)
Nitrogen	2.9
Oxygen	2.8
Helium	3.0
Argon	1.9
Krypton	11,000
Xenon	93,000

[4] G. Weir and T. Muneer, 'Energy and environmental impact analyses of double-glazed windows', *Energy. Convers. Manag.,* 1998, **39**, 243–56.

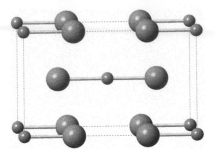

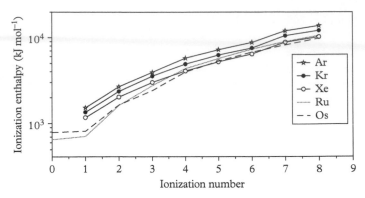

Figure 4.5 *The unit cell of XeF₂.*

Figure 4.4 *Comparison of the atomization and ionization enthalpies (kJ mol⁻¹) of elements in groups 8 and 18.*

Figure 4.6 *Linking of XeF₄ molecules.*

Table 4.6 *Annual production (Mt) of group 1 and 17 elements.*

Element	Production
hydrogen	50
fluorine	0.017
chlorine	65
bromine	0.35
iodine	0.031

With an excess of fluorine, and a catalyst of NiF_2, the reaction can be driven to XeF_4 and even XeF_6. XeF_4 is also a molecular solid (Fig. 4.6), with a square planar geometry at xenon. The molecular structures of XeF_2 and XeF_4 can both be rationalized by the **valence shell electron pair repulsion** (VSEPR) model. For Xe(II) and Xe(IV) there are three and two non-bonding electron pairs, respectively. So the total electron pair counts will be five and six. Hence the geometry of XeF_2 can be described as a trigonal bipyramid with three equatorial lone pairs; for XeF_4 the base structure is an octahedron with *trans* lone pairs. There are F–F interactions which link the individual molecules. These links are very complex in XeF_6. The oxidation state of +VI is also attainable in the trioxide, and the maximum oxidation state of +VIII can be attained with XeO_4; both of these are very endothermic compounds and can reform the elements explosively.

As can be anticipated from Fig. 4.4, oxidation of krypton and argon become increasingly difficult. KrF_2 is well established. It is a very powerful fluorinating agent and will oxidize Xe to XeF_6 and gold to Au(V). KrF_4 has been identified under very high-pressure conditions. ArF_2 is not isolable. H–Ar–F has been observed under cryogenic conditions, but was observed to decompose by 27 K.

4.1.2 Group 1 and 17 Elements

Hydrogen and chlorine are produced on a multi-million-tonne scale annually (Table 4.6). Considering hydrogen first, the most abundant element in the Universe, its atmospheric composition (Table 4.3) is very low, about 50% that of krypton per volume. Partially, this is due to its ready diffusion, like helium, but is largely because it is chemically fixed. Two of the most obviously available hydrogen sources are water and fossil fuels, and the thermodynamics of these reactions are presented in Table 4.7. As expected, the splitting of water is highly

Table 4.7 *Prediction of the thermodynamics for the production of hydrogen under standard conditions.*

Reagents	Products	$\Delta_r H^\circ$ /kJ mol^{-1}
$H_2O_{(l)}$	$H_{2(g)} + 0.5\ O_{2(g)}$	285.8
$CH_{4(g)} + H_2O_{(l)}$	$3\ H_{2(g)} + CO_{(g)}$	250.1
$CO_{(g)} + H_2O_{(l)}$	$H_{2(g)} + CO_{2(g)}$	2.8
$CH_{4(g)} + H_2O_{(l)} + O_{2(g)}$	$3\ H_{2(g)} + CO_{(g)}$	−35.7
$3\ C_{(s)} + O_{2(g)} + H_2O_{(l)}$	$H_{2(g)} + 3\ CO_{(g)}$	−45.7

endothermic and so energy needs to be provided, perhaps electrochemically or photochemically. The current major source of hydrogen is from fossil fuels. The second line of Table 4.7 describes **steam methane reforming** (SMR), which is carried out at high temperature (750–1,100 K). It affords a particular mixture of H_2 and CO, which, in general, is called **synthesis gas**. It is also a highly endothermic process, even when the water is considered from its gas phase as the starting point. The next reaction, the **water gas shift reaction** (WGSR), is exothermic when considering the water in the vapour phase ($\Delta_r H^\circ$ −41.5 kJ mol^{-1}), and is used at somewhat lower temperatures (450–790 K) to increase the yield of H_2. The CO_2 is removed from the mixture by a pressure-swing adsorption process thus affording H_2. Partial oxidation (line 4) is a variation of steam reforming with O_2 within the reagent stream, shown here for methane. This is exothermic and viable for a selection of raw materials, including biomass and plastics. Finally, coal gasification is an extreme variant of this, which is also exothermic. It also affords the highest proportion of CO_2. The embodied energy of H_2 produced by the SMR process has been estimated as *c.*180 MJ kg^{-1}.

The majority of the hydrogen gas produced industrially is used to fix nitrogen as ammonia ($\approx$ 55%) and this is described in Section 4.1.3. Hydrogenation as part of petrochemical processing at refineries is the second highest use ($\approx$ 25%), and the conversion of synthesis gas to methanol consumes half of the remainder ($\approx$ 10%). Currently little is used as a fuel.

Chlorine is also a highly abundant element, with NaCl as the major source. The production of salt in 2017 is estimated as 280 Mt. The use for de-icing on highways is similar to that for all chemical industry ($\approx$ 40% each). Nearly 90% of the chemical use is routed by the conversion to chlorine gas and caustic soda. One method of Cl_2 production is as a side product for the electrochemical production of sodium and lithium using a Downs cell (Section 3.1.1). However, generation of Cl_2 does not for itself require the high-temperature, non-aqueous conditions demanded by the alkali metals. The majority of production involves the **chlor-alkali** process.[5] This involves cathodic reduction of protons (Eq. 4.11) and anodic oxidation of chloride (Eq. 4.12). The result is that H_2 and NaOH are co-products

[5] S. Lakshmanan, T. Murugesan, 'The chlor-alkali process: Work in Progress', *Clean Techn. Environ. Policy*, 2014, **16**, 225–34.

with Cl_2 (Eq. 4.13). The production capacity of chlor-alkali cells was 62.8 Mt y^{-1} in 2008.

$$Cathode: \ 2\,H^+_{(aq)} + 2\,e \rightarrow H_{2(g)} \tag{4.11}$$

$$Anode: \ 2\,Cl^-_{(aq)} \rightarrow Cl_{2(g)} + 2\,e \tag{4.12}$$

$$Overall: \ 2\,NaCl + 2\,H_2O \rightarrow Cl_2 + H_2 + 2\,NaOH. \tag{4.13}$$

The separation of the electrochemical cells has been a major factor of reactor design. In a diaphragm cell, developed from 1851, materials such as asbestos and PTFE (polytetrafluoroethylene) have been employed as the divider. An alternative approach was the Castner–Kellner cell (1892), in which a pool of mercury acts as the cathode. In this case sodium is reduced and the metal generated dissolves into the mercury to form an amalgam (Eq. 4.14). Hence the overall cell reaction becomes Eq. 4.15.

$$Cathode: \ 2\,Na^+_{(aq)} + 2\,Hg + 2\,e \rightarrow 2Na - Hg \tag{4.14}$$

$$Overall: \ 2\,NaCl + 2\,Hg \rightarrow Cl_2 + 2\,Na - Hg \tag{4.15}$$

In a second compartment, called the denuder, the amalgam is reacted with water (Eq. 4.16). Overall, this process generates the same co-products as the diaphragm cell, and the mercury is returned to the electrochemical cell. Environmental pressures to reduce mercury emissions[6] have curtailed new installations of this type of cell. As can be seen from Figure 4.7, the aerial emissions of mercury

[6] European Environment Agency, http://www.eea.europa.eu.

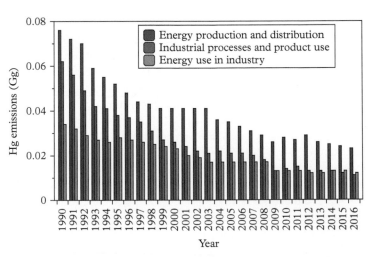

Figure 4.7 *Emissions of mercury to the air by the three major contributions in the EU33 region.*

from industrial activity in the EU area were reduced by a factor of six from 1990 to 2016. The dominant sources are now considered to be energy related.

$$Overall: \quad 2\,Na-Hg+2\,H_2O \rightarrow H_2+2\,NaOH+2\,Hg \qquad (4.16)$$

In large measure due to mercury poisoning, Japan has closed all of its mercury cells, and switched to membrane cells. The principle is very similar to that of a diaphragm cell, but a sodium-permeable ion-exchange polymer membrane is employed to separate the cathodic and anodic environments. Anodes are often titanium-based, and can include activating catalysts such as RuO_2; the hydrogen-evolving cathodes are generally fabricated from nickel. Membranes are bilayered with a perfluorosulphonic acid polymer as the core structure and a poly-perfluorocarboxylic acid as a thin layer on the cathodic side. The perfluoralkene provides a very chemically resistant framework for the polymer, required for the aggressive nature of the electrolytes. The sulphonic acid groups are generally supplied as their sodium salts. They are strong acids, and thus bind relatively weakly to the sodium ions and allow a high ion permeability. The carboxylic function is a weaker acid, and hence will coordinate more strongly to the sodium. It will thus draw the sodium ions towards the cathodic chamber of the cell. Generic structures of these two polymers are represented in Fig. 4.8.[7]

A schematic of the overall chlor-alkali cells is shown in Fig. 4.9, based on a design published by Euro Chlor.[8] On the anodic side of the membrane, the saturated brine solution enters and is diluted within the cell. Chlorine gas is formed at the electrode according to Eq. 4.12, and the excess sodium ions migrate across the ion-exchange polymers. The input reagent on the cathodic side is pure water which is mixed into a caustic soda solution (33%). Hydrogen is formed at the

[7] M. Paider, V. Fateev, K. Bouzek, 'Membrane electrolysis—History, current status and perspective', *Electrochim. Acta*, 2016, **209**, 737–56.

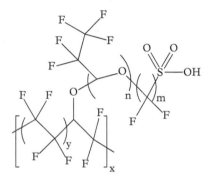

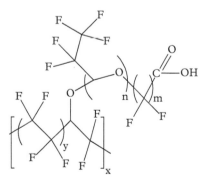

Figure 4.8 *Example of the sulphonic (top) and carboxylic (bottom) acid sites in perfluoroalkene polymers.*

[8] http://www.eurochlor.org.

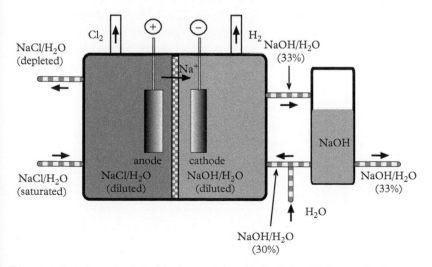

Figure 4.9 *Schematic of the diaphragm chlor-alkali cell for chlorine production.*

cathode according to Eq. 4.11. The sodium transferred across the membrane balances the charge of the hydroxyl ions formed after the removal of protons at the cathode. Thus caustic soda solution of higher concentration (33%) is transferred for extraction.

The major uses of chlorine are in the chemical industry: the manufacture of polyvinylchloride (PVC), as reagents for other high-volume organic materials (e.g. isocyanates for polyurethanes), and of inorganic materials (e.g. to form $TiCl_4$ in the manufacture of the pigment TiO_2). The co-location of chlorine and caustic soda also provides the opportunity for the manufacture of sodium hypochlorite, NaOCl, or, as an aqueous solution, chlorine bleach, by Eq. 4.17.

$$2\,NaOH + Cl_2 \rightarrow NaClO + NaCl + H_2O. \tag{4.17}$$

Like the group 18 elements, the members of this group have electronegativities which decrease significantly down the group (Fig. 2.12). Accordingly, their ionization enthalpies also decrease with period (Fig. 4.10). As the most electronegative of the halogens, fluorine is formally considered to have an oxidation number of −1 in all compounds. The electronegativity of oxygen is higher than that of chlorine so in the hypochlorite anion, chlorine is ascribed the oxidation number of +1. It is evident from Fig. 4.10 that the ionization energies of chorine and bromine are similar, and those of iodine rather lower. As a result it can be anticipated that the stability of the higher oxidation sates of the halogens will increase with period number.

Even for chlorine, oxidation states up to +7 are attainable. In the manufacture of sodium hypochlorite, the chlorination of NaOH is carried out in an electrochemical cell to maximize the conversion to NaOCl. That requires a low-temperature process (<40 °C). At higher temperature the electrochemical oxidation proceeds further to sodium chlorate (Eq. 4.18). The chlorate is oxidized further electrochemically to Cl(VII) in sodium perchlorate (Eq. 4.19).

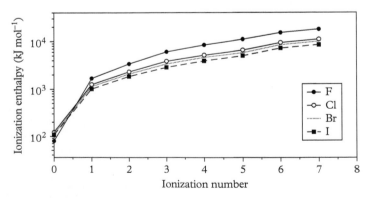

Figure 4.10 *Comparison of the atomization and ionization enthalpies ($kJ\ mol^{-1}$) of elements in group 17.*

$$NaCl + 3\,H_2O \rightarrow NaCl^{V}O_3 + 3\,H_2 \tag{4.18}$$

$$NaCl^{V}O_3 + H_2O \rightarrow NaCl^{VII}O_4 + H_2. \tag{4.19}$$

By comparison to chlorine (10 ppm), fluorine (0.4 ppm) is much less abundant in the Universe, but has a higher estimated crustal abundance (553 rather than 145 ppm). The mine production of *fluorite* or *fluorspar*, CaF_2, was estimated as 6 Mt in 2017, with known reserves of 270 Mt.

The main uses of fluorine derived from calcium fluoride are summarized in Fig. 4.11. A small proportion of fluorite is used directly, but the majority is directed towards metal smelting (mostly iron), and the manufacture of HF by the reaction in Eq. 4.20:

$$CaF_{2(s)} + H_2SO_{4(l)} \rightarrow 2\,HF_{(g)} + CaSO_{4(s)}. \tag{4.20}$$

The two major uses of hydrofluoric acid, HF, are in the manufacture of *cryolite*, Na_3AlF_6, used as solvent in the electrochemical manufacture of aluminium (Section 3.2.2), and for the production of fluorocarbons. The type of fluorocarbons produced is varying in response to the demands of the Montreal Protocol (Section 1.7) to reduce ozone-depleting gases in the stratosphere. Hence the production of chlorofluorcarbons (CFCs) is very low, and the majority of fluorocarbons are now hydrofluorocarbons (HFCs). The manufacture of hydrochlorofluorocarbons (HCFCs) is still significant. About

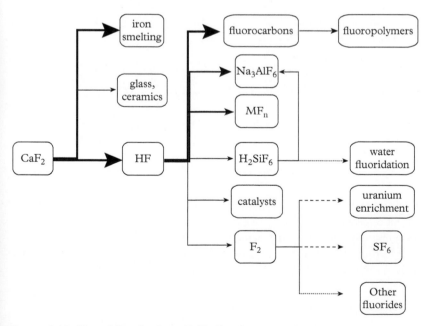

Figure 4.11 *Uses of fluorine from CaF_2, fluorite.*

16% of all fluorocarbons are converted into polymers, such as PTFE. A significant proportion of HF is utilized in a variety of metal fluoridations ($\approx$ 16%). Minor applications in terms of quantity include the production of fluorosilicic acid, about 5% of which is employed in water fluoridation, and in the manufacture of catalysts for the petrochemical industry. Only $\approx$ 1% of CaF_2 is converted into F_2. This is then used to make fluoride compounds, the majority of which are UF_6, for uranium enrichment (Section 3.4.2), and the inert, electrically insulating gas sulphur hexafluoride (see Section 1.7).

Bromine reserves are extremely large. Production of its brine in 2017 was in the region of 350 kt, considerably less than that of the lighter halogens, and future supply seems secure. Elemental bromine is generally produced by oxidation with chlorine (Eq. 4.21), and then extracted by steam distillation. The two major uses are in flame-retarding materials and, as salts of group 2 elements, as drilling fluid lubricants.

$$Cl_{2(g)} + 2\,Br^-_{(aq)} \rightarrow 2\,Cl^-_{(aq)} + Br_{2(g)}. \tag{4.21}$$

Production of iodine materials in 2017 was an order of magnitude lower than than of bromine (31 kt), with substantial known reserves (6.4 Mt). The manufacture of I_2 is commonly the direct analogue of bromine (Eq. 4.21). Purification is either by blowing the volatile iodine out of solution with air, trapping it in HI/H_2SO_4 and reducing it with SO_2 to iodide, or by reaction with iodide to form $I_3{}^-_{(aq)}$ and separation by ion exchange. A second chlorine oxidation stage generates the iodine product. In Chile, the sodium nitrate ore contains 5% of sodium iodate, NaI^VO_3. After crystallization of $NaNO_3$, a solution enriched in iodate is reacted with SO_2 to form iodine initially (Eq. 4.22), and then again to generate iodide (Eq. 4.23):

$$2\,IO^-_{3(aq)} + 5\,SO_{2(aq)} + 4\,H_2O_{(l)} \rightarrow I_{2(aq)} + 8\,H^+_{(aq)} + 5\,SO^{2-}_{4(aq)} \tag{4.22}$$

$$I_{2(aq)} + SO_{2(aq)} + 4\,H_2O_{(l)} \rightarrow 2\,I^-_{(aq)} + 4\,H^+_{(aq)} + SO^{2-}_{4(aq)}. \tag{4.23}$$

Finally, mixing of the iodate solution with this solution of iodide results in a conproportionation reaction to afford elemental iodine as a solid (Eq. 4.24):

$$IO^-_{3(aq)} + 5\,I^-_{(aq)} + 6\,H^+_{(aq)} \rightarrow I_{2(s)} + 3\,H_2O_{(l)}. \tag{4.24}$$

The main uses of iodine are in pharmaceuticals, principally as disinfectant and antiseptic agents, and as X-ray imaging contrast agents, exploiting the high-X-ray absorption cross-section of the element.

Unfortunately, obtaining estimates of embodied energy (EE) and carbon (EC) for the halogens from the open literature has proved problematical.

4.1.3   Group 15 and 16 Elements

The two most abundant gases in the atmosphere, N_2 and O_2 (Table 4.3), are clearly in great abundance. However, these are not immutable entities,

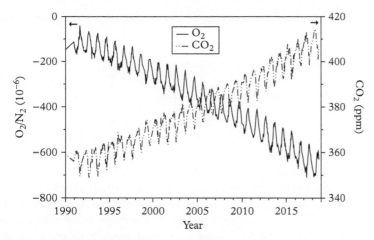

Figure 4.12 *The O/N ratio (right axis) and CO_2 concentration (left axis) recorded at the Scripps Institution pier, La Jolla in California.*

both elements having cycles across the spheres of chemical and biochemical transformations on Earth. An example of this is presented in Fig. 4.12. This shows the O_2/N_2 ratio relative to a reference mixture and multiplied by 10^6 (per meg abbreviated to pmeg) (Eq. 4.25);[9,10] the reference was sampled from the atmosphere in the mid 1980s. In Fig. 4.12, this is compared with the concentration of atmospheric CO_2 at the same location.

$$\delta(O_2/N_2) = \left(\frac{(O_2/N_2)_{(sample)}}{(O_2/N_2)_{(ref)}} - 1 \right). \tag{4.25}$$

Both show seasonal variation, with the O_2/N_2 ratio high in August, and the partial pressure of CO_2 low at that time. The seasonal changes might be expected to be ascribed to differences in the rate of photosynthesis (Eq. 4.26) between summer and winter. Photosynthesis can be expected to have a 1:1 stoichiometry between CO_2 lost and O_2 produced as carbon is fixed as sugars. Respiration is the reverse (dark) reaction.

$$6\,CO_2 + 6H_2O + light \rightarrow C_6H_{12}O_6 + 6O_2. \tag{4.26}$$

The change in $\delta(O_2/N_2)$ could occur though changes in concentration of both gases, and this gives us Eq. 4.27 for changes in that measurement:

$$\Delta\big(\delta(O_2/N_2)\big) = \left(\frac{\Delta(O_2)}{X_{O_2}} - \frac{\Delta(N_2)}{X_{N_2}} \right) x \frac{1}{M_{air}}. \tag{4.27}$$

Here $\Delta(O_2)$ and $\Delta(N_2)$ are changes in the quantity of these gases (in moles) in a sample of M_{air} moles of air; the X values are the mole fractions of the

[9] R. F. Keeling, S. R. Shertz, 'Seasonal and interannual variations in atmospheric oxygen and implications for the global carbon cycle', *Nature*, 1992, **358**, 723–7.

[10] Given the concentration of O_2 in the atmosphere (Table 4.3), the unit relationship is 4.8 pmeg = 1 ppm volume.

[11] A. C. Manning, R. F. Keeling, 'Global oceanic and land biotic carbon sinks from the Scripps atmospheric oxygen flask sampling network', *Tellus*, 2006, **58B**, 95–116.

[12] M. Battle, S. Mikaloff Fletcher, M. L. Bender, R. F. Keeling, A. C. Manning, N. Gruber, P. P. Tans, M. B. Hendricks, D. T. Ho, C. Simonds, R. Mika, and B. Paplawsky, 'Atmospheric potential oxygen: new observations and their implications for some atmospheric and oceanic models', *Global Biogeochem. Cycles*, 2006, **20**, GB1010.

gases (Table 4.3). Fluxes of N_2 are orders of magnitude smaller than those of O_2,[11] and thus the oxygen changes dominate $\delta(O_2/N_2)$. The change in the CO_2 concentration is given directly by Eq. 4.28:

$$\Delta X_{CO_2} = \frac{\Delta CO_2}{M_{air}}. \tag{4.28}$$

As a result, the relative speed of the photosynthesis/respiration cycle of processes, atmospheric CO_2 can be considered as proxy-O_2, and part of the count of **atmospheric potential oxygen** (APO) within Eq. 4.29.[12] Here the factor α_B represents the mean O_2 to CO_2 stoichiometry of 1.1 estimated from the biosphere. The seasonal extremes are in the region of 16 ppm for CO_2 and 100 pmeg for O_2/N_2. Applying Eq. 4.29 alone would predict a variation of 85 pmeg for this change in CO_2 concentration ($\delta[CO_2]$). The major factor in the seasonal variation then appears to be the balance between photosynthesis and respiration.

$$APO = \delta(O_2/N_2) + \frac{\alpha_B}{XO_2}\delta[CO_2]. \tag{4.29}$$

The change in APO (moles) is thus given as Eq. 4.30:

$$\Delta APO = \Delta O_2 + \alpha_B.\Delta CO_2. \tag{4.30}$$

In addition, the two measurements show an annual trend. The mean CO_2 level increased from ≈ 355 ppm in the early 1990s to ≈ 410 ppm in the late 2010s. In the same time, the O_2/N_2 has dropped from ≈ -100 pmeg to ≈ -650 pmeg, with a mean annual reduction of 19 pmeg, or ≈ 4 ppm. Measurements at other stations around the globe indicate this is a global trend. Overall, changes in CO_2 and O_2 concentrations in the atmosphere can be ascribed to three factors: F, the source of CO_2 from fossil fuel combustion and cement manufacture; B, the land biotic CO_2 sink; and O the ocean CO_2 sink, giving Eq. 4.31:

$$\Delta CO_2 = F - O - B. \tag{4.31}$$

$$\Delta O_2 = -\alpha_F F + \alpha_B B + Z. \tag{4.32}$$

Just as the biotic oxygen change can be given a stoichiometric factor (α_B), there can be a fossil fuel factor, α_F. Z in this case is the net exchange of O_2 between the oceans and atmosphere. Hence the oxygen changes can be expressed as Eq. 4.32. The empirical value for α_F is 1.39. Such a value approximates to the burning of a hydrocarbon of formula CH_x, where $x \approx 2$ (Eq. 4.33):

$$CH_2 + 1.5\,O_2 \rightarrow CO_2 + H_2O. \tag{4.33}$$

This averaged figure will be a mixture of processes with O_2/CO_2 ratios from 0 (calcination of carbonates) to 2 (combustion of methane). The effect of

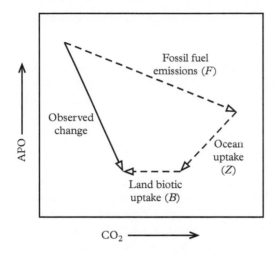

Figure 4.13 *Contributions to the change in APO and CO$_2$ concentration in the atmosphere.*

these components is shown qualitatively in Fig. 4.13. If fossil fuel emissions increase the atmospheric CO$_2$ concentration but reduce the O$_2$ concentration by a larger extent, then the APO will also reduce (Eq. 4.19). Given that Eq. 4.29 includes a term for the respiration–photosynthesis interconversion, the APO does change by this process, although atmospheric CO$_2$ is mitigated. The additional reduction of CO$_2$ and APO is assigned to ocean uptake; Z has been refined as 0.45×10^{14} mol yr^{-1}.

The industrial productions of nitrogen and oxygen gases are clearly linked. The long-standing method for their separation has been by fractional distillation, and this is favoured for high-purity and large-scale production. Newer methods include **pressure-swing adsorption** (PSA) using zeolite adsorbers (e.g. Ag in zeolite 5A), and polymer film or ceramic membrane separation. The major use of O$_2$ gas is in the steel industry, with medical and chemical industries also being large outlets. Production was estimated (in 2016) as 1.2 Mt per day. In addition to being the second most abundant atmospheric element, it is the most abundant in the crust, and hence supply is not an issue. It is literally our bedrock.

The element of oxygen pervades a high proportion of essential materials. The vast majority have the oxidation state of -2, as oxides, hydroxides, and neutral molecules of which water is the archetype. Peroxides (O$_2^{2-}$) and superoxides (O$_2^{-}$) exhibit oxidation states of -1 and -0.5, respectively. There are much fewer compounds of positive oxidation states, ranging from $+0.5$ (O$_2^{+}$ and O$_2$F) to $+1$ (FOOF) and $+2$ (OF$_2$).[13]

Nitrogen is much less prevalent in the Earth's crust (19 ppm, similar to La and Nd). Sodium nitrate is the major mineral, mined from deposits in Chile and Peru. Its major use is as fertilizer, but there is an increasing demand for use in batteries in solar power plants. Peak production from the mines was $\approx$ 3 Mt in the

[13] R. Marx and K. Seppelt, 'Structural investigations on oxygen fluorides', *Dalton Trans.*, 2015, **44**, 19659–62.

1910s, but in very large measure nitrogen fertilizers are now manufactured via the Haber–Bosch process (Eq. 4.34). The production of ammonia as fixed nitrogen in 2017 was $\approx$ 150 Mt; worldwide production of fertilizers was estimated to be 200 Mt in 2018, of which 119 Mt was nitrogen-based. This can be compared to the estimates of natural annual fixation of nitrogen prior to the industrial revolution of 203 Mt.[14] Thus the industrial production is a significant perturbation, and the overall result is considered to be a transfer of 50–70 Mt of nitrogen to the oceans.

Ammonia formation is exothermic ($\Delta_f H^\circ$ −45.9 kJ mol^{-1}), but a catalyst is required for this reaction to proceed. This is generally an iron catalyst preconditioned in a way to have a core–shell structure of a metallic surface associated with a layer of FeO over a core of *magnetite*, Fe$_3$O$_4$, with main group oxides as promoters. The slow step is involved in cleaving the strong N–N triple bond of the adsorbed N$_2$ molecule. The catalyst requires elevated temperatures, 400–500 °C, for a reasonable reaction rate. However, the reaction has a high negative entropy given that it is a conversion of four molecules to two. As a result, the equilibrium constant, K_p, decreases with temperature and is in the range of 10^{-4} to 10^{-5} over this operating temperature range. To partially offset that, a high pressure is employed (150–250 bar), which, according to the Le Châtelier Principle, will cause the reaction to move to the lower number of molecules. In the reactors conversion is *c.*15%, and this is condensed out of the gas stream in a cooler vessel, allowing the reagents to cycle round.

$$N_{2(g)} + 3\, H_{2(g)} = 2\, NH_{3(g)}. \tag{4.34}$$

The estimates for the embodied energy and carbon are shown in Table 4.8.[15] These are similar to those for steel (Table 3.13). Given the scale of production this will form a significant fraction of the environmental impact of industrial activity; it is estimated to be 1% of all CO$_2$ emissions.

Much of the ammonia is converted into nitric acid, HN^{V}O$_3$, substantially to combine with ammonia to provide the nitrogen-rich fertilizer ammonium nitrate (N^{-III}H$_4$)(N^{V}O$_3$); nitric acid production in 2016 was estimated as 62.2 Mt. (NH$_4$)(NO$_3$) is a highly disproportionate salt. Nitrogen reaches its highest positive oxidation state in ions such as nitrate and (NF$_4$)$^+$.

Oxidizing nitrogen through eight oxidation numbers is not performed in one step. The first is the highly exothermic oxidation ($\Delta_r H^\circ = -$ 900 kJ mol^{-1}) to nitric oxide, NO, by Eq. 4.35:

$$4\, NH_{3(g)} + 5\, O_{2(g)} \rightarrow 4\, NO_{(g)} + 6\, H_2O_{(g)}. \tag{4.35}$$

The oxidation by O$_2$ is carried out under moderate pressure ($\approx$ 12 bar) and high temperature (*c.*930 °C). Again, a catalyst is required and this is generally a gauze of two PGMs, platinum (90%) and rhodium (10%). This formally raises the nitrogen oxidation state to +2 in the odd electron compound NO, unusual in the *s* and *p* block chemistry. After mixing with air, the stream is compressed to *c.*10 bar, and the exothermic ($\Delta_r H^\circ = -116$ kJ mol^{-1}) reaction in Eq. 4.36 occurs

14 D. Fowler, M. Coyle, U. Skiba, M. A. Sutton, J. N. Cape, S. Reis, L. J. Sheppard, A. Jenkins, B. Grizzetti, J. N. Galloway, P. Vitousek, A. Leach, A. F. Bouwman, K. Butterbach-Bahl, F. Dentener, D. Stevenson, M. Amann, and M. Voss, 'The global nitrogen cycle in the twenty-first century', *Phil. Trans. R. Soc.,* B, 2013, **368**, 20130164.

Table 4.8 *Embodied energy (EE) and carbon (EC) of ammonia production.*

Method	EE	EC
	MJ/kg	kgCO$_2$e/kg
Haber–Bosch	41	1.92

15 L. F. Razon, 'Life Cycle Analysis of an Alternative Haber-Bosch Process: Non-renewable Energy Usage and Global Warming Potential of Liquid Ammonia from Cyanobacteria', *Environ. Prog. Sustain Energy*, 2014, **33**, 618–24.

Table 4.9 *Annual emissions of nitrogen-containing gases into the atmosphere, and their lifetimes.*

Gas	Sources	Emissions (Mt)	Lifetime
NH_3	terrestrial	60	hours
	oceans, volcanoes	9	hours
NO_x	combustion, fossil fuel	30	days
	combustion, biomass	4	days
N_2O	terrestrial	13	years
	oceans	5.5	years
NO	terrestrial	5	days
NO_2	combustion, oxidation		days
HNO_3	hydroxylation		hours

Note: NO_2 and HNO_3 are predominately formed in the atmosphere.

with nitrogen now in the +4 oxidation state. It is then cooled to $\approx 40\,°C$, allowing the equilibrium Eq. 4.37 to move to the right ($\Delta_r H° = -55$ kJ mol^{-1}).

$$2\,NO_{(g)} + O_{2(g)} \rightarrow 2\,NO_{2(g)}, \qquad (4.36)$$
$$2\,NO_{2(g)} = N_2O_{4(g)}. \qquad (4.37)$$

The gas mixture is then intercepted by a stream of water, which disproportionates to oxidation states +2 and +5 (Eq. 4.38) affording nitric acid and nitric oxide ($\Delta_r H° = -104$ kJ mol^{-1}):

$$3\,N_2O_{4(g)} + 2\,H_2O_{(l)} \rightarrow 4\,HNO_{3(aq)} + 2\,NO_{(g)}. \qquad (4.38)$$

The principal use of nitric acid is in fertilizers (80%); the vast majority of the fertilizers (96%) are in the form of ammonium and calcium ammonium nitrates. Half of the remainder is utilized in the manufacture of organic materials: dyes (from nitrobenzene), polyamides (from hexanedioic acid), and polyurethanes (from toluidine di-isocyanate).

4.2 Atmospheric Components

In Chapter 1, the build-up of natural (e.g. CH_4) and synthetic (e.g. NF_3) gases in the atmosphere was described, together with their radiative forcing effects.

Generally these are molecules with long lifetimes in the atmosphere. The compounds of nitrogen discussed in Section 4.1.3 generally differ in regard to their lifetimes (Table 4.9). Apart from nitrous oxide, N_2O, these are all short-lived but are part of a large natural flux. That flux has been considerably disturbed by human activity.[14] It is estimated that $\frac{2}{3}$ of the evolution of ammonia to the atmosphere is anthropogenic, and virtually all of the NO/NO_2 mixture, labelled as NO_x, is also our own cause. These two gases interconvert rapidly in the troposphere. NO is oxidized by ozone (Eq. 4.39), and, in the day, NO_2 is photolysed back to NO (Eq. 4.40), and an oxygen atom, which combines with O_2 to reform ozone.

$$NO + O_3 \rightarrow NO_2 + O_2. \tag{4.39}$$

$$NO_2 + h\nu \rightarrow NO + O. \tag{4.40}$$

Nitric oxide is also oxidized by hydroperoxy and organic peroxy radicals (Eq. 4.41). Taken together, when NO_x is $>$ 30 ppt, these processes can lead to an increase in tropospheric ozone.

$$RO_2 + NO \rightarrow NO_2 + OR, R = H \text{ or organic } R. \tag{4.41}$$

Nitrogen dioxide is removed from the troposphere by reaction with hydroxyl radicals (Eq. 4.42), which requires a third body to mediate it. This can occur in free air or in aerosols. If ammonia is present this can combine with HNO_3 to form some ammonium nitrate in the aerosols (Eq. 4.43). Depending upon the atmospheric composition in the area, clouds can be acidic, basic, or even sources of nitrogen fertilizer!

$$NO_2 + OH \rightarrow HNO_3 \tag{4.42}$$

$$HNO_3 + NH_3 = (NH_4)(NO_3). \tag{4.43}$$

It is estimated that in the thirty-three countries of the European area emissions of NO_x reduced by 41% from 1990 to 2011. In 2011, the major cause was due to road transport (41%), with the sector of energy production and distribution the second largest source (23%). The reduction in road transport NO_x emission over that time period of 48% occurred in a time of expanding activity. Urban areas have been slower to show reductions. The cause of the reduction in road traffic emissions can be ascribed to the developments in exhaust catalysts, such as three way catalysts.

4.2.1 Exhaust Gas Remediation

The ideal conversion of fuel in an internal combustion engine would be for the mixture of hydrocarbons to be oxidized completely to water and carbon dioxide.

This would provide the maximum energy return, and afford two non-toxic products, though admittedly one is a greenhouse gas. However, imperfect combustion leads to the formation of CO and also of carbonaceous particles. Some oxidation of N_2 may also occur to form NO_x. Any organic nitrogen in the fuel can also be oxidized to NO_x and sulphur to SO_x. This last problem has been addressed to a significant extent during the refining of fuels to reduce the organic heteroatoms.

The **two way catalyst** tackles the problem of having products of incomplete combustion in the exhaust stream. The first way is with a catalyst to mediate the oxidation of carbon monoxide (Eq. 4.44), and the second to fully burn unreacted or partially oxidized fuel. This is exemplified with octane in Eq. 4.45. Gasoline (petrol)[16] is a mixture of hydrocarbons, not pure octane. The stoichiometry of combustion in Eq. 4.45 would imply the ideal O_2:fuel molar ratio to be 12.5:1. The air:fuel mass ratio is labelled lambda (λ). A stoichiometric ratio is ascribed a λ value of 1. This has an air:fuel mass ratio of 14.7:1; this would correspond to an O_2:octane ratio of c.12.2. Lean and rich fuel mixes have λ values which are >1 or <1, respectively.

$$2\,CO + O_2 \rightarrow 2\,CO_2 \qquad (4.44)$$
$$e.g.\ C_8H_{18} + 12.5\,O_2 \rightarrow 8\,CO_2 + 9\,H_2O. \qquad (4.45)$$

The **three way catalyst** (TWC) also mediates reactions the removal NO generated in the combustion cylinder. This can occur by three different reactions, with the reductant being a hydrocarbon (Eq. 4.46), H_2 (Eq. 4.47), and CO (Eq. 4.48), and the products are water, and/or CO_2 and N_2.

$$e.g.\ CH_4 + 4\,NO \rightarrow CO_2 + 2\,H_2O + 2\,N_2 \qquad (4.46)$$
$$2\,H_2 + 2\,NO \rightarrow 2\,H_2O + N_2 \qquad (4.47)$$
$$2\,CO + 2\,NO \rightarrow 2\,CO_2 + N_2. \qquad (4.48)$$

It is very beneficial to avoid a reaction such as Eq. 4.49 in which the reduction of nitrogen is incomplete and results in the evolution of N_2O, which has a long lifetime in the atmosphere.

$$H_2 + 2\,NO \rightarrow 2\,H_2O + N_2O. \qquad (4.49)$$

The demands on an exhaust catalyst are very high. The reaction conditions are provided by the operating conditions, which will vary from a cold start to a high cruising speed. The temperature along the converter will also vary across the manifold, reducing with increasing distance from the engine. There are also three classes of reactions, and it may not be easy to optimize all of them.

This is evident from the behaviour of an engine under differing air/fuel ratios (Fig. 4.14).[17] In this example, engine power is maximized under fuel-rich conditions, but in this regime, removal of hydrocarbon (HC) and CO from the

[16] The test fuel is a mixture of *i*-octane and *n*-heptane. In the octane number definition, the former is allocated 100% and the latter 10%.

[17] J. Kašpar, P. Fornasiero, and N. Hickey, 'Automotive catalytic converters: current status and some perspectives', *Catal. Today*, 2003, **77**, 419–49.

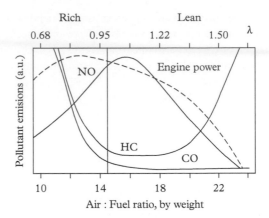

Figure 4.14 *The effect of air/fuel ratio on emissions. Reproduced from Kašpar, Fornasiero, and Hickey (n 17).*

exhaust stream is far from optimized. That can occur under lean-burn conditions. As the fuel concentration becomes further reduced at high air/fuel ratios, the engine temperature drops, fuel is unburnt, and the engine power falls away. However, the efficiency of NO_x removal is poor in this example, and is highest under fuel-rich conditions. Under very lean conditions the reduction in engine temperature also reduces the oxidation of nitrogen and so NO_x emission also drops away.

The catalyst structures themselves are complex and are on different length scales. There is a ceramic honeycomb (or monolith) which allows the exhaust gases to pass through and come into contact with the active components of the surface. The catalyst itself is based on oxide particles, generally an alumina stabilized with lanthanum or barium to prevent conversion to α-Al_2O_3 under high temperatures (up to $c.1,000\,^{\circ}C$). On these alumina particles are the active metals, generally Rh, Pd, and Pt. These have co-promoters such as cerium–zirconium mixtures which have oxygen storage capabilities. The promoters can lower the operating temperature and so assist with the conversions under cold start conditions. Cerium we have discussed (Section 3.4.1) as having both Ce^{III} and Ce^{IV} available; zirconium further increases the oxygen availability. Thus oxygen is available from these oxides as the oxidation state of the metal is reduced. They can be reoxidized at the higher gas temperatures provide by a warm engine.

The reactions at the PGM can in principle follow one of two classical pathways, for example, for CO oxidation.[18] In the **Eley–Rideal** mechanism, mobile CO reacts with an immobile surface oxide site. However, kinetic data on rhodium surfaces favour the alternative **Langmuir–Hinshelwood** mechanism in which a surface site, labelled M^* in Eq. 4.50, can bind CO and molecular oxygen (Eq. 4.51), which compete for metal sites. CO binds the more strongly and thus under a fuel-rich environment, there be few adsorption sites possible for the rather sparse gaseous O_2 molecules to adsorb. Under leaner conditions, the O_2/CO ratio will

[18] P. Granger and V. I. Parvulescu, 'Catalytic NO_x Abatement Systems for Mobile Sources: From Three-Way to Lean Burn after-Treatment Technologies', *Chem. Rev.,* 2011, **111**, 3155–207.

be higher, and so the probability of Eq. 4.51 is considerably increased. Once adsorbed, the O=O bond can be cleaved to afford two surface oxygen sites (Eq. 4.52). When in close proximity adsorbed CO can react with an adsorbed O site to liberate CO_2 (Eq. 4.53).

$$CO_{(g)} + M^* = M^*(CO) \qquad (4.50)$$

$$O_{2(g)} + M^* = M^*(O_2) \qquad (4.51)$$

$$M^*(O_2) + M^* \rightarrow 2\,M^*(O) \qquad (4.52)$$

$$M^*(O) + M^*(CO) \rightarrow 2\,M^* + CO_{2(g)}. \qquad (4.53)$$

With NO also present in the gas stream, additional adsorption processes can occur: (Eqs 4.54–4.58), generally mediated by rhodium. Competition for adsorption sites is now between NO (Eq. 4.54) and CO (Eq. 4.50), a much more even contest. Dissociation of NO can occur (Eq. 4.55) and N_2 formation can occur by two alternative reactions of adsorbed species (Eqs 4.56 and 4.57). Finally, these same sites in Eq. 4.56 can also result in the formation of nitrous oxide by Eq. 4.58. The optimum efficiency for removal of NO from the exhaust stream is just under the stoichiometric air/fuel ratio. Initially a compromise mix was adopted minimize the emissions of NO, CO, and hydrocarbons. However, engine management systems equipped with electronic fuel injection and gas sensors mean that the gas composition injected into a cylinder is rapidly modulated (sub-second repetition rates) so that a spread of compositions around the lambda point is created and thus more optimized compositions for the removal of each gas are passed through.

$$NO_{(g)} + M^* = M^*(NO) \qquad (4.54)$$

$$M^*(NO) + M^* \rightarrow M^*(N) + M^*(O) \qquad (4.55)$$

$$M^*(NO) + M^*(N) \rightarrow M^* + M^*(O) + N_{2(g)} \qquad (4.56)$$

$$2\,M^*(N) \rightarrow 2\,M^* + N_{2(g)} \qquad (4.57)$$

$$M^*(NO) + M^*(N) \rightarrow 2\,M^* + N_2O_{(g)}. \qquad (4.58)$$

The interplay between supports, metals, and the reacting gases are complex and intertwined. *In situ* IR and X-ray spectroscopic studies[19] show that the structure changes are rapid and the rapid gas composition changes help avoid the metal particles from sintering into larger aggregates. In that case the ratio of active surface to bulk atoms would deteriorate thus increasing the proportion of the PGM atoms that are merely spectators.

As noted in Table 3.18, the environmental demands of producing PGMs are very high. Development of TWC materials and operating processes has greatly reduced the amount of these metals required for a working catalyst. The manufacture of a TWC unit has been estimated to result in the emission of 0.60 kg of NO_x.[20] However, it has been estimated that over the lifetime of a vehicle, the catalyst reduces those emissions by *c.*300 kg. Indeed the break-even driving distance was only $\approx$ 250 km.

[19] J. Evans, *X-ray Absorption Spectroscopy for the Chemical and Materials Sciences*, Wiley, Chichester, 2018.

[20] T. Bossi and J. Gediga, 'The Environmental Profile of Platinum Group Metals', *Johnson Matthey Technol. Rev.*, 2017, **61**, 111–21.

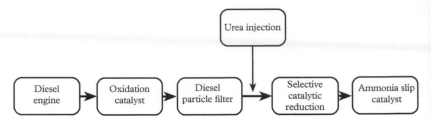

Figure 4.15 *A schematic of exhaust treatment from a diesel-powered vehicle engine.*

[21] This is also called the flame equivalence ratio.

[22] T. Johnson, 'Diesel Engine Emissions and their Control', *Platinum Metals Rev.*, 2008, **52**, 23–37.

[23] E. Borfecchia, P. Beato, S. Svelle, U. Olsbye, C. Lamberti, and S. Bordiga, 'Cu–CHA—a model system for applied selective redox catalysis', *Chem. Soc. Rev.*, 2018, **47**, 8097–133.

[24] D. Fino, S. Bensaid, M. Piumetti, and N, Russo, 'A review on the catalytic combustion of soot in Diesel particulate filters for automotive applications: From powder catalysts to structured reactors', *Applied Catal.* A, 2016, **509**, 75–96.

Diesel engines operate under varying and high λ ratios (varying between ≈ 2 to 10). The two-way (oxidation) catalysts perform well under these conditions, and so are employed in diesel vehicles. However, NOx removal is not effective under such lean compositions. Hence rhodium would not be employed, and the PGMs used are platinum and palladium. There are regimes of temperature and fuel-to-air stoichiometry[21,22] under which a diesel engine can create significant quantities of soot (fuel/air > 2 and 1,200–2,300 °C) and NO$_x$ (fuel/air < 2 and $>1,900$ °C). Pre-mixing of gases prior to injection can help minimize the production of both pollutants, but this cannot currently be relied on for emission control. As a result, additional stages are incorporated after the two-way oxidation catalyst reactor. The designs of these stages vary, but a representative option is shown in Fig. 4.15.[23]

In this scheme, removal of the particulates from incomplete reduction of fuel occurs in the diesel particle filter. Ceramic materials like silicon carbide and derivatives of the mineral *cordierite*, $Mg_2Al_4Si_5O_{18}$, have been used for this purpose. Cordierite is commonly employed as the basis of the support monolith. The structure (Fig. 4.16) is based upon aluminosilcate tetrahedra, with magnesium ions in a distorted octahedral site. Down the centre of the unit cell there is a cavity with transannular O...O distances of 5 Å. It has a melting point of 1,435 °C, sufficient for exhaust gas conditions within all of the stages. The deposition of soot into the particle filter will cause a deterioration in effectiveness and so oxidation of the particles is utilized to continually strip them away. Under the conditions of a diesel exhaust (200–500 °C), a catalyst is required for the purpose.[24] It is evident that NO$_2$ contributes to the combustion of the carbon in soot in addition to O$_2$. The positioning of the filter after the two-way oxidation catalyst before the NO$_x$ removal chamber is advantageous in this regard. The mobile gas phase oxidants can migrate to carbon particles and effect oxidation. The values of E_a are lower for CO$_2$ formation than CO, and much lower for oxidation by NO$_2$ (39 kJ mol^{-1}) (Eq. 4.59) than by O$_2$ (127 kJ mol^{-1}) (Eq. 4.60):

$$C + 2\,NO_2 \rightarrow CO_2 + 2\,NO \tag{4.59}$$

$$C + O_2 \rightarrow CO_2. \tag{4.60}$$

Oxidation catalysts mediate the regeneration of NO$_2$ within the particle filter, a role that can be effected by platinum and CeO$_2$. A wide variety of other catalysts

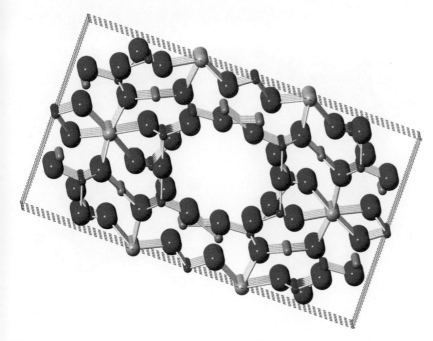

Figure 4.16 *View down the c axis of cordierite, $Mg_2Al_4Si_5O_{18}$ (five unit cells); red = O, yellow = Mg, light blue = Al, dark blue = Si.*

have been described based upon *perovskites* (e.g. $LaMnO_3$), oxides (e.g. La_2O_3), and alloys (e.g. $Pb_{10}La_5Co_{85}O_x$). This remains a field of high activity.

One approach to continual regeneration of the particle filter system is by **fuel-borne catalysts** (FBCs).[25] Additives can either be hydrocarbon-soluble materials (e.g. ferrocene, $Fe(C_5H_5)_2$), microemulsions of metal salts or complexes (e.g. $Cu(NO_3)_2$ and $Ce(CH_3CO_2)_3 \cdot H_2O$), or nanoparticles (e.g. TiO_2 and CeO_2). These additives may be injected upstream of the diesel particle filter carried by fuel, or prior to the combustion chambers in which case they will form fine particles mainly of oxides during combustion.

The removal of NO_x is effected by a variation of Eq. 4.46, and is called a **selective catalytic reduction** (SCR). Rather than use a hydrocarbon as the reductant, ammonia is often employed (Eq. 4.61):

$$e.g.\ 2\,NH_3 + 2\,NO + \frac{1}{2}O_2 \rightarrow 3\,H_2O + 2\,N_2. \tag{4.61}$$

The ammonia is generated on-site by decomposition of urea (Eq. 4.62) and hydrolysis of the isocyanic acid (HNCO) so formed (Eq. 4.63):

$$H_2N - C(O) - NH_{2(solv)} \rightarrow HNCO_{(g)} + NH_{3(g)} \tag{4.62}$$

[25] J. J. Jelles, M. Makkee, and J. A. Moulijn, 'Ultra low dosage of platinum and cerium fuel additives in diesel particulate control', *Top. Catal.*, 2001, **16**, 269–73.

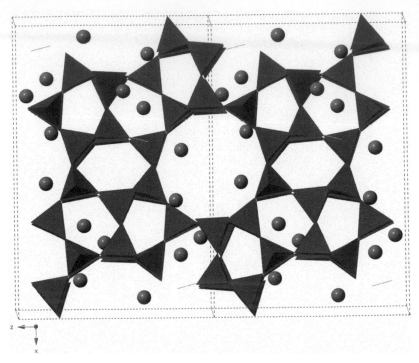

Figure 4.17 *View down the b axis (two unit cells) of a ZSM-5 sample: $H_{41.61}Al_{1.78}$ $Si_{94.22}O_{212}$, showing the framework of tetrahedra with locations of water molecules.*

$$HNCO_{(g)} + H_2O_{(g)} \rightarrow CO_{2(g)} + NH_{3(g)}. \tag{4.63}$$

The first catalysts used in transportation diesel engines were based upon those already developed for fixed-location plants. Typically this was a mixture of early d-block oxides of the type V_2O_5(2 wt%)-WO_3 (3 wt%)/TiO_2 on a honeycomb structure.[18] However, new catalysts needed to be sought to extend the operable temperature range.[26] It is also apparent that with the order of the processes in Fig. 4.15, the regeneration processes in the particle filter can deactivate vanadium-based SCR catalysts.[27]

As introduced in Section 3.2.3, aluminosilicates can be synthesized with a range of three-dimensional architectures. Extrapolating from silica (SiO_2) with Si–O–Si linked tetrahedra, substitution of silicon by aluminium requires a charge-balancing proton. The Al–OH–Si units so formed can have high acidity, and derive some of their applications as solid acid catalysts. One such zeolite is labelled ZSM 5 (Fig 4.17). The structure is depicted in a polyhedral mode showing the locations of the aluminosilicate tetrahedra. An important feature of the structure is the ten-ring with ten tetrahedral sites and ten bridging oxygen atoms, with a trans-annular O..O distance of 8.2 Å. The acid sites in conjunction with the channels

[26] M. Piumetti, S. Bensaid, D. Fino, and N. Russo, 'Catalysis in Diesel engine NO_x aftertreatment: a review', *Catal. Struct. React.*, 2015, **1**, 155–73.

[27] B. Guan, R. Zahn, H. Lin, and Z. Huang, 'Review of state of the art technologies of selective catalytic reduction of NO_x from diesel engine exhaust', *Appl. Therm, Eng.*, 2014, **66**, 395–414.

catalyse the dehydration of methanol to gasoline hydrocarbons. An alternative channel structure is provided by *chabazite* (Fig. 4.18). This exhibits an eight-ring with transannular O..O distances of $c.6.5$ Å. A similar structure is adopted by the **silicoaluminophosphate** SAPO-34. In such a structure there are normally alternating aluminium and silicon/phosphorus tetrahedral sites. There can be acid sites in SAPOs; in the absence of the silicon, we would have an aluminium phosphate material, which, like silica, does not have attendant hydroxyl groups.

The acidic sites can be ion exchanged with metal cations to create specific sites near pores of different sizes. An example of this is shown in Fig. 4.19 with copper as the metal. At this exchange level, five different copper sites could be identified. Three of the sites were located in relatively small channels, but the other two are located near the walls of the large channels in ZSM-5; these latter two were also populated when the exchange level was 30% lower. It was the site in which the copper was held towards the centre of the six-ring that was absent. Thus the metal-loaded microporous solids can be synthesized with a variety of steric and electronic influences which can affect catalytic performance.

Both copper and iron exchanged zeolites such as ZSM-5 do catalyse the ammonia-SCR of NO; however, their utility is compromised by poor hydrothermal stability above 700 °C.[28] The chabazite class of zeolites provide higher chemical resistance and can maintain isolated copper(II) sites in the catalytic cycle (Fig. 4.20), avoiding the formation of CuO crystallites which reduces catalytic activity.

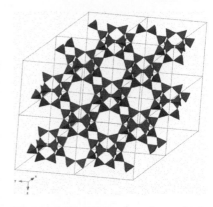

Figure 4.18 *Diagonal view (eight unit cells) of a chabazite sample, $H_{2.2}$ $Al_{2.2}Si_{33.8}O_{72}$, showing the framework of tetrahedra.*

[28] S. J. Schmeig, S. H. Oh, C. H. Kim, D. B. Brown, J. H. Lee, C. H. F. Peden, and D. H. Kim, 'Thermal durability of Cu-CHA NH_3-SCR catalysts for diesel NO_x reduction', *Catal. Today*, 2012, **184**, 252–61.

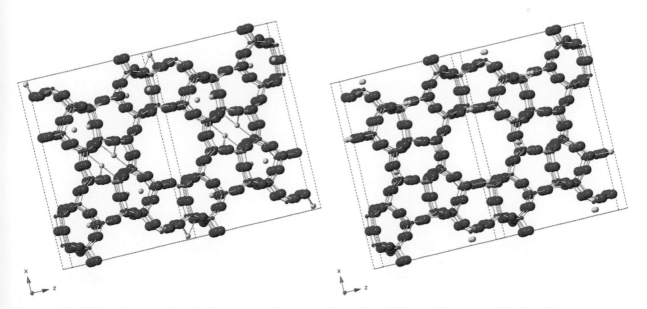

Figure 4.19 *Views of the ZSM-5 derivative $H_{1.33}Al_{3.7}Cu_{2.37}Si_{92.3}O_{192}$; left: sites Cu(1,4,5); right: sites Cu(2,3); red = O, blue = Si/Al, green = Cu.*

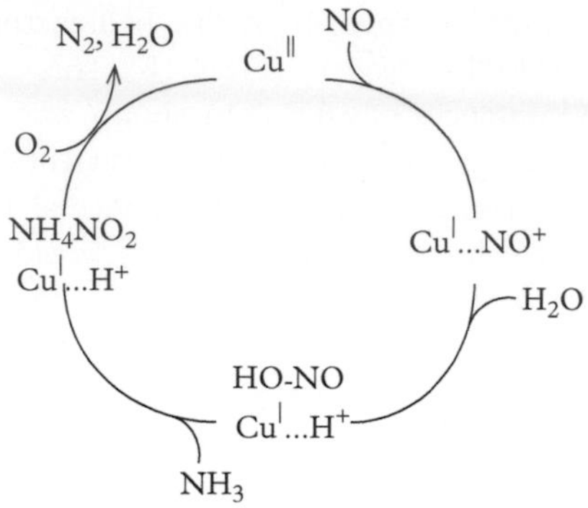

Figure 4.20 *Reaction scheme for the copper-catalysed standard NH₃ SCR of NO.*

[29] A. M. Beale, F. Gao, I. Lezcano-Gonzalez, C. H. F. Peden, and J. Szanyi, 'Recent advances in automotive catalysis for NO$_x$ emission control by small-pore microporous materials', *Chem. Soc., Rev.,* 2015, 52, 7371–405.

[30] J. H. Kwak, J. H. Lee, S. D. Burton, A. S. Lipton, C. H. F. Peden, J. Szanyi, 'A Common Intermediate for N$_2$ formation in Enzymes and Zeolites: Side-On Cu-Nitrosyl complexes', *Angew. Chem. Int. Ed.,* 2013, **52**, 9985–9.

[31] M. J. Hazlett and W. S. Epling, 'Heterogeneous catalyst design: zoned and layered catalysts in diesel vehicle aftertreatment monolith reactors', *Canad. J. Chem. Eng.,* 2019, **97**, 188–206.

The catalytic properties of these catalysts are not fully understood, in spite of very considerable efforts.[23,29] The reaction scheme in Fig. 4.20 outlines one possibility, which allows for the bifunctional properties of a metal-in-zeolite catalyst (acid–base and redox).[30] Alternative schemes involve monomeric and dimeric copper sites. This scheme, for the standard NH₃-SCR reaction (Eq. 4.61), starts from a Cu(II), which can be reduced to a transient Cu(I) complex of NO⁺. This in turn can be hydroxylated leaving a proton near the Cu(I) site. The nitrous acid so formed is neutralized by ammonia to form ammonium nitrite. Heated under the oxygen-rich conditions, the nitrite salt conproportionates to evolve N₂ and water. The copper is oxidized back to Cu(II) and water is also formed in this way to complete the catalytic cycle. Catalysts of this type have been adopted in more recent light-duty diesel engines for transportation. There are limitations caused by thermal instability >800 °C, sensitivity to the build-up of sulphate from fuel-borne sulphur, and low activity at low exhaust gas temperatures (≈150 °C).

The final step is to remove the unreacted ammonia after the SCR-NH₃ step (Eq. 4.61). Ammonia is in excess in the SCR module to minimize the escape of NO$_x$ to the atmosphere, but may itself slip through. However, it is an undesirable pollutant and the normal removal process is effected by an **ammonia slip catalyst** (ASC). The ideal reaction is shown in Eq. 4.64, but further oxidation to N₂O and NO$_x$ is clearly to be avoided. One catalyst adopted is Pt/Al₂O₃, added to the catalyst monolith surface as a washcoat. In Fig. 4.15 it is represented in a separate box, but many architectures have been tried, such as a slice at the end of the SCR monolith or a double layer.[31] The platinum catalyst in question will catalyse the further oxidation to nitrogen oxides at high temperatures. This problem has been reduced with formulations very dilute in platinum.[27]

$$2\,NH_3 + 1.5\,O_2 \rightarrow N_2 + 3H_2O. \tag{4.64}$$

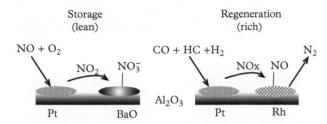

Figure 4.21 *Reaction schemes for a Pt–Rh–Ba/Al$_2$O$_3$ catalyst in a lean NO$_x$ trap (LNT).*

An alternative approach is often called the **lean NO$_x$ trap** (LNT) catalyst.[26,27,31] This is a composite catalyst material that can store NO$_x$ under lean-burn conditions and then reduce the stored nitrogen oxides, generally using a PGM catalyst on alumina, under rich conditions. For a diesel engine this cannot be achieved under normal operating conditions and so a pulsed fuel injection is required for the reduction/regeneration step. The storage of the acid gases is carried out by a basic metal oxide, typically of potassium or barium. Focusing on the example of barium, in a Pt–Rh–Ba/Al$_2$O$_3$ catalyst BaO will react with water to form its hydroxide under normal ambient conditions, and dehydrate at elevated temperatures (Eq. 4.65). The NO$_x$ gases will adsorb onto the oxide/hydroxide which will result in a build-up of Ba(NO$_2$)$_2$ and Ba(NO$_3$)$_2$ in the storage step. The platinum will catalyse the oxidation of NO, and so the nitrate would predominate (Eq. 4.66). In the reduction period, the stored nitrate/nitrite may migrate from the barium particles to the PGMs and then be reduced to N$_2$, as summarized in Fig. 4.21.

$$BaO_{(s)} + H_2O_{(g)} = Ba(OH)_{2(s)} \tag{4.65}$$

$$BaO_{(s)} + NO_{2(g)} = Ba(NO_3)_{2(s)}. \tag{4.66}$$

As the target for increasingly clean exhaust emissions grows, further developments in the chemical processing in the exhaust train will be essential. One key target is an efficient deNO$_x$ process for low temperatures.

4.2.2 Carbon Capture and Storage (CCS)

As described in Section 1.3, there is a close link between the concentration of atmospheric CO$_2$ and mean global surface temperature. Indeed there is a strong correlation between them over millions of years of the Earth's history. Sometimes temperature has led the curve, as in the start of ice ages, and at other times the changes in CO$_2$ concentration have caused temperature change. The introduction of what is now tens of Gt per year by fossil fuel burning (Fig. 1.7) makes CO$_2$ concentration the leading factor at present.

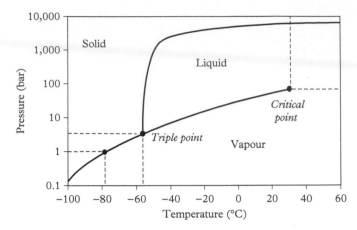

Figure 4.22 *Phase diagram of CO_2.*

Mitigation of this by CO_2 removal could involve both physical and chemical methods. The physical parameters are presented in the phase diagram of CO_2 in Fig. 4.22. At 1 bar pressure and ambient temperatures it is the vapour phase which is stable, and lowering the temperature to $-78\,°C$ passes the transition to solid (**dry ice**). Liquid CO_2 can only be accessed above the **triple point** at which solid, liquid, and vapour are in equilibrium (5.1 bar, $-56.6\,°C$). The liquid–vapour transition curve follows a path to the **critical point** at 73.8 bar and $31.1\,°C$. The pressures for these points correspond to water depths of 53 m and 570 m, respectively. Options for storage locations are outlined in Table 4.10.[32] An option which is excluded is storage as a liquid pool under deep water. The density of liquid CO_2 (0.77 g cm^{-3}) is substantially less than that of water and so this is intrinsically unstable. There are two major risks: dissolution lowering the pH of the water, and leakage into the atmosphere. This would result in a low-lying layer of CO_2-rich and O_2-deficient air, with potentially tragic results.[33]

The risk of storage in oceans is similar to those in land-enclosed water systems, and indeed is happening to a great extent by dissolution from the atmosphere. Underground voids or reservoirs are more viable and have a considerable capacity, but none are risk-free and considerable care is required in assessing sites. Mineralization is an option with high potential capacity and greatest permanence, provided the geochemistry and geology are both favourable.

At higher temperatures and pressures than the critical point the **supercritical** fluid exists. The properties of a supercritical fluid (like sc-CO_2) are somewhat between a liquid and a gas, for example in density, viscosity, and diffusivity. The critical density of CO_2 is 0.469 g cm^{-3}. Supercritical fluids have no surface tension and can diffuse through solids. Thus the volume of CO_2 can be compressed by a factor of $c.370$ by liquifying at elevated pressures, but under supercritical conditions diffusion through solid materials is enhanced. In pipelines, pressures range between 85 and 210 bar, which will either be liquid or sc-CO_2, depending

[32] J. M. Allwood and J. M. Cullen, *Sustainable Materials: With Both Eyes Open*, UIT Cambridge, 2012, Cambridge UK.

[33] Such an event occurred in August 1986 when a large bubble from a naturally occurring reservoir of CO_2 was released from Lake Nyos in Cameroon.

Table 4.10 *CO_2 storage options.*

Storage	Capacity Gt(CO_2)	Risk
Oil/gas reservoirs	675–900	low
Coal seams	20–900	medium
Saline aquifers	10^3–10^4	medium
Oceans	$>10^3$	high
Mineralization	large	?

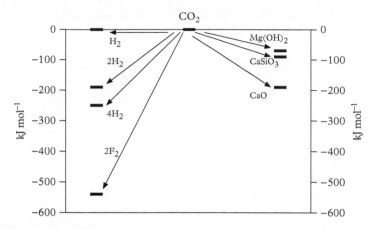

Figure 4.23 *Options for chemical fixing of CO$_2$.*

upon the temperature. On ships CO$_2$ is transported as a cryogenic liquid close to the triple point conditions. Carbon dioxide from natural sources is saturated with water, but otherwise generally has just traces of H$_2$S as an impurity. However, industrially generated CO$_2$ has a wider range of impurities from operations (CO, O$_2$, H$_2$S, SO$_x$, NO$_x$, H$_2$) and may require a variety of capture methods (amines, NH$_3$, solvents).[34]

Even though CO$_2$ is sometimes dubbed a thermodynamic sink, there are several types of reaction which are exothermic and thus could, on that criterion, be used as a method of chemical fixation (Fig. 4.23). Due to the normally high bond energy of element-to-fluorine bonds, conversion of CO$_2$ to CF$_4$ is highly exothermic. However, this is very unlikely to be an appropriate method. As discussed in Section 4.1.2, production of F$_2$ is highly energy intensive and carried out on only a minor scale (Fig. 4.11). In addition, the product, CF$_4$, is possibly the most long-lived greenhouse gas with a very high global warming potential (Sections 1.8, 1.9) and so its production should be reduced rather than enhanced. The stoichiometric reaction of CO$_2$ with H$_2$ is the water gas shift reaction (Section 4.1.2), and the carbon product, CO, is not attractive for very long-term storage. With excess H$_2$, carbon can be reduced to elemental carbon and with a 4:1 ratio methane would result. In all cases, water is the co-product, and so for the reaction in Eq. 4.67, all products are safe and storable. However, at present production of hydrogen is predominantly from fossil fuel sources. Hence adoption of this as a storage method will require a switch to photo- and electro-chemical methods of hydrogen generation.

$$CO_{2(g)} + 2\,H_{2(g)} \rightarrow C_{(s)} + 2\,H_2O_{(l)}. \tag{4.67}$$

On the right-hand side of Fig. 4.23, there are three examples of gas–solid reactions. The most exothermic of these is the reverse of calcination of carbonates, as in the manufacture of cement. The carbonate is the sole product and is

[34] M. E. Boot-Handford, J. C. Abanades, E. J. Anthony, M. J. Blunt, S. Brandani, N. MacDowell, J. R. Fernández, M.-C. Ferrari, R. Gross, J. P. Hallett, R. S. Haszeldine, P. Heptonstall, A. Lyngfelt, Z. Makuch, E. Mangano, R. T. J. Porter, M. Pourkashanian, G. T. Rochelle, N. Shah, J. G. Yao, and P. S. Fennell, 'Carbon capture and storage update', *Energy. & Environ. Sci.*, 2014, 7, 130–89.

also safe and storable. However, since CaO has to be produced by the reverse reaction, there is no evident gain in CO_2 storage in the overall process. It is, however, utilized in a scheme called calcium looping (CaL) for carbon capture. Carbonation is performed at $c.650$–$700\,°C$ and the reverse, calcination, at $900\,°C$. $Mg(OH)_2$ and it calcium analogue both occur naturally as minerals, *brucite* and *portlandite*, respectively. The conversion to the carbonate is favoured by the higher charge/volume ratio of the carbonate anion over hydroxyl, and the by-product, again, is water (Eq. 4.68). This is a more viable option for storage. However, these are relatively uncommon minerals and, again, the scale of its applicability does not match that of the problem to be solved.

$$CO_{2(g)} + Mg(OH)_{2(s)} \rightarrow MgCO_{3(s)} + H_2O_{(l)}. \tag{4.68}$$

The final example is the exchange of silicate for carbonate (Eq. 4.69). Again, the small size of the carbonate anion contributes significantly to the driving force for this reaction (see Eqs. 2.75–2.77). The products are both storable, and indeed are minerals themselves. In terms of scale, the fixing of 1 Gt of CO_2 would require 2.6 Gt of $CaSiO_3$.

$$CO_{2(g)} + Ca(SiO_3)_{(s)} \rightarrow CaCO_{3(s)} + SiO_{2(s)}. \tag{4.69}$$

A wide spectrum of methodologies for carbon capture have been investigated. One is using ethanolamine, which can be manufactured from the addition of ammonia to ethylene oxide (Eq. 4.70), and derivatives of it. The capture and recovery is effected by the equilibria in Eqs 4.71 and 4.72. The amines are often supported on high-area materials like activated charcoal in beds used to extract CO_2 from flue gases, for example in power stations. Beds can then be heated to allow collection of the gas, with reactivation of the absorbent.

$$C_2H_4O + NH_3 \rightarrow HOCH_2CH_2NH_2 \tag{4.70}$$
$$CO_2 + HOCH_2CH_2NH_2 = HOCH_2CH_2NH_2^+ - CO_2^- \tag{4.71}$$
$$CO_2 + HOCH_2CH_2NH_2 + H_2O = HOCH_2CH_2NH_3^+ + HCO_3^-. \tag{4.72}$$

Abstraction methods like this can be utilized in a precombustion mode as well as postcombustion. For example, coal gasification under air and water followed by a water gas shift reaction can provide a mixture of CO_2, H_2, and N_2. Abstraction of CO_2 leaves what is then a carbon-free fuel.

A reagent can be grafted onto the surface of a support like silica. The surface contains silanol groups and these can substituted for Si–OMe groups. The number of surface links varies with preparation method, but the disubstitution shown in Fig. 4.24 is typical. This leaves a diamine chemically bonded to the silica surface which should be resistant to leaching from the surface.[35] This material

[35] J. A. Wurzbacher, C. Gebald, and A. Steinfeld, 'Separation of CO_2 from air by temperature-vacuum swing adsorption using diamine-functionalized silica gel', *Energy Environ. Sci.*, 2011, **4**, 3584–92.

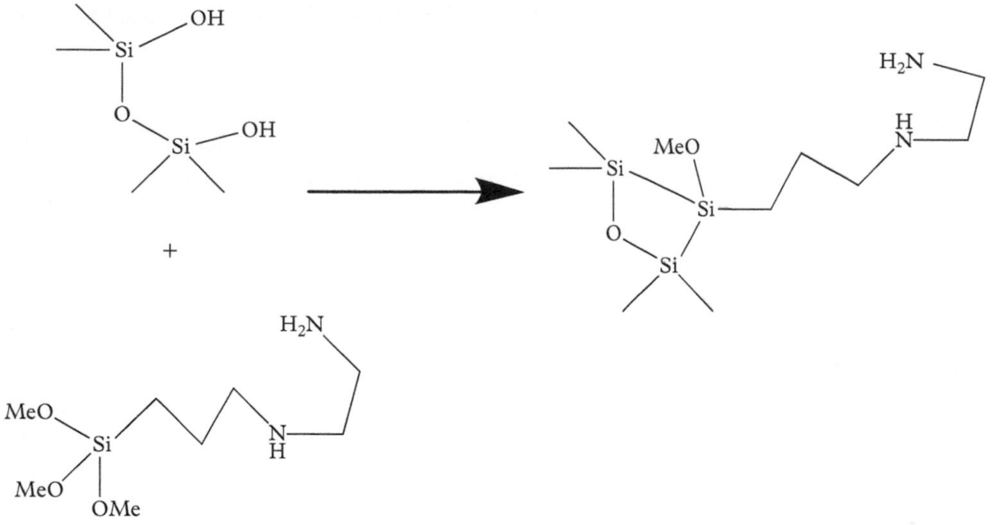

Figure 4.24 *Functionalization of a silica surface by amines.*

absorbs CO_2 from air at concentrations of 400–440 ppm, and the gas can be desorbed by an argon flow. It proved to be stable over a significant number of adsorption/desorption cycles (forty). Using a related silylating reagent,[36] $Si(OMe)_2MeSiCH_2CH_2CH_2NH_2$, a cellulose-based adsorbent was synthesized to allow CO_2 capture under ambient conditions, and desorption at low pressure (e.g. 62 mbar) and elevated temperature (90 °C). From this a small-scale installation has been developed in which the captured CO_2 was utilized in a market garden.[37]

An alternative approach is related to Eq. 4.69, using high lattice energy of carbonates of small counter ions. Lithium orthosilicate, Li_4SiO_4, was found to absorb CO_2 from ambient air at room temperature.[38] This is a material of interest in the nuclear industry, including a potential use in deuterium–lithium nuclear fusion reactors. The reaction in Eq. 4.73 is exothermic at 298 K ($\Delta_r H°$ −134 kJ mol^{-1}). The performance of this material is attractive as absorption is rapid at 500 °C, and regeneration can be carried out <750 °C, considerably lower than for calcium looping (CaL) in both directions. The rate of the chemical reaction in Eq. 4.73 is only the limiting step for the outer layer of the solid absorbents; deeper into the particles it is the rate of diffusion into the solid that will become limiting. Synthesizing materials which have structural porosity and can retain that during many recycling steps is a challenge. These materials will also tie up lithium, and thus research is ongoing to improve material synthesis, enhance rates with additives, and seek alternative absorbents for locations such as flues in power stations.[39] This would effect capture but not storage.

$$Li_4SiO_{4(s)} + CO_{2(g)} = Li_2SiO_{3(s)} + Li_2CO_{3(s)}. \qquad (4.73)$$

[36] J. A. Wurzbacher, C. Gebald, S. Brunner, and A. Steinfeld, 'Heat and mass transfer of temperature-vacuum swing desorption for CO_2 capture from air', *Chem. Eng. J.*, 2016, **283**, 1329–38.

[37] http://www.climeworks.com.

[38] M. Kato, S Yoshikawa, and K. Nakagawa, 'Carbon dioxide absorption by lithium orthosilicate in a wide range of temperature and carbon dioxide concentrations', *J. Mater. Sci. Lett.*, 2002, **21**, 485–7.

[39] Y. Hu, W. Liu, Y. Yang, M. Qu, and H. Li, 'CO_2 capture by Li_4SiO_4 absorbents and their applications: Current developments and new trends', *Chem. Eng. J.*, 2019, **359**, 604–25.

[40] H. T. Schaef, B. P. McGrail, and A. T. Owen, 'Carbonate mineralization of volcanic province basalts', *Int. J. Greenh. Gas Con.*, 2010, **4**, 249–61.

[41] I. Gunnarsson, E. S. Aradóttir, E. H. Oelkers, D. E. Clark, M. T. Arnarson, B. Sigfússon, S. Ó. Snæbjörnadóttir, J. M. Matter, M. Stute, B. M. Júlíusson, and S. R. Gíslason, 'The rapid and cost-effective capture and sub-surface mineral storage of carbon and sulphur at the CarbFix2 site', *Int. J. Greenh. Gas Con.*, 2018, **79**, 117–26.

[42] A. Sanna, M. Uibu, G. Caramanna, R. Kuusik, and M. M. Maroto-Valer, 'A review of mineral carbonation technologies', *Chem. Soc., Rev.*, 2014, **43**, 8049–80.

[43] Chapter 1 lead author R. Millar, Technical Annex 1A; Figure 11, IPCC report SR15, 2018, https://www.ipcc.ch/sr15/.

[44] N. von der Assen, P. Voll, M. Peters, and A. Bardow, 'Life cycle assessment of CO_2 capture and utilization: a tutorial review', *Chem. Soc., Rev.*, 2014, **43**, 7982–94.

[45] P. Markewitz, W. Kuckshinrichs, W. Leitner, J. Linssen, P. Zapp, R. Bongartz, A. Schreiber, and T. E. Müller, 'Worldwide innovations in the development of carbon capture technologies and the utilization of CO_2', *Energy Environ. Sci.*, 2012, **5**, 7281–305.

[46] T. Wilberforce, A. Baroutaji, B. Soudan, A. H. Al-Alami, and A. G. Olabi, 'Outlook of carbon capture technology and challenges', *Sci. Total Environ.*, 2019, **657**, 56–72.

Samples of basalt from different sites across the world have been tested for carbon storage by mineralization.[40] Generally calcium, magnesium, and iron ($\approx \frac{1}{4}$ of the mass) could be liberated from the silicaceous rock samples and then mineralized, but this was subject to the variation with site. A consequence of the presence of H_2S could be coatings of iron sulphide which inhibits carbonate formation. An interesting approach has been the combined capture and storage of effluent gas from a geothermal power plant in Iceland, using an injection site called CarbFix2.[41] The exhaust gas, which is rich in H_2S as well as the CO_2, is mixed with condensate water from the plant, and the gas-charged water piped to the injection site, at which point it is sprayed onto hot rocks (240–250 °C) which start at a depth of *c.*750 m. The aquifer reservoir at the depth of *c.*2,000 m is very much larger than the volumes of the minerals created, which can thus be accommodated. In terms of scaling up such a project it is worth noting that at a pressure of 25 bar, 27 t of water is required to dissolve 1 t of CO_2. However, it has been noted that about half of the carbon pool on Earth is in the form of carbonate minerals (>60,000,000 Gt), with a potential change from CCS by mineralization of 10,000 Gt.[42] This can be compared with estimates for 2018 of annual and cumulative emissions of CO_2 of *c.*40 Gt and 2,200 Gt, respectively.[43]

An alternative approach is **carbon capture and utilization** (CCU). An example of the utilization of CO_2 is methanol synthesis by reduction with H_2, described in Chapter 5 (Fire). It is estimated that the potential for utilization may reach 2 Gt per year, under 5% of that currently emitted to the atmosphere; life-cycle assessment (LCA) shows that CCU can have significant beneficial environmental effects and so can make a contribution to reducing net CO_2e emissions.[44] However, for utilization, CO_2 purity must be higher than normal anthropogenic emissions.

The most developed process for post-combustion treatment is Oxyfuel (Fig. 4.25).[45,46] This scheme depicts a coal-powered power station, and demonstration plants of this nature have been validated to a 30 MW scale; they are expected in future to be scaled up to have a capacity for 100–500 MW production. The sequence picks up processes from much of this chapter. There is an air separation unit to produce oxygen (95–97%) for high-temperature combustion of the fuel, in this case pulverised coal. This gas is mixed with recycled flue gas to maximize

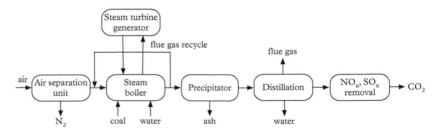

Figure 4.25 *Schematic of an Oxyfuel system for a coal-fired power station.*

the consumption of O_2. Particulates, water, and sulphur and nitrogen oxides are removed to provide CO_2 that can be utilized or stored.

4.3 Atmospheric Solution: Wind Turbines

In terms of carbon dioxide, a broad conclusion of this chapter is that there is scope for storage of liquid CO_2 and sequestering by mineralization in carefully chosen reservoirs in the Earth's crust, but this will also reduce the available potential oxygen (APO) at the surface. Reduction of CO_2 emissions will reduce the requirement for CCS operations, and one energy source for doing so is that provided by solar energy and the Earth's rotation, namely wind. This is a physical rather than a chemical process, but there are some key chemical elements in its efficient operation. Between 1980 and 2016 the use of electricity worldwide has increased by a factor of three, and the generation by non-fossil fuel sources has only just exceeded this, with the proportion rising from 30 to 35%. The source with the greatest increase has been wind power.

The power potentially provided by wind (kinetic energy ($1/2mv^2$) per unit time (t) is based upon Eq. 4.74, where ρ is the density of air (1.3 kg m^{-3}) and A is the area covered by the rotation of the blade.[47] Hence, the best location of wind turbines is in regions of high probability of appropriate wind speeds (≈ 5 to ≈ 25 m s^{-1}). This is assisted by increased height above the ground, and this also allows an increase in area with longer rotor blades. As a result the power density increases by a factor of 3–4 with an increase in height from 10 m to 100 m.

$$\frac{\frac{1}{2}mv^2}{t} = \frac{1}{2}\rho Av^3. \tag{4.74}$$

Effective wind turbines are therefore substantial, with the tower head of a substantial 6 MW peak-power offshore installation having a mass of $\approx$360 t. Approximate material requirements per unit power for a smaller, onshore wind turbine are presented in Table 4.11.[48] The major materials are structural: concrete and steel. But key functional components are composites for the blades (CFRP—carbon fibre reinforced polymer) and rotor (GFRP—glass fibre reinforced polymer), steel for drives, aluminium and copper for motors and neodymium for magnets in the generator. Indeed, a 2 MW wind turbine contains $\approx$ 25 kg of neodymium.[49] The energy return on energy invested (EROI) of installed wind power systems has been estimated on operational studies as 19.8, with this figure rising to 25.2 when conceptual studies were also included.[50] This is a favourable measure, but it is clear that it is dependent upon location. Another important operation factor is the ratio of output to the area of land use. Estimates for this also vary widely with installation, with means of *c.*2 or 3 MW km^{-2} being published.[47,51]

[47] D. J. C. MacKay, *Sustainable Energy—Without the Hot Air*, UIT Cambridge, Cambridge, UK, 2009.

Table 4.11 *Onshore wind turbine materials.*

Material	kg/kW
aluminium	0.8–3
CFRP	5–10
concrete	380–600
copper	1–2
GFRP	5–10
steel	85–100
neodymium	0.04
plastics	0.2–10

[48] M. F. Ashby, *Materials and the Environment*, Butterworth-Heinemann, Elsevier, Oxford, 2nd edn, 2018.

[49] M. F. Ashby, *Materials and Sustainable Development*, Butterworth-Heinemann, Elsevier, Oxford, 2016.

[50] I. Kubiszewski, C. J. Cleveland, and P. K. Endres, 'Meta-analysis of the net energy return for wind power systems', *Renew. Energ.*, 2010, **35**, 218–25.

[51] P. Denholm, M. Hand, M. Jackson, and S. Ong, *Land-Use Requirements of Modern Wind Power Plants in the United States*, NREL Technical Report NREL/TP-6A2-45834, 2009.

Availability of sufficient land area in suitable areas may be a limit for onshore installations, with offshore sites extending the option. However, the use of f-block elements must also be considered.

4.3.1   High-Performance Magnets

As described in Section 3.3.3, magnetic materials depend upon having unpaired electrons, as observed largely in transition and rare-earth elements. Permanent magnets may either be ferromagnets, in which magnetic moments from a single type of site remain aligned, or ferrimagnets in which different types of sites can be opposed (an antiferromagnetic coupling) but there is a net formula magnetic moment (Fig. 3.32). The simplest formula for estimating the contribution from an iron atom is via the spin-only formula, Eq. 3.28. The example of Nd of the lanthanides is chosen because of its application in wind turbines (Table 4.11). Its sites are likely to be in the +III oxidation state which, from Table 3.20, has a valence configuration of $4f^3$, that is $S = \frac{3}{2}$. The spin-only formula assigns a value of 3.87 μ_B for S number. However, the experimental values for two $4f$ ions with $S = \frac{3}{2}$ are 3.62 and 9.34 μ_B for Nd(III) and Er(III) fluorides ($4f^{11}$), respectively. Maximizing the values of M_S and M_L in that order according to Hund's rules (Section 2.8), both ions will also have the same M_L value of 6 (Fig. 4.26), giving them a term symbol of ^4I. However, the coupling between the spin and orbital moments is energetically important for the $4f$ elements and is generally larger than ligand–field splittings. That provides the quantum number J, which can take values from L-S in unit steps to L+S. In this case, Hund's third rule must also be considered:

For configurations which are less than half-filled the J=L-S state is the most stable, and for configurations which are more than half-filled the J=L+S is the most stable (Fig. 4.27).

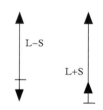

Figure 4.27 *L-S and L+S coupling.*

[52] J. H. Van Vleck, Nobel Prize for Physics, 1977.

Hence for Nd(III) and Er(III), the most stable states are labelled as $^4I_{\frac{9}{2}}$ and $^4I_{\frac{15}{2}}$, respectively. For ions such as these the magnetic moment can be calculated using a similar formula to that of the spin-only formula, but based upon the quantum number coming from the strong coupling of S and L (Eq. 4.75), as proposed by Van Vleck.[52] The factor g_J, called the **Landé g-factor**, must also be calculated from all three quantum numbers (Eq. 4.76). The calculated values from Van Vleck's equations are 3.62 and 9.57 μ_B for f^3 and f^{11}, respectively, close to experimental values for NdF$_3$ and ErF$_3$.

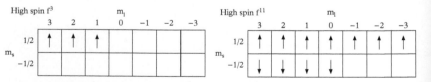

Figure 4.26 *The microstates of f^3 and f^{11} ground states.*

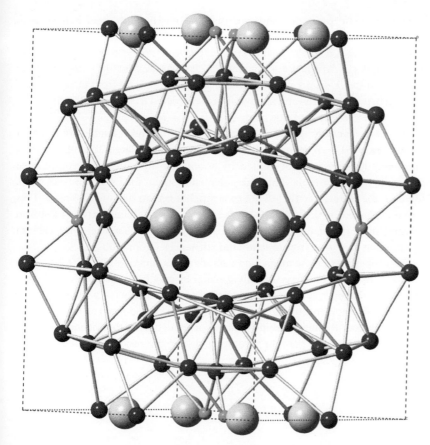

Figure 4.28 *The unit cell of $Nd_2Fe_{14}B$; yellow = Nd, red = Fe, green = B.*

$$\mu_{eff} = g_{\mathcal{J}}\sqrt{\mathcal{J}(\mathcal{J}+1)} \tag{4.75}$$

$$g_{\mathcal{J}} = \frac{3}{2} + \frac{S(S+1) - L(L+1)}{2\mathcal{J}(\mathcal{J}+1)}. \tag{4.76}$$

When L = 0, as for Gd(III), $4f^7$, this simplifies to the value 2. The gadolinium has a large magnetic moment due to having seven unpaired electrons (μ_{eff} calculated to be 7.79 μ_B).

A key class of materials is of the type $Nd_2Fe_{14}B$, in which differing lanthanides, magnetic $3d$ elements, and early main group component can be substituted. The three components are heated in a vacuum-induction furnace to create the alloy. The unit cell of this neodymium iron boride is shown in Fig. 4.28. It is a complicated structure, but can be seen to have three features. First, the boron is in a six-coordinate site within a trigonal prism of iron atoms (Fe–B 2.10, 2.12 Å). Second, the neodymium sites are quite isolated from each other (Nd..Nd 3.48, 3.76 Å), with the closest approaches to them being by iron atoms (to each Nd there are sixteen Fe atoms being 3.06 and 3.40 Å). Finally, the iron atoms form layers with strong Fe–Fe interactions.

The magnetic properties indicate that the neodymium does behave as a tripositive ion, and thus the expected component of the magnet moment of the formula is ≈ 7.2 μ_B. The moment expected per iron atom can be estimated from that of $La_2Fe_{14}B$ which has a diamagnetic lanthanide ion ($4f^o$), and provides that value of ≈ 2.1 μ_B per iron atom. The resulting moment from ferromagnetic coupling between the iron and neodymium is estimated to be $2 \times 3.62 + 14 \times 2.1 = 36.6\mu_B$, the experimental value being 37.7 μ_B at 4 K.

The ferromagnetic coupling between the iron and lanthanide magnet moments in $Nd_2Fe_{14}B$ opens up the prospect of gaining a larger formula moment if the early lanthanide is substituted by a late one for which the ground state follows the J=L+S formula. Thus the potential moment for $Er_2Fe_{14}B$ is 48.5 μ_B. However, the moment at 4K (12.9 μ_B) is much less than that of $Nd_2Fe_{14}B$. This can be explained by the switch to antiferromagnetic coupling between the iron and lanthanide moments. This may be related to the smaller size of the erbium ion, resulting in shorter iron-to-lanthanide distances (from 3.00 Å).

The properties of these magnetic materials are outlined in Fig. 4.29. The loop is called a **hysteresis** curve. Initially a material may have no magnetization as the individual moments are randomly oriented. Application of a magnetic field (H) causes the moments to align until saturation is reached when all the moments are fully aligned. Reducing the field causes a reduction in the magnetization of the material, and at the point of zero applied field the residual magnetization is

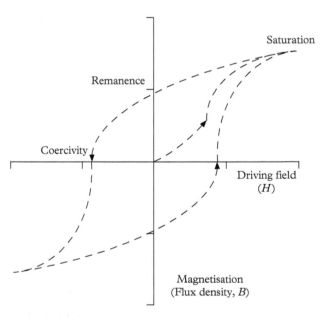

Figure 4.29 *Hysteresis loop showing the magnetization of a material under a driving field (H).*

Table 4.12 *Properties of different classes of permanent magnets.*

Material	Moment (μ_B)	T_C (°C)	BH_{max} (kJ m^{-3})	Saturation (T)	Coercivity (kA m^{-1})
Fe_3O_4	4.1	575	28	0.048	≈ 28
$SrFe_{12}O_{19}$	20	452	≈ 28	0.4	≈ 200
$SmCo_5$	8.9	720	160	1.14	≈ 700
$Nd_2Fe_{14}B$	37.7	320	450	1.61	≈ 900

called the **remanence** (B_r). The magnetization can be reduced to zero which requires a field in the opposite direction which measures the resistance of the material to having its magnetization removed, called **coercivity** (H_c). Continuing to increase the field in this negative direction will achieve saturation when the magnetic moments are all now aligned in the opposite direction. A figure of merit of magnets is called the **maximum energy product** (BH_{max}), given by Eq. 4.77:

$$BH_{max} = B_r \; x \; H_c. \tag{4.77}$$

The properties of different classes of permanent magnets are summarized in Table 4.12. The first item, *magnetite*, is a ferrimagnetic *spinel* with both Fe(II) and Fe(III) sites, and was discussed in Section 3.3.3. $SrFe_{12}O_{19}$ is an example of a mixed metal *ferrite*, in which all magnetic sites are Fe(III). There are five different iron sites in the unit cell, and again this is a ferrimagnet. This and its barium analogue are in wide use in recording media, loudpeakers, and magnetic stripe cards. Having higher coercivity than Fe_3O_4 renders these ferrites their value as permanent magnets. The samarium cobalt magnets have higher BH_{max} values with increased coercivity than the ferrites, but suffer in comparison with the neodymium–iron magnets because cobalt is less abundant than iron and samarium is less abundant than neodymium. Accordingly, the Nd–Fe magnets are widely employed and it was estimated that in 2007 63,000 t of Nd and 15.7 t each of Pr and Dy were in use, largely in computers, audio systems, wind turbines, and automobiles.[53] The relatively low Curie temperature for $Nd_2Fe_{14}B$ points to its tendency to reduce coercivity at high temperatures. Addition of other lanthanides such as dysprosium and terbium mitigate against this, and as a result partial substitution of Nd by them (3–10%) is generally employed, particularly for motors in automobiles. Dy (0.002 ppm) are Tb (0.0005 ppm) are both considered to be in much lower crustal abundance than Nd (0.01 ppm), and thus are considered to be at great risk in terms of supply against demand in clean energy application (Fig. 3.51). Accordingly, research into alternatives to these rather rare earths has a strong impetus to reduce a restriction of the capacity for non-fossil-fuel energy generation and transportation.

[53] X. Du, T. E. Graedel, 'Global Rare Earth In-Use Stocks in NdFeB Permanent Magnets', *J. Ind. Ecol.*, 2011, **15**, 836–43.

4.4 Questions

1. Construct a diagram of the type in Fig. 4.2 to plot the decay processes from ^{232}Th to ^{220}Rn, and its decay to the stable isotope ^{208}Pb. The data required is in Table 4.13. Comment on which isotopes may be observable in samples from the Earth's crust.

2. Propose molecular structures for the following positive oxidation state compounds and ions of elements of groups 17 and 18:
XeO_3F_2, BrF_5, $ClO_3{}^-$, $I_3{}^-$, IF_7, $[IO_3(OH)_3]^{2-}$.

3. From May 1993 to May 2018, measurements of the atmosphere at the Scripps Institution pier indicated that the O_2/N_2 ratio had changed from -163 to -691 pmeg and the CO_2 concentration had risen from 362 to 413 ppm.

   (i) Calculate the change in the atmospheric potential oxygen (APO) concentration over this period of time, assuming a stoichiometry of 1.1 for O_2/CO_2.

   (ii) Account for the difference between the change in APO as compared to the variation in O_2/N_2 ratio.

   (iii) The annual change in O_2 concentration is considerably larger than that of N_2. Estimate the mean annual change in O_2 concentration at this site and compare that with the values of CO_2.

4. Calculate the standard Gibbs free energies for three stages in diesel exhaust treatment using the data provided below in the square brackets, and comment on the results.

Table 4.13 *Decay characteristics.*

Isotope	Decay	Half life
^{232}Th	α	1.4×10^{10} y
^{228}Th	α	1.9 y
^{228}Ac	β	6.1 min
^{228}Ra	β	5.75 y
^{224}Ra	α	3.6 d
^{220}Rn	α	55 s
^{216}Po	α	0.145 s
^{212}Po	α	299 ns
^{212}Bi	β	60.6 min, 64%
^{212}Bi	α	60.6 min, 36%
^{212}Pb	β	10.6 h
^{208}Tl	β	3 min

(i) Consider particulate carbon removal by direct reaction with O_2 and with NO_2.

(ii) Consider the ideal equation for NO_x removal starting with an NO/O_2 mixture and selective chemical reduction (SCR) with ammonia.

(iii) Consider the oxidation of ammonia in the ideal ammonia slip process.

[$\Delta_f G^\circ$ kJ mol^{-1}: $CO_{2(g)}$ -394.4; $NH_{3(g)}$ -16.4; $NO_{(g)}$ 87.6; $NO_{2(g)}$ 51.3; $OH_{2(g)}$ -228.6.]

5. Adsorbents such as zeolites have been investigated for application in CCS (carbon capture and storage). In the gas phase, CO_2 has a larger molecular diameter than N_2: 5.1 and 4.3 Å, respectively. However, CO_2 has the larger charge variation (the quadrupole moment) due to the different Z_{eff} of C and O. Hence it would be expected to adsorb more strongly. It has the smaller effective kinetic diameter in zeolites (3.3 Å as compared to 3.6 Å for N_2).

(i) What properties would be important for an effective absorbent for carbon capture and for carbon storage? What structural features might achieve these properties?

(ii) Zeolite structures include apertures with six-rings (2.3 Å, sodalite), eight-rings (4 Å, zeolite A and chabazite), ten-rings (5.5 Å, silicilite and ZSM-5), and twelve-rings (7.4 Å, faujasite, zeolites X and Y). Which may have use in CCS?

(iii) How might changing the counter ions in aluminosilicates be used to improve CCS properties?

6. In rare-earth iron garnets, $RE_3Fe_5O_{12}$, the iron ions are in two different sites, tetrahedral and octahedral, of ratio 3:2, in which the spins orient antiparallel. The magnetic properties indicate that the three rare-earth ions have moments antiparallel to the net iron moment.

(i) Which type of magnetism does this describe?

(ii) Calculate the magnetic moments that you would expect from the individual rare-earth sites (La and Dy) and the Fe sites.

(iii) Estimate the magnetic moment from the formula unit for RE = lanthanum (La) and dysprosium (Dy) that could be exhibited at saturation magnetism.

Magnetic moments (μ_B) can be calculated from:

$$\mu_S = 2\sqrt{S(S+1)}, \text{ or} \tag{4.78}$$

$$\mu_J = 2\sqrt{J(J+1)} \tag{4.79}$$

where

$$g_J = \frac{3}{2} + \frac{S(S+1) - L(L+1)}{2J(J+1)}. \tag{4.80}$$

5

Fire

[1] https://www.un.org/sustainable development/.

[2] P. T. Anastas and J. C. Warner, *Green Chemistry: Theory and Practice*, Oxford University Press, Oxford, 1998.

[3] http://www.rsc.org/journals-books-databases/about-journals/green-chemistry.

[4] M. F. Ashby, *Materials and Sustainable Development*, Butterworth-Heinemann, Oxford, 2016.

Much of the consequence of reducing the dependence on fossil fuels for energy use falls within the scope of this chapter. Many of our commonplace materials of life result from conversion of crude oil into petrochemicals. A simplified extract of this is given in Fig. 5.1, which incorporates petroleum and natural gas streams. The end points shown are predominantly synthetic polymers including thermosetting resins from phenols and formaldehyde (methanal), thermoplastics, rubbers, and foams.

This is a sideline of the conversion to fuel, amounting to only 4–5% of the total. Petrochemical processes thus gain from the economies of scale associated with the energy industry. There is a complex web of downstream processes between the crude sources and industrial products, few of which can be reinvented rapidly. Therefore ways to provide primary and intermediate reagents are required which accord with the Sustainable Development Goals of the United Nations[1]

The terms **Green Chemistry** and **Sustainable Chemistry** are often used in this venture. Green Chemistry[2] has been defined as: 'the utilization of a set of principles that reduces or eliminates the use of hazardous substances in the design, manufacture and application of chemical products.' Sustainable chemistry has a broader scope which is recognized as an extension green chemistry.[3] Fig. 5.2 provides some definition to this scope, based upon an assessment of sustainable materials.[4] In this book, it is the upper half of the hexagon that is discussed, with factors such as embodied energy (EE), embodied carbon (EC), and water (virtual water, VW). A complete assessment of the sustainability of a chemical process will also involve the socio-economic factors in the lower half of Fig. 5.2. This figure is general enough to be versatile and can also serve to assess the sustainability of an energy-generating process.

5.1 Petrochemicals from Carbon Oxides

There are routes to synthesis gas (or syngas), CO and H_2, from a variety of fossil fuels and also from biomass and plastic waste. In Section 4.1.2, the manufacture of hydrogen gas was described, using a combination of steam reforming or partial oxidation and the water gas shift reaction. Much of this hydrogen is used as a component of ammonia (Section 4.1.3). The CO_x fraction of syngas can be utilized as a single carbon atom building block for organic compounds. Syngas routes into petrochemicals represent a reductionist approach: everything has been converted to the lowest common denominator.

Elements of a Sustainable World. John Evans, Oxford University Press (2020). © John Evans.
DOI: 10.1093/oso/9780198827832.003.0005

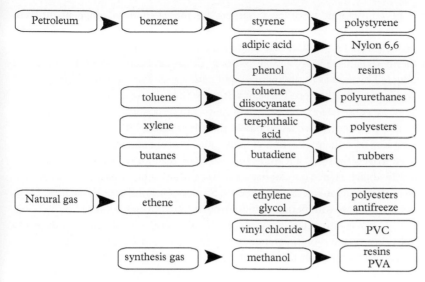

Figure 5.1 *Examples of petrochemical sequences.*

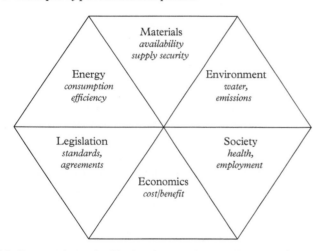

Figure 5.2 *Sectors of sustainable chemistry and energy, based on Fig. 2.5 in Ashby,* Materials and Sustainable Development *(n 4).*

Some of the boiling-point cuts from petroleum refining are shown in Table 5.1. This shows a little of the breadth of chemical targets. The two lightest cuts, gases and LPG (liquified petroleum gas), are relatively simple mixes. However, for all the heavier cuts, the overall compositions have wider carbon number coverages than the peaks presented in Table 5.1. Light naphtha is a liquid fraction which has a boiling-point range of 30–90°C, and the boiling-point ranges increase with carbon number: heavy naphtha 90–200°C, kerosene 150–270°C, diesel 200–350°C.

[5] G. C. Bond, *Heterogeneous Catalysis: Principles and Application*, Oxford Chemical Series 18, Clarendon Press, Oxford, 1974.

Table 5.1 *Petrochemical fuels.*

Class	Composition
gases	CH_4, C_2H_6
LPG	C_3H_8, C_4H_{10}
light naphtha	C_5–C_7
heavy naphtha	C_6–C_{10}
gasoline	C_7–C_{11}
kerosene	C_{10}–C_{14}
diesel	C_{14}–C_{18}

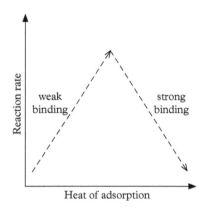

Figure 5.3 *A volcano plot for catalysed reactions.*

[6] D. Sheldon, 'Methanol Production: A Technical History', *Johnson Matthey Technol. Rev.*, 2017, **61**, 172–82.

The conversions to different targets are mediated by a range of different catalysts comprising mostly *d*-block metals supported on a high-area oxide. The choice of catalyst involves the right balance of adsorption energies to allow a thermodynamically favourable pathway down an energy slope to the target product. This is akin to finding the right pathway down a mountain slope to get to the next overnight stop. This can be represented in a volcano plot like Fig. 5.3. The syngas conversion processes require CO to be adsorbed to reaction sites on the catalyst surface, called **associative chemisorption**.[5] If the heat of adsorption is very low then the proportion of active sites bearing an activated CO molecule will be extremely small and thus any catalysed reaction will proceed with very low probability. Hence developing a catalyst with a higher binding energy should increase the reaction rate. As discussed in Section 3.3, bond energies in the *d*-block generally increase with period number from 4 (*3d*) through 5 (*4d*) to 6 (*5d*) (Fig. 3.18). There is also a tendency for the bond energies to decrease with group number across the *d*-block. Hence strong adsorption of CO will be expected for a heavy, early *d*-block element like tungsten (period 6, group 6). However, in that case the bonding is so strong that the absorbed CO blocks the reactive sites and thus acts as a poison by preventing any further CO conversions. Thus the reaction rate can drop after a critical value of the heat of adsorption, $\Delta_{ad}H°$ (Fig. 5.3).

A further factor is the effect on reaction selectivity. On highly reactive metal surfaces like those of tungsten, the chemisorption of CO can proceed further to **dissociative chemisorption**[5] involving the cleavage of the CO bond to create W–C and W–O sites. Such sites could react with absorbed hydrogen to form an alkane. If there is a sufficiently high concentration of surface sites, then migration of carbon can lead to catenation, and then hydrogenation will afford longer-chain alkanes. This gives some guidance for selecting potential catalysts. To achieve oxygenated products with high selectivity then relatively weak CO adsorption is required which avoids dissociative chemisorption. Thus the *d*-block elements of choice would be late in period 4. For hydrogenation to low-carbon-number hydrocarbons a longer retention time for CO is required to allow a sequence of hydrogenation steps to occur, and thus the favoured elements would be from slightly lower group numbers. To generate longer-chain hydrocarbons there needs to be stronger binding of CO to allow a build-up of (hydro)carbonaceous species that can migrate and catenate before being liberated as an alkane or alkene. To achieve this, metals are likely to be earlier in the *d*-block. However, this will be limited by the tendency for site poisoning as in the right-hand side of the plot in Fig. 5.3.

5.1.1　Methanol Synthesis

Methanol is an important chemical intermediate with wide application in the manufacture of chemicals, materials, and fuels.[6] Annual production of methanol occurs in over ninety world-scale plants and is in excess of 110 Mt. A direct reaction between CO and H_2 to form methanol in the vapour phase is exoergic

(Eq. 5.1), and indeed the hydrogenation of CO_2 (Eq. 5.2) is only slightly endoergic; it is clear that the condensation of the two liquid products will make even this second reaction favourable. All materials excluding methanol are related by the water gas shift reaction (WGSR) (Eq. 5.3). Both routes to methanol reduce the number of gaseous species and thus become less favourable with increasing temperature. As a result moderate temperatures (up to 250 °C and pressures (50–100 atm) tend to be employed.

$$CO_{(g)} + 2\,H_{2(g)} = CH_3OH_{(g)}, \; \Delta_f G° - 25.1 \; kJ\,mol^{-1} \qquad (5.1)$$

$$CO_{2(g)} + 3\,H_{2(g)} = CH_3OH_{(g)} + H_2O_{(g)}, \; \Delta_f G° \; 3.5 \; kJ\,mol^{-1} \qquad (5.2)$$

$$CO_{(g)} + H_2O_{(g)} = CO_{2(g)} + H_{2(g)}, \; \Delta_f G° - 28.6 \; kJ\,mol^{-1}. \qquad (5.3)$$

The most important catalyst type includes two elements in the expected region of the *d*-block: copper and zinc, generally written as Cu/ZnO, in a ratio of ≈70:30 with a smaller amount of Al_2O_3 (5–20%).[7] Other elements such as magnesium have also been added to the catalyst. The preparation route is critical for achieving high catalytic activity with *zincian malachite* $(Cu,Zn)_2(OH)_2(CO_3)$ identified as a key precursor. The catalyst is then activated with hydrogen at 190–230 °C, at which point the copper oxide is reduced to metal crystallites with ZnO and Al_2O_3 particles; these oxides can prevent sintering of the copper to larger particles and can also have a synergic enhancement of reaction rate.

Although it has been viewed as a synthesis gas reaction, carbon-14 tracer studies showed that the majority or all of the catalytic flux emanates from CO_2, depending upon the CO_2/CO ratio (of 0.02 to 1).[8] Identifying the active sites generated is still an ongoing topic.[9] It is clear that the form of the high-area copper nanocrystallites is important. A cut along the unit cell face, which has the Miller (*h,k,l*) indices of (100) would leave a surface atom with a coordination number of 8 since four of the twelve atoms in the bulk are removed (Fig. 5.4). The most stable surface of the *fcc* metal is the close-packed plane, Cu(111), which cuts across the body diagonal of the unit cell giving Miller index (*h,k,l*) values of unity. That provides a Cu–Cu coordination number of 9, having had a set of three atoms

[7] M. Behrens and R. Schlögl, 'How to Prepare a Good Cu/ZnO Catalyst or the Role of Solid State Chemistry for the Synthesis of Nanostructured Catalysts,' *Z. Anorg. Allg. Chem.*, 2013, **639**, 2683–95.

[8] G. C. Chinchen, P. J. Denny, D. G. Parker, M. S. Spencer, and D. A. Whan, 'Mechanism of methanol synthesis from CO2/CO/H2 mixtures over copper/zinc oxide/alumina catalysts: use of ^{14}C-labelled reactants', *Appl. Catal.*, 1987, **30**, 333–338.

[9] M. Behrens, F. Studt, I. Kasatkin, S. Kül, M. Hävecker, F. Abild-Pedersen, S. Zander, F. Girgsdies, P. Kurr, B.-L. Kniep, M. Tovar, R. W. Fischer, J. K. Nørskov, and R. Schlögl, 'The Active Site of Methanol Synthesis over Cu/ZnO/Al2O3 Industrial Catalysts', *Science*, 2012, **336**, 893–8987.

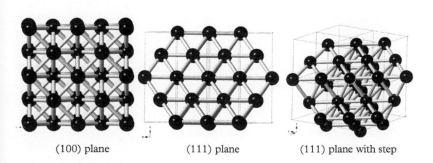

    (100) plane   (111) plane   (111) plane with step

Figure 5.4 *Surface planes of copper.*

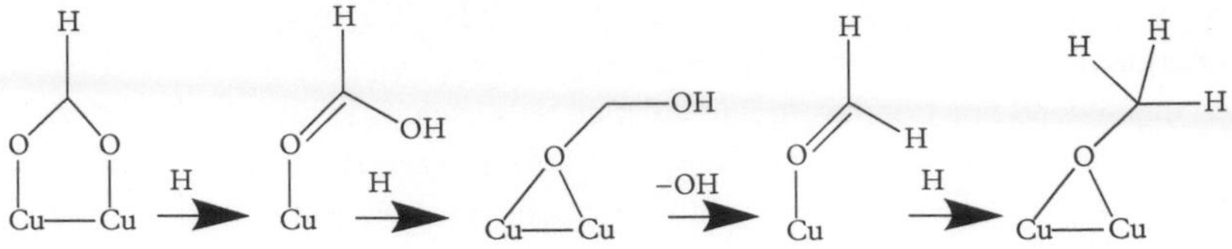

Figure 5.5 *Steps in methanol synthesis from CO_2.*

[10] P. Wu and B. Yang, 'Significance of Surface Formate Coverage on the Reaction Kinetics of Methanol Synthesis for CO_2 Hydrogenation over Cu', *ACS Catal.*, 2017, **7**, 7187–95.

removed by the surface cut. If perfect, such a surface would be infinitely long in two dimensions. If the cut in one direction is halfway along the unit cell, giving a Cu(211) plane, then a monatomic step is introduced, which removes two more neighbours from the atom at the top of the step. This seven-coordinate atom is likely to have a higher binding energy to adsorbates. This may help reduce energy barriers in the catalytic reactions and/or generate stable spectator ligands which could modify the reactivity of neighbouring atoms in a beneficial way.[10]

The sequence of adsorbates formed from CO_2 and H_2 is shown in Fig. 5.5. Rapid reaction of CO_2 with Cu–H surface sites affords formate, which is a prevalent surface species. Stepwise incorporation of hydrogen atoms from those adsorbed on the metal surface gives rise to formic acid and the methanediolate OCH_2OH. At this point a C–O bond is cleaved as a hydroxyl group migrates to the surface leaving adsorbed methanal (formaldehyde). It is at this juncture that the cycle becomes common with that from CO. A further hydrogen transfer results in another stable surface species, methoxide, which can be hydrogenated once more to afford methanol. These recent studies do indeed indicate that adsorption and the barriers in these steps are made more favourable by moving from Cu(111) to Cu(211), by incorporating a zinc at the step site, and also by the presence of spectator adsorbates like formate and CO.[11]

[11] L. C. Grabow and M. Mavrikakis, 'Mechanism of Methanol Synthesis on Cu through CO_2 and CO Hydrogenation', *ACS Catal.*, 2011, **1**, 365–84.

5.1.2 Methanol Carbonylation

Ethanoic (acetic) acid is a significant derivative from methanol, with annual production approaching 16 MT. The major production route is via a **homogeneous catalysis** process. This implies that all reagents and the catalyst are in the same phase. The majority of industrial catalysts are **heterogeneous** with the reagents and products generally being in fluid phases and the catalysts incorporated into a solid phase. This generally improves product separation. However, the complexity of having a multiplicity of sites in a solid host has disadvantages. A single site in a solution can be investigated to a higher precision and once generated may have higher specific activity and selectivity. Such sites are exemplified by the carbonylation of methanol by the Monsanto (rhodium based) and Cativa (iridium based) processes which sit within a sequence of organic reactions in Fig. 5.6. An organic halide is utilized for the metal catalysed stage, and the preferred iodomethane can be formed from methanol with hydriodic acid, HI. This stage inserts a CO into the C–I bond. Acetyl iodide is rapidly hydrolysed to ethanoic

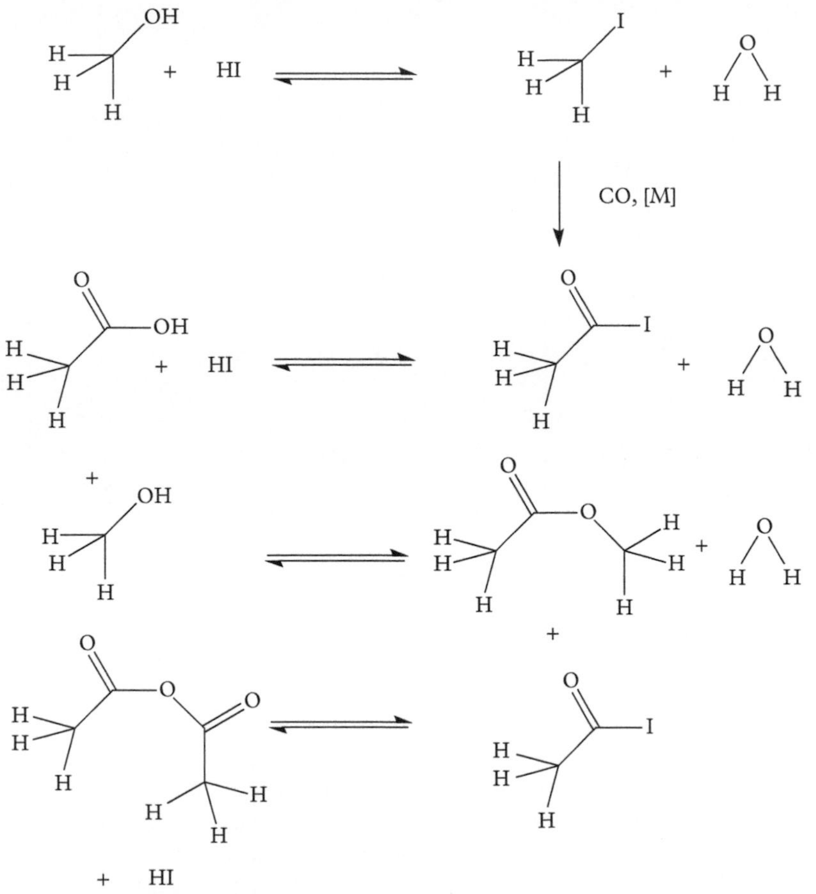

Figure 5.6 *Sequence from methanol carbonylation.*

acid. Under relatively anhydrous conditions ethanoic acid will be esterified by methanol. The acetate ester too can react with acetyl iodide and thus afford acetic anhydride. This C_4 molecule is formed of CO_2, CO, and H_2 with high specificity and has water as the side product.

The iodide is recycled around the sequence but is required to create the active species $[RhI_2(CO)_2]^-$ in the Monsanto Process. This is a Rh(I) complex which will have a $4d^8$ electronic configuration. That favours a square planar geometry, as presented in Fig. 5.7. The iodomethane undergoes an **oxidative addition** reaction with the rhodium complex via an ionic (S_N2) nucleophilic substitution. The iodide anion liberated coordinates to what is now a Rh(III) centre; this has an eighteen-electron valence count and thus is likely to be stable by the **inert gas rule**. This new complex will undergo a **migratory insertion** reaction in which the methyl group migrates from the rhodium to an adjacent carbonyl ligand thus forming an acetyl group. Coordination of CO reforms an

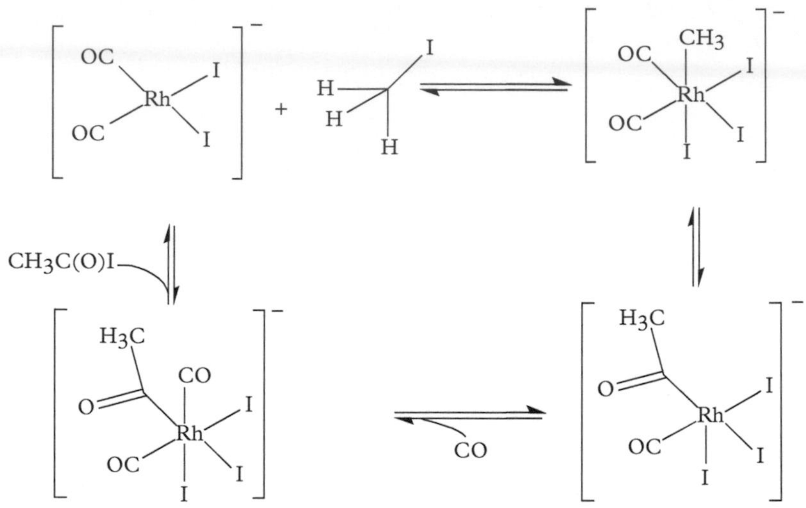

Figure 5.7 *Rhodium catalysed carbonylation.*

inert gas configuration. The acetyl group can exit by bonding to an adjacent iodide by a **reductive elimination** reaction, regenerating the original Rh(I) complex and thus completing the catalytic cycle. The Monsanto process requires a high concentration of iodide to avoid precipitation of RhI$_3$ within the reaction vessels. It also affords some propanoic acid as a byproduct, which is difficult to separate from ethanoic acid. Both of these problems can be much alleviated by changing to iridium, as in the Cativa process.[12] The higher bond energies of the *5d* metal render its complexes more stable in general, and it can operate at much lower iodide levels. The methyl complex formed after the oxidative addition is the longest-lived member of the cycle and so following the CO insertion step the cascade is quicker to the end product. Hence there is no acetyl species present sufficiently long for it to be reduced and undergo a further CO insertion to form the C$_3$ acid.

[12] A. Haynes, P. M. Maitlis, G. E. Morris, G. J. Sunley, H. Adams, P. W. Badger, C. M. Bowers, D. B. Cook, P. I. P. Elliott, T. Ghaffer, H. Green, T. R. Griffin, M. Payne, J. M. Pearson, M. J. Taylor, P. W. Vickers, and R. J. Watt, 'Promotion of Iridium-Catalysed Methanol Carbonylation: Mechanistic Studies of the Cativa Process', *J. Am. Chem. Soc.*, 2004, **126**, 2847–61.

5.1.3   Methanol Dehydration

A less selective, but valuable, type of conversion is via the dehydration of methanol, mediated with solid acid catalysts of the type H-ZSM-5 (Fig. 4.17). This was developed by Mobil in the 1970s and a plant for methanol to gasoline (MTG) established in New Zealand in 1985; this operated to manufacture LPG and gasoline until 1997. This is part of a family of processes including methanol-to-hydrocarbons (MTH), -olefins (MTO) and -aromatics (MTA). What is to be minimized is the deposition of coke which both reduces the conversion into desirable product and blocks pores in the catalyst.[13] The initial stage is a partial dehydration to dimethylether (DME, CH$_3$OCH$_3$), as shown in Eq. 5.4, which

[13] U. Olsby, S. Svelle, K. P. Lillerud, Z. H. Wei, Y. Y. Chen, J. F. Li, J. G. Wang, and W. B. Fan, 'The formation and degradation of active species during methanol conversion over protonation zeotype catalysts', *Chem. Soc. Rev.*, 2015, **44**, 7155–7176.

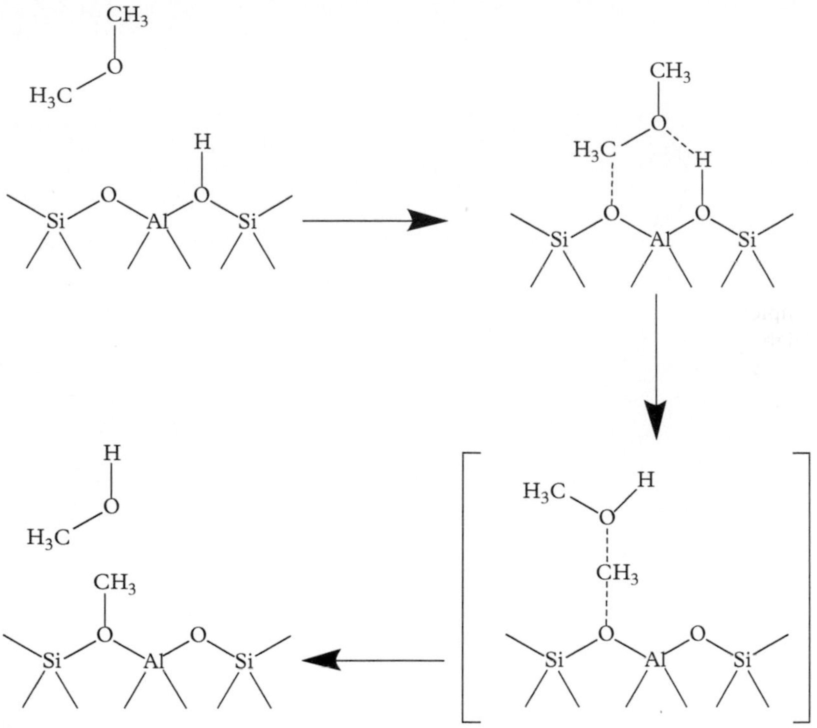

Figure 5.8 *Direct methylation of a zeolitic acid site.*

is exothermic in the gas phase. The subsequent feed to other reactors is a mix of methanol and DME. A plausible mechanism for the initiation of the carbon chains is the direct methylation at an acidic aluminosilicate site, as in Fig. 5.8, and possibly also assisted by another methanol molecule. The DME molecule is shown to be adsorbed primarily by the hydrogen bonding between the acidic proton and the electron-rich oxygen of the ether. There can be a weaker hydrogen bond between the aluminosilicate oxygen and a CH group of a methyl.

$$2\,CH_3OH_{(g)} \;\rightarrow\; CH_3OCH_{3(g)} \;+\; H_2O_{(g)}, \; \Delta_r H^\circ \; -23.6 \; kJ \; mol^{-1}. \qquad (5.4)$$

There is a substantial initiation time for this process, carried out from 400–450 °C. A scheme showing a proposed route to the initial C–C formation step is shown in Fig. 5.9, starting from the methylated surface in Fig. 5.8, an incipient methyl cation.[13] A hydride transfer to this electron-deficient carbon would liberate the lowest hydrocarbon, methane, and convert the new DME molecule to an oxonium cation. This leaves the surface site negatively charged which will promote a C–H interaction with the third DME molecule. This in turn will activate a nucleophilic addition to the $(CH_2=OCH_3)^+$ cation resulting in the formation of dimethoxyethane and regeneration of the original active acidic site. The new ether

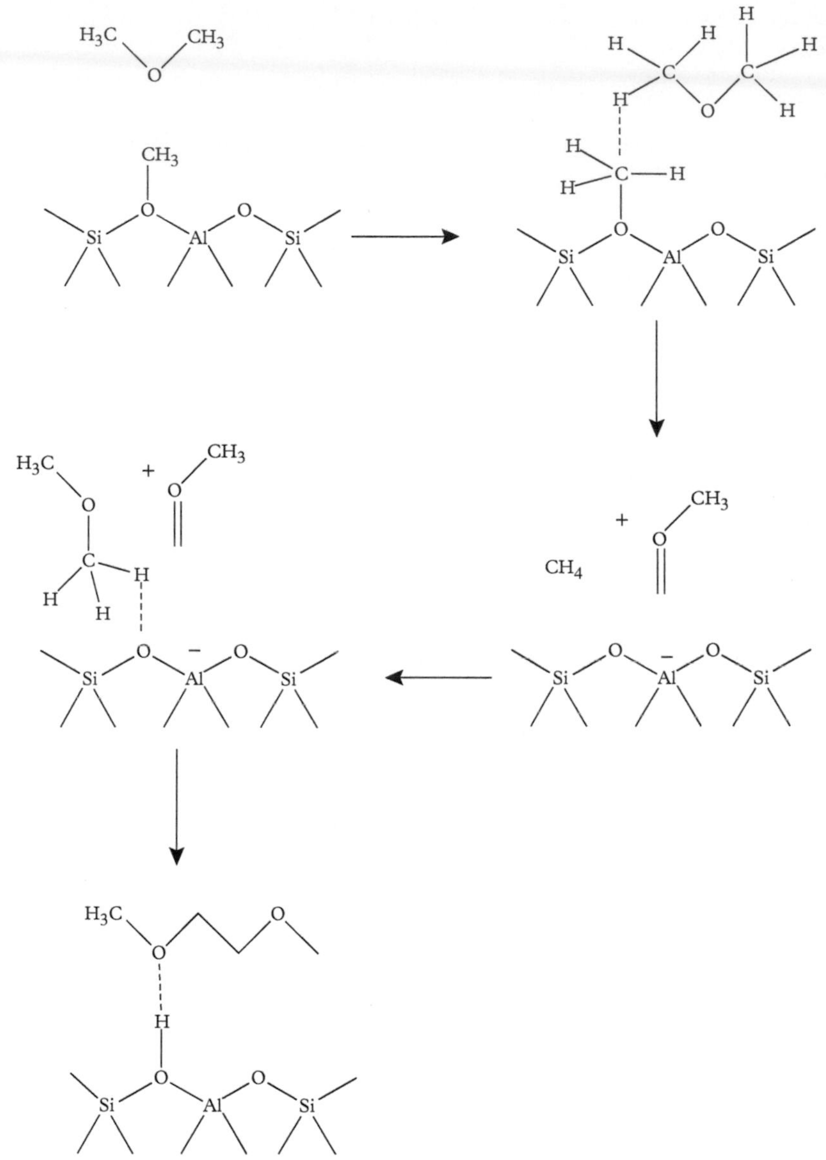

Figure 5.9 *Initial C–C formation from DME a zeolitic acid site.*

can then enter into these schemes again and further homologate to longer chain ethers and alkanes.

As the reaction enters a steady state, a hydrocarbon pool (HCP) mechanism becomes established, and most products are from secondary processes involving this pool of material. The zeolite catalyst bears a series of organic intermediates that can be identified by ^{13}C NMR and which react further with the DME/MeOH feed to afford isolated products. Such intermediates include the species in Fig. 5.10. These include methylcyclopropanes, cyclopentadiene, cyclopentenium, and cyclohexadienium ions. Most of these have been observed in eight-ring windowed structures like the acid form of SAPO-34 (Fig. 4.18), whilst the larger carbenium ions tend to be found in H-ZSM-5, which has ten-ring windows running in two dimensions. These are two of the bases of commercially utilized catalysts, and the products are influenced by the morphology of the catalyst pores: H-ZSM-5 is favoured for gasoline production (MTG) and H-SAPO-34 is used for olefin (alkene) production (MTO). The development of methylated- and poly-aromatics is hinted at by the structural progression in Fig. 5.10. Once stable, bulky polyaromatic species like methylnaphthalenes, phenanthrenes, and pyrenes are formed they block the process and deactivate the catalyst. In practice the spent catalysts are regenerated by heating in air to burn off these residues.

Application of methanol conversion has increased again over the last decade. In 2009 the first coal-to-liquid plant using a MTG process (ExxonMobil) began in Shanxi Province in China, on a 100 kt per year scale. Since then fixed-bed installations of over 1 Mt per year have been licensed in China and the USA. These facilities have been established for natural gas or coal to gasoline conversions, but other precursors for the synthesis gas stage might be utilized in future.

5.1.4   Methanation

Moving to the left of copper in the d-block would be expected to move upwards on the volcano plot in Fig. 5.3. Indeed, nickel is well known to have a strong affinity for CO and will react directly with it to form $Ni(CO)_4$. This interconversion is the basis of the process (Eq. 5.5) developed in 1890 for the extraction and purification of nickel metal. At 50–60 °C, $Ni(CO)_4$ can be extracted as a gas in this favourable reaction. The reaction has a large negative standard entropy $\Delta_r S°$, -507 J mol^{-1} K^{-1}, and can be reversed at 220–250 °C to provide nickel in high purity.

$$Ni_{(s)} + 4\,CO_{(g)} \rightarrow Ni(CO)_{4(g)}, \ \Delta_r G° - 39.5 \ kJ \ mol^{-1}. \quad (5.5)$$

Hence it might be expected that, under different conditions, nickel would adsorb CO and might allow a catalysed conversion to occur. As shown in Fig. 4.23, the reduction of CO_2 to CH_4 is exothermic, as it is for CO (Eqs 5.6, 5.7). At 1 bar pressure and a ratio of H_2/CO_2 of 4:1, CH_4 is the predominant carbon-containing gas up to ≈ 400 °C, and at 30 bar this preference extends to ≈ 600 °C.

$$CO_{2(g)} + 4\,H_{2(g)} \rightarrow CH_{4(g)} + 2\,H_2O_{(g)}, \ \Delta_r H° - 164 \ kJ \ mol^{-1} \quad (5.6)$$
$$CO_{(g)} + 3\,H_{2(g)} \rightarrow CH_{4(g)} + H_2O_{(g)}, \ \Delta_r H° - 206 \ kJ \ mol^{-1}. \quad (5.7)$$

As discussed in Section 4.2.2, production of methane from the carbon oxides is not presently likely to displace the supply from natural gas. However, there are situations when this is an important process, including the removal of CO from

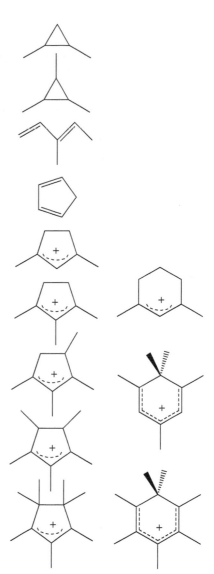

Figure 5.10 *Organic intermediates in methanol dehydration on zeolites.*

[14] S. Rönsch, J. Schneider, S. Matthis-chke, M. Schlüter, M. Götz, J. Lefebvre, P. Prabhakaran, and S. Bajohr, 'Review on methanation: from fundamentals to current projects'. *Fuels*, 2016, **166**, 276–96.

hydrogen coming as a by-product of ethene manufacture. In the manufacture of hydrogen for ammonia synthesis it can be used to remove CO_x which are poisons for the catalysts in that process. Also it can provide a route to a **synthetic natural gas** (SNG) from coal, wood, biomass or waste.[14] Methanation processes from CO and CO_2 were first reported by Sabatier and Senderens in 1902. Many d-block elements will act as catalysts, with ruthenium having the highest activity but poorer selectivity than nickel. Nickel is generally employed for a compromise combination of selectivity, activity, availability and cost. The most commonly used support is alumina. As in many catalyst examples, the alumina is not corundum (α-Al_2O_3), but a high area material such as γ-Al_2O_3. This is a disordered material with the Al ions in both octahedral and tetrahedral holes in a *ccp* type of oxide lattice. The NiO on the surface can react with γ-Al_2O_3 to form some $NiAl_2O_4$, a spinel (see section 3.3.3). Indeed MgO can be used as a promoter, and can also react with γ-Al_2O_3 to form *spinel* itself, $MgAl_2O_4$, which is also utilised as a support. For reduction of CO_2, silica has advantages as a support as it has less affinity for CO_2. The materials are then activated under hydrogen to form nickel clusters on the oxidic support. Under differing operating conditions NiO and Ni_3C clusters may also be present.

Adsorption studies indicate that, once adsorbed, CO_2 rapidly dissociates with adsorbed CO remaining adsorbed on nickel (Equ. 5.8). In model studies on a closed-packed plane of a single crystal of nickel, Ni(111), in the absence of H_2, CO_2 caused the formation of NiO, which adsorbed more CO_2 to form a carbonate (Equ. 5.9); CO desorbed from these new surfaces.[15] The adsorbed oxide reacts with H_2 to liberate water and the carbonate is also lost. The CO is absorbed onto metal surface sites.

[15] C. Heine, B. A. J. Lechner, H. Bluhm, and M. Salmeron, 'Recycling of CO_2: Probing the Chemical State of the Ni(111) Surface during Methanation Reaction with Ambient-Pressure X-Ray Photoelectron Spectroscopy'. *J. Am. Chem. Soc.*, 2016, **138**, 13246–52.

$$CO_{2(g)} + [Ni_a] \rightarrow [Ni_a] - CO_2 \rightarrow [Ni_a] - CO + [Ni_a] - O \quad (5.8)$$
$$[Ni_a] - O + CO_{2(g)} \rightarrow [Ni_a] - CO_3. \quad (5.9)$$

A simplified scheme of the steps in the catalytic process is provided by Eqs 5.10–5.14.[16] When the syngas feed includes $^{13}C^{16}O$ and $^{12}C^{18}O$ there is no scrambling of the labels which points to the cleavage of the C–O bond of the adsorbed CO being the rate determining step.

[16] J. Sehested, S. Dahl, J. Jacobsen, and R. Rostrup-Nielsen, 'Methanation of CO over Nickel: Mechanism and Kinetics at High H_2/CO ratios', *J. Phys. Chem. B*, 2005, **109**, 2432–38.

$$CO_{(g)} + [Ni_a] = [Ni_a] - CO, K_{CO}, \Delta H \approx -122 \ kJ \ mol^{-1} \quad (5.10)$$
$$0.5 \ H_{2(g)} + [Ni_a] = [Ni_a] - H, K_H, \Delta H \approx -43 \ kJ \ mol^{-1} \quad (5.11)$$
$$[Ni_a] - CO + [Ni_b] \rightarrow [Ni_a] - C + [Ni_b] - O, E_a \approx 97 \ kJ \ mol^{-1} \quad (5.12)$$
$$[Ni_a] - C + 2 H_2 \rightarrow CH_{4(g)} + [Ni_a], \ fast \quad (5.13)$$
$$[Ni_b] - O + H_2 \rightarrow OH_{2(g)} + [Ni_b], \ fast. \quad (5.14)$$

The first two equations show the absorption of CO and H_2 as exothermic pre-equilibria. The rate-determining step (Eq. 5.12) requires a second nickel adsorption site as the CO bond is cleaved. The bond energies provided by the new carbide and oxide must be considerable as the adsorption of CO provides only a small proportion of the 1072 kJ mol^{-1} required to cleave the CO bond in the gas phase. The hydrogenation steps for both of these species are considered to be fast, and no intermediate hydrogenation stages (methylidyne, CH, methylidene, CH$_2$, methyl, CH$_3$) are identified.

5.1.5 Fischer–Tropsch Synthesis (FTS)

The generic reaction for the Fischer–Tropsch synthesis is shown in Eq. 5.15. As for methanation (Eq. 5.7), this is exoergic at low temperatures, but becomes increasingly unfavourable at higher temperatures due to reduction in the number of gaseous molecules. It was reported in 1923 by Franz Fischer and Hans Tropsch of the Kaiser Wilhelm Institute in Berlin. The first commercial plant was established in Germany in 1936. It was preferred over direct hydrogenation as a route to the conversion of coal to liquid hydrocarbon fuels since it provided better aviation fuel. It was generally developed where oil supply was limited and coal accessible. Accordingly it was also developed in the Republic of South Africa and from 1955 became the core feature of the country's fuel and chemical industry.

$$(2n+1)\, H_2 + n\, CO \rightarrow C_nH_{2n+2} + n\, H_2O. \qquad (5.15)$$

As compared to methanation, the intermediates between the carbide and methane, CH$_x$, will be required to have a longer residence time to allow their migration and catenation. This will be favoured by elements with higher heats of adsorption (Fig. 5.3). Accordingly cobalt, iron, and ruthenium are effective catalysts. On the basis of cost and availability, iron is the preferred element, but there are important roles for cobalt catalysts for particular products.

Iron catalysts are typically formed from Fe(II) salts by precipitation and drying to form Fe$_2$O$_3$, haematite. Under hydrogen this reduces to Fe$_3$O$_4$, magnetite (cf. Section 3.3.3), and on to α-Fe at higher temperatures. The process operates close to the equilibrium in Eq. 5.16:

$$\frac{3}{4}\, Fe_{(s)} + H_2O_{(g)} = \frac{1}{4}\, Fe_3O_{4(s)} + H_{2(g)}. \qquad (5.16)$$

In addition, carbide phases such as Fe$_5$C$_2$ and Fe$_7$C$_3$ are formed. Under some conditions, activity increases with the concentration of the latter carbide. It may be that Fe$_7$C$_3$ and α-Fe are active phases,[17] but opinion is divided with other authors finding correlations between activity with the presence of Fe$_5$C$_3$ and its surface area.[18,19] The presence of magnetite has been related to an oxidative deactivation. Particle sizes in the nm regime affect both the reaction rate and the product distribution, with 2 nm particles being especially effective. Theoretical studies indicate that carbide phases chemisorb less strongly than metals, but also offer a lower barrier to dissociation. It appears that iron catalysts may benefit from being later in the volcano plot (Fig. 5.3) and having the chemisorption energies reduced by carbide formation. This is in contrast to the next element in the row, cobalt.

[17] J. Gaube and H.-F. Klein, 'The promoter effect of alkali in Fischer-Tropsch iron and cobalt catalysts', *Appl. Catal., A,* 2008, **350**, 126–132.

[18] H. M. Torres Galvis, J. H. Bitter, T. Davidian, M. Ruitenbeek, A. I. Dugulan, and K. P. de Jong, 'Iron Particle Size Effects for Direct Production of Lower Olefins from Synthesis Gas', *J. Am. Chem. Soc.,* 2012, **134**, 16207–15.

[19] V. R. R. Pendyala, U. M. Graham, G. Jacobs, H. H. Hamdeh, and B. H. Davis, 'Fischer-Tropsch Synthesis: Deactivation as a Function of Potassium Promoter Loading for Precipitated Iron Catalyst', *Catal. Lett.,* 2014, **144**, 1704–16.

[21] J. P. den Breejen, P. B. Radstake, G. L. Bezemer, J. H. Bitter, V. Frøseth, A. Holmen, and K. P. de Jong, 'On the Origin of the Cobalt Size Effects in Fischer-Tropsch Catalysis', *J. Am. Chem. Soc.*, 2009, **131**, 7197–7203.

[22] A β hydride elimination process.

[23] A reductive elimination process.

Cobalt catalysts are generally supported on alumina or silica. Carbide phases, such as Co_2C, are present, but the activity is related to the surface area of the metal. The turnover frequency[20] increases with average particle size up to 6 nm, and is constant above that figure.[21] Below 6 nm it is thought that the increased number of low-coordination metal sites adsorb CO too strongly and block the sites. For the smaller particles there is higher hydrogen coverage and this may account for the higher proportions of methane. The residence time of CH_2 species also increases with particle size up to 6 nm, which will also favour higher homologation.

An outline scheme for the Fischer–Tropsch synthesis is shown in Fig. 5.11. Under CO/H_2, the CO is cleaved and reduced resulting in a CH_x chain growth point such as the methylidene depicted here. The catalyst surface can also have a growing chain and a surface hydride. Three possible reactions of this mixture of sites are shown. The top one is migration of the alkyl group to the methylidene, resulting in chain growth, ascribed with a probability of α. The other two processes lead to chain termination. The elimination of an alkene can occur by transfer of a hydrogen from the β carbon of the chain to the metal.[22] This is the normal primary reaction pathway. The alternative involves transfer of a hydrogen from the metal to the α carbon of the alkyl chain[23] generating the alkane.

This polymerisation type of process can be simulated by an Anderson–Schulz–Flory (ASF) distribution (Eq. 5.17), where W_n is the weight fraction of hydrocarbons of n atoms.[17] This implies that all chain growth processes have the same probability, whatever the site or the chain length. This is a considerable simplification and it can be alleviated by increasing the number of α values, as in Eq. 5.18, with a second site of fraction f_2:

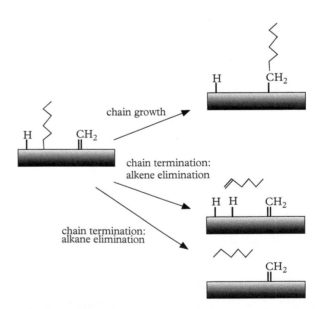

Figure 5.11 *Chain growth and termination in Fischer–Tropsch synthesis.*

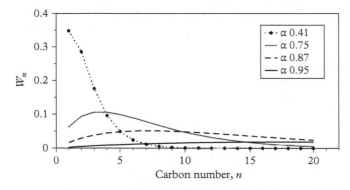

Figure 5.12 *Anderson–Schulz–Flory distributions for FTS with four α values.*

$$W_n = n(1 - \alpha)^2 \alpha^{(n-1)} \tag{5.17}$$

$$W_n = (n(1 - \alpha_1)^2 \alpha_1^{(n-1)})(1 - f_2) + (n(1 - \alpha_2)^2 \alpha_2^{(n-1)})(f_2). \tag{5.18}$$

Predictions based on Eq. 5.17 are shown in Fig. 5.12. For the lowest of the α values chosen (0.41), over 60% of the weight fraction of the products would be SNG and ≈ 25% LPG. Increasing α to 0.75 moves the peak to LPG, but this now amounts to 21% of the weight fraction. The distribution is wider and provides a better option for naphtha/gasoline cuts (Table 5.1). Kerosene and diesel are provided by α values near 0.87, and very long molecules (waxes) by an α of 0.95. Clearly a wide range of hydrocarbon mixes can be generated by an FTS process depending upon the properties of the catalytic site under the operating conditions.

Alkali metal salts are commonly utilized as prospective modifiers of activity and selectivity. A model for how this may operate is given in Fig. 5.13. The synthesis can be considered as in a surface science experiment where potassium vapour is deposited on a metal surface. The *s*-block element will transfer its valence electron to the metal, rendering it electron rich, and the potassium provides the charge balance. With the promoted metal more electron rich, the backbonding to the π^* orbitals of the CO will be enhanced thus strengthening the M-CO chemisorption. In contrast, the more electron-rich metal will accept less electron density from the adsorbed hydride. Hence the promotion would increase the surface coverage of carbon over hydrogen, and this would be expected to favour chain growth (i.e. increase α).

Addition of 0.5% K_2CO_3 to an unsupported cobalt catalyst reduced the rate of catalysis by more than a factor of 2,[17] indicating that this is an undesirable modification for this high-demand metal. The selectivity for the primary 1-alkene product also reduces, and this effect increases considerably with chain length. Just as CO chemisorption should be enhanced by alkali metal promotion, then the backbonding to the π^* orbital of an alkene will also be enhanced thus increasing the residence time on the surface and enhancing the probability of isomerization to the 2-alkene or hydrogenation to the alkane.

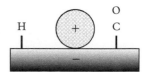

Figure 5.13 *A model for promotion by alkali metals.*

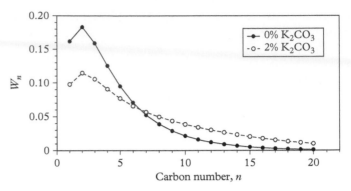

Figure 5.14 *Change in Anderson–Schulz–Flory distributions on addition of K_2CO_3 to an iron catalyst for FTS (data from Gaube and Klein, n 17).*

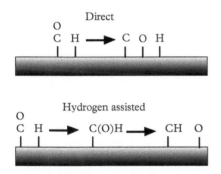

Figure 5.15 *Direct and hydrogen-assisted initial steps in FTS.*

[24] H.-J. Li, C.-C. Chang, and J.-J. Ho, 'Density Functional Calculations to Study the Mechanism of the Fischer-Tropsch Reaction on Fe(111) and W(111) Surfaces', *J. Phys. Chem. C*, 2011, **115**, 11045–55.

[25] Y. He, P. Zhao, J. Yin, W. Guo, Y. Yang, Y.-W. Li, C.-F. Huo, and X.-D. Wen, 'CO Direct versus H-Assisted Dissociation on Hydrogen Coadsorbed χ-Fe$_5$C$_2$ Fischer-Tropsch Catalysts', *J. Phys. Chem. C*, 2018, **122**, 20907–17.

In contrast, the addition of 2% K_2CO_3 to an unsupported iron catalyst causes a much smaller reduction (10%) in the reaction rate.[17] Smaller loadings of potassium can enhance the reaction rate and also prolong the catalyst lifetime.[19] The effect of the addition of potassium can be seen from Fig. 5.14. Under particular conditions (293 K, H_2/CO = 1), the carbon number distribution of the iron catalyst could be modelled with two sites according to Eq. 5.18 with f_2 0.47 and α_1 and α_2 values of 0.52 and 0.71, respectively. For the catalyst with added K_2CO_3, f_2 increased to 0.67 and α_2 to 0.82. The increase in the proportion of the higher α site, and its probability value, has significantly enhanced the gasoline and diesel components of the liquid products.

Steps to understand the reaction mechanism through theoretical methods, generally density functional theory have required the choice of active sites. The Fe(111) surface of iron is not a close-packed plane, since α-Fe has a bcc structure. Instead it provides a puckered surface.[24] Fig. 5.15 provides two options for the activation of CO. In the top version this is via direct cleavage of the C–O bond to form carbide and oxide adsorbates. The alternative is by a hydrogen-assisted step in which a formyl group is formed which splits into an oxide and a methylidyne. It is this which was found to be preferable in this iron surface. Alternatively the active sites may emanate on a carbide, such as Fe$_5$C$_2$. Calculations on one phase of this carbide show that the preferred route could be either of these alternatives, depending upon the nature of the exposed crystal plane. Highly energetic sites will favour a direct mechanism and those with moderate stability favoured the hydrogen-assisted mechanism. The most stable plane displayed little activity.[25] As can often be the case, the most promising catalytic sites can be the most tantalizing to obtain!

Fischer–Tropsch synthesis is presently included in the portfolio of hydrocarbon interconversions starting from coal and also from natural gas, as in the PetroSV refinery in Mossel Bay, South Africa. Cracking plants (generally using solid acid catalysts) are used to convert a proportion of high-carbon-number waxes into diesel and other liquid fuels. Alternative, possibly more sustainable, non-fossil fuels are being assessed. A life-cycle assessment of the conversion of biogas (from anaerobic digesters) to gasoline, diesel, and electricity has shown the complexity of the issue.[26] Compared to a fossil-fuel-derived product stream, the footprint of diesel from biogas was indeed assessed to have lower environmental impact in terms cumulative non-renewable energy demand (CED_{nr}) and ozone-layer depletion (ODP). However, in terms of global warming potential (GWP), acidification (AP), and eutrophication (EP, the excess soil nutrients such a phosphates), it was assessed to be more environmentally intrusive. The overall environmental profile was assessed to be poorer than the conventional FTS route to diesel with the main cause of this being within the biogas production. Improvements are needed to reduce methane leakage and the heat demand in that step. Reductions of NO and NH_3 emissions in the biogas-to-liquid stage are also required.

However, compared to landfill and incineration, gasification of waste biomass (and municipal waste) will have higher energy recovery, and means should also be developed to lower the emission of environmental pollutants. In an assessment of the conversion of forestry residues in Brazil,[27] woody residues were identified as having relatively low nitrogen, sulphur, and chlorine content. The partial oxidation of these residues produces liquid (tar) and solid (char) residues in addition to the syngas stream of H_2, CO, CO_2, H_2O, and CH_4. In this model, a water wash is included to remove traces of HCl and NH_3; H_2S is trapped in an absorber (e.g. ZnO) and CO_2 is removed by compression. In the model local CCS modules were also included, providing strong advantages in emission reduction with a minor increase on the expected cost of the product. Rather more difference was reported from the choice of wood, with pine expected to afford $\approx 4\%$ higher yield than eucalyptus. Residues from both though could be viable and utilized according to their geographic distribution. At present the process is not considered to be economically viable but it remains a future possibility. The biochar residues can also be utilized, with applications like solid additives promoting carbon sequestration, precursors for carbon absorbents, and catalysts for tar removal.[28] Hence these can further reduce the net carbon emissions of these processes. It is to be expected that pilot plants will be established for alternative sources of syngas.

5.2 Polymers

5.2.1 Polyalkenes

The worldwide production of plastics in 2017 has been estimated at 348 M t, and this figure has been rising at an annual rate of 3.7%. Of this the major product is polyethylene (PE) (Table 5.2, with data from n[29]). Approximately 60% of the ethene manufactured is converted into PE. This is achieved by the exothermic conversion of C=C bonds into a pair of single bonds (Eq. 5.19). The first process developed was via radical polymerization, which uses initiators such

26 Z. Navas-Anguita, P. L. Cruz, M. Martín-Gaboa, D. Iribarren and J Dufour, 'Simulation and life-cycle assessment of synthetic fuels produced via biogas dry reforming and Fischer-Tropsch synthesis', *Fuel*, 2019, **235**, 1492–1500.

27 I. S. Tagomori, P. R. R. Rochedo, and A. Szklo, 'Techno-economic and geo-referenced analysis of forestry residues-based Fischer-Tropsch diesel with carbon capture in Brazil', *Biomass Bioenergy*, 2019, **123**, 134–148.

28 S. You, Y. S. Ok, D. C. W. Tsang, E. E. Kwon, and C.-H. Wang, 'Towards practical application of gasification: a critical review form syngas and biochar perspectives', *Crit. Rev. Environ. Sci. Tec.*, 2018, **48**, 1165–1213.

29 The appendix in M. F. Ashby, with D. Ferrer Balas and J. Segalas Coral, *Materials and Sustainable Development*, Butterworth-Heinemann, Oxford, 2016.

Table 5.2 *Estimated annual production and environmental data for polymers.*

Polymer	Production Mt yr^{-1}	EE MJ kg^{-1}	EC kgCO$_2$e kg^{-1}	Water usage L kg^{-1}
polyethylene (PE)	82	81	2.8	58
polyvinylchloride (PVC)	51	58	2.5	210
polypropylene (PP)	43	80	3.1	39
polyester	40	40	3	200
polystyrene (PS)	12	97	3.8	140
phenolic resins	10	79	3.6	52
butyl rubber (IIR)	10	120	6.6	110
polyethylene terephthalate (PET)	9.1	85	4	130
acrylonitrile butadiene styrene (ABS)	5.6	95	3.8	180
polyamides (PA)	3.7	120	8	180
epoxy resins	0.12	130	7.2	28

as organic peroxides such as *t*-BuOOBu-*t*, with the polymerization performed at high pressures (1,000–3,000 bar) and elevated temperatures (150–300 °C). The radical process has low selectivity and so the resulting material will have a variety of side chain lengths and positions. The material has a relatively low density (0.930 g cm^{-3}), and is termed LDPE. Annual production by this route is ≈ 19 Mt. The majority ($c.\frac{3}{4}$) is converted into films for uses like cling-film and milk carton linings, although smaller proportions are shaped into common items such as buckets (injection moulding), squeezable bottles (blow moulding), and flexible water pipes (extrusion).

$$n\,CH_2 = CH_2 \rightarrow [-CH_2CH_2-]_n\,,\, \Delta_r H° - 92\,k\mathcal{J}\,mol^{-1}. \quad (5.19)$$

The extreme conditions required for the radical polymerization prompted research into alternative, catalytic methods. Two methods based on *d*-block metals are Ziegler–Natta (titanium on magnesium chloride), and Phillips (chromium on silica), and they operate at much lower pressures (10–80 bar) and somewhat lower temperatures (80–150 °C). The catalytic reactions occur at particular classes of single-metal sites and thus can provide much more control of the polymerization process. For example, the probability of chain growth over termination may be controlled by the ethene/H$_2$ ratio in the supply. There are very few side chains and thus the linear chains of comparable length can pack more closely (density

0.94–0.965 g cm^{-3}) and provide a stronger, more rigid material called high-density polyethylene (HDPE). Annual production of this material is in the region of 24 Mt. Substantially less of this version of PE is converted into film (*c.*16%), and this provides the base for shopping bags. The destinations of the products of moulding methods are for uses which require higher strength and/or rigidity, like dustbins (injection moulding), detergent bottles (blow moulding), and water pipes (extrusion).

The catalyst for the Zeigler–Natta process is considered to be α-TiCl₃, the structure of which is shown in Fig. 5.16. The Ti(III) centres are octahedrally coordinated to chlorides. These octahedra form layers and there is a much weaker interlayer interaction between chloride ions. Thus the structure is prone to cleavage under mild conditions. The titanium (III) chloride is supported on MgCl₂, which has a similar, layered structure. Having a fragile support is important since the catalyst remains in the final material. Its break-up into extremely small particles avoids compromising the appearance of the material formed. The defects in the surface sites can provide five-coordinate centres, and an ethene molecule can in principle coordinate to that site though its π orbital. This a high oxidation state for an alkene complex and it has a d^1 configuration. Hence it will have little propensity for gaining any extra stability from backbonding into the ethene π* orbital. Weak binding to one site alone will not afford polymerization activity. Alkylaluminium regents are used as activators, and it is sufficient for one on the chloride ligands to be substituted by an alkyl group to provide a plausible polymerization mechanism (Fig. 5.17). The alkyl group transferred from the aluminium acts a the initial growth point of the polymer chain. After coordination of the ethene to a *cis*

Figure 5.16 *The structure of TiCl₃; large = Cl, small = Ti.*

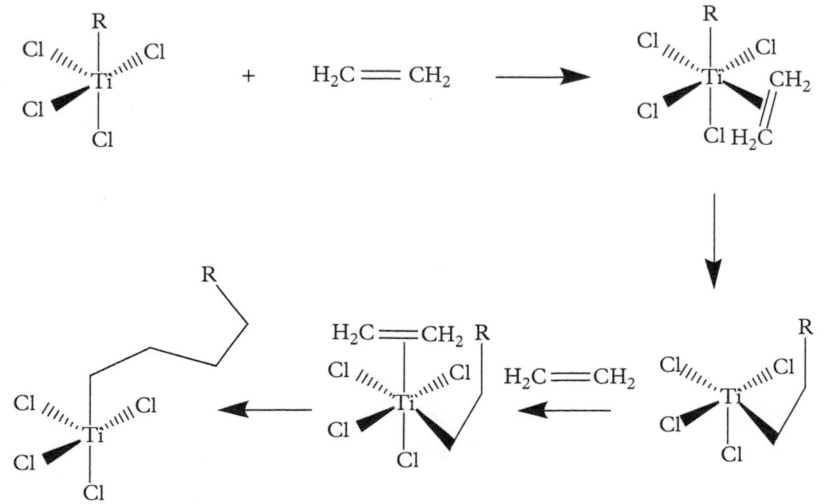

Figure 5.17 *Schematic mechanism of the early steps of the Ziegler–Natta polymerization of ethene.*

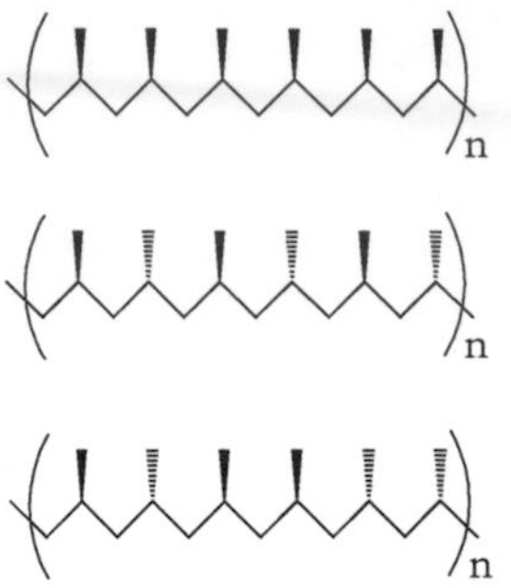

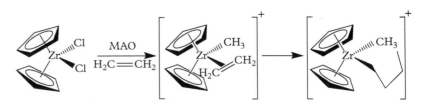

Figure 5.18 *Tacticity of polypropene: isotactic (top), syndiotactic (middle), atactic (bottom).*

[30] M. F. Ashby, *Materials and the Environment*, Second Edition, Butterworth-Heinemann, Waltham, USA, 2013.

[31] A metallocene, or sandwich compound, includes the unit [M(C₅H₅)₂].

position, a migratory insertion can move the alkyl group onto what become the β carbon of the new chain. The new vacant site can also bind an ethene, and then the migration of the longer alkyl group results in the polymer chain returning to the the initiating site. The growing alkyl chain would have a windscreen wiper type of motion at the titanium reaction centre. The alternative mechanism of ethene inserting into the Ti–C bond would not create this motion.

This more controlled process allowed the development of side-chain substituted PE, the simplest of which is polypropylene (PP). For head-to-tail polymerization, there are three general classes of chain stereochemistry (Fig. 5.18). In an **isotactic** polymer all the substituents are on the same side of the polymer chain, while they alternate in a **syndiotactic** polymer. Alternatively, the polymer may not be stereoregular, and **atactic** polymers have side chains in random orientations. Such materials are rubbery at room temperature, and have a practical temperature range of up to 120 °C. Commercial PP made from Ziegler–Natta catalysts is predominantly isotactic with some atactic components (1–5%). The side chains in the isotactic polymer lower the density as compared to PE (0.98–0.91 g cm⁻³) and also cause the chain to adopt a helical conformation which leads to a stiffer material with a much higher melting point ($\approx$ 185–220 °C). Its properties as a light, stiff material give rise to it predominant uses in rigid packaging (28%), carpet textiles (21 %), pipes and automobile plastics (20%), food packaging films (16%), and consumer products (15%) such a furniture and toys. By comparison, syndiotactic PP can now be synthesized and has a melting point range of 161–186 °C. It is worth noting that the EE, EC, and water usage values of PP and HDPE are much lower than those of aluminium alloys (EE 200–220 MJ kg⁻¹, EC 11–13 kg kg⁻¹, water usage 495–1,490 L kg⁻¹).[30] It is also of note that water is not used for the dyeing of PP, in contrast to cotton.

Further control of the polymer structure has been achieved by use of molecular catalysts based upon metallocene structures,[31] or other ligand systems building on from this chemistry known as post-metallocene catalysts. The archetype for this chemistry is [Zr(C₅H₅)₂Cl₂] (Fig. 5.19). This formally is a Zr(IV), $4d^0$, centre, since the aromatic cyclopentadienyl (cp) ligand (with six π electrons) is negatively charged. This can be converted into [Zr(C₅H₅)₂(CH₃)₂], which is formally coordinatively unsaturated, having sixteen valence electrons in the zirconium coordination sphere. In principle this can provide a vacant site for an ethene to coordinate and initiate polymerization. However, the complex is

Figure 5.19 *Schematic on the onset of polymerization using a metallocene catalyst.*

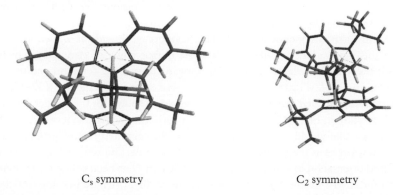

C$_s$ symmetry   C$_2$ symmetry

Figure 5.20 *Catalyst structures for tacticity selectivity in polypropylene: left for syndiotactic, right for isotactic; grey = C, white = H, purple = Zr, yellow = π-bonded ligands.*

unreactive to ethene and it requires further activation to create a more reactive vacant site. This was achieved by a partially hydrolysed sample of Al(CH$_3$)$_3$ called methylaluminoxane (MAO). This complicated material acts in at least three ways: (1) it exchanges methyl groups for chloride on zirconium, (2) it acts as a Lewis acid to remove a (CH$_3$)$^-$ ligand, and (3) it provides a large non-coordinating counter-ion. It is employed in large excess (*c.* 1,000×), and other promoters have been developed which provide all these functions. The activity of these catalysts is very high, possibly because the back-bonding from a *4d*° centre to the ethene is negligible and the alkene rapidly undergoes the migratory insertion. This also inhibits the chain termination by a β-hydride elimination mechanism (Eq. 5.20) and thus chain lengths are extremely long. These catalysts are used either in solution as homogeneous catalysts or in a heterogeneous system when bound to a support.

$$[M - CH_2CH_2R] \rightarrow [M - H(CH_2 = CH_2R)]. \tag{5.20}$$

Such catalysts have allowed the manufacture of a more controlled version of low-density PE (LDPE) called linear low-density PE (LLDPE). This is a copolymer of ethene and a higher α-alkene, typically butene, hexene, or octene, chosen to optimize the required properties. In the absence of long side chains present in LDPE, the chains in LLDPE can slide against each other and achieve a much higher strength. Accordingly, this material is much used in plastic bags as less material is required to provide viable strength in use.

Organometallic complexes such as the zirconocenes can be fine-tuned for enhanced selectivity and activity. Linking the two cyclopentadienyl rings with a -CR$_2$- or -SiR$_2$ forms a more constrained structure which increases the tilt of the two spectator aromatic rings; this is called an *ansa* ligand. This provides a more open space at the reaction centre, and increases the polymerization rate.

Selectivity can be enhanced by introducing steric constraints into the catalytic region of the complex, as shown in Fig. 5.20. In the complex on the left the two cyclopentadienyl rings are linked by a -CMe$_2$- bridge. Two benzo groups were utilized to form an extended ligand plate (a fluorenyl ligand) as shown on the top of this figure. This fragment has a single mirror plane of symmetry, which gives it the symmetry label C_s. Coming forward from the zirconium centre on the left is an *iso*-butyl ligand that would be formed after a migration of a Zr–CH$_3$ group to a coordinated propene, as in the mechanism shown for the Ziegler–Natta polymerization in Fig. 5.17. In Fig. 5.20, a second propene has been coordinated into the site that would have been occupied by the migrated methyl group. The interaction between the propene methyl group and the fluorenyl ligand gives a preferred orientation in which the alkene double bond has been rotated from that shown for ethene in Fig. 5.19 with the propene methyl oriented into the stericly least hindered side of the complex. The effect of this on the propene is depicted in Fig. 5.21. The terminal methylene is significantly closer to the metal than the central CH group and bears a higher negative charge. So it has substantial character like the lower of the two resonance forms. The migration of the alkyl group to the β carbon can be viewed as an intramolecular nucleophilic attack. The windscreen-wiper motion of the steps of the polymerization process then will create head-to-tail polymers with the methyl groups on alternating sides of the backbone, namely a syndiotactic polymer.

A different symmetry is required to afford an isotactic polymer. An example of this is shown in the right-hand complex in Fig. 5.20. This also has an *ansa* ligand with a linking -CMe$_2$- bridge but now the two benzo groups are split between the two sides of the bridge, creating two coordinated indenyl groups in which there is a bulky *tertiary*-butyl substituent. Overall this fragment approximates to having a two-fold axis of symmetry, which gives it the label C_2. This provides an important steric driver for the orientation of the growing chain and the propene. As shown in Fig. 5.20, with the ligand symmetry the propene remains oriented in the same plane as depicted in Fig. 5.19. The methyl group of the propene can slot between the *t*-butyl group on the bottom left of the drawing and the benzo group. As a result every time the chain migrates to a propene, it does so with the propene methyl oriented in that direction generating an isotactic polymer.

In terms of potential sustainability, **bio-based** monomers can be derived from biomass by fermentation methods (ethene, butene) as well as by gasification to synthesis gas (also a source of propene). Thus dehydration can afford ethene according to Eq. 5.21. This is an endothermic reaction but can be driven forward at high temperatures (330–480 °C) over an oxide catalyst such as alumina or H-ZSM5. This can also form a route to the glycol portion of PET, 1,2-dihydroxyethane. Sometimes the 'bio' label is attached to the alkene and resulting polymer, which should be identical to their fossil-fuel-derived equivalents.

$$C_2H_5OH_{(g)} \rightarrow C_2H_{4(g)} + H_2O_{(g)}, \quad \Delta H° = 49 \, kJ \, mol^{-1}. \qquad (5.21)$$

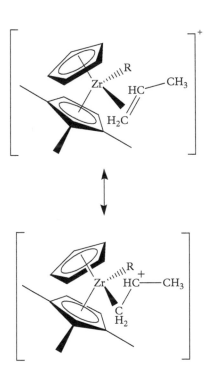

Figure 5.21 *Resonant forms of zirconocene propene complexes.*

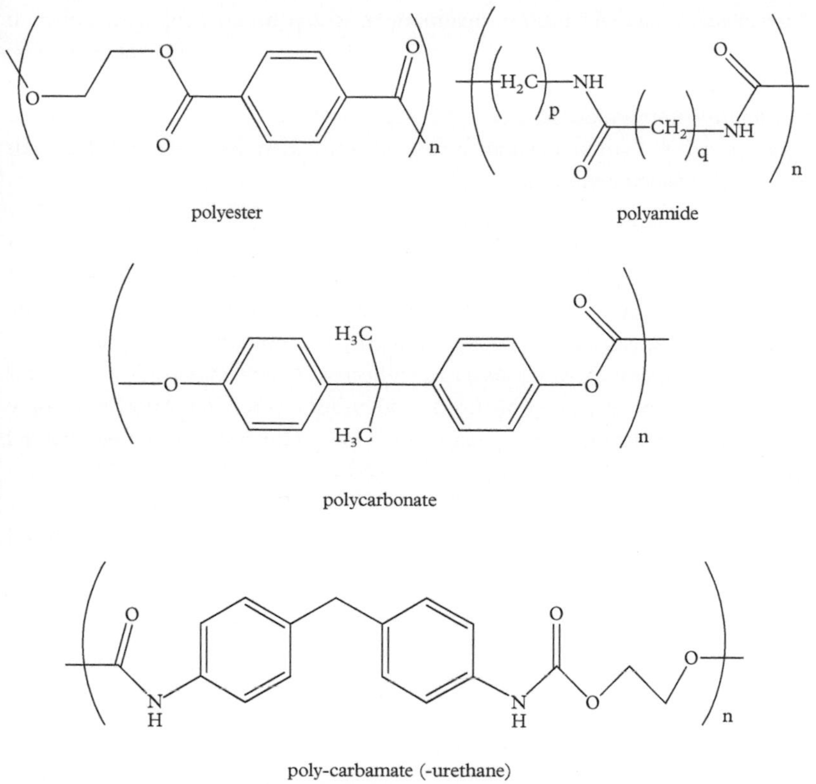

Figure 5.22 *Polymer links using functional groups.*

The effect of the methyl substitution is also to provide a higher density of *tertiary* carbons than in PE. These afford lower-energy sites for radicals. Thus PP is more susceptible to UV-initiated oxidation. Degradation can occur microbially and in implants in the human body, but these are predominantly surface effects and breakdown of the polymers is very slow. This carbon chain backbone structure is very common in polymer materials, for example in all polyvinyls, such as PVC, polystyrene (Table 5.2), and PVA, with side chains of chloride, phenyl, and acetate, respectively.

5.2.2 Polymers Linked by Functional Groups

There are many polymers produced with functional group links on a very large scale (Table 5.2). These are headed by polyesters, which include PET (poly-ethyleneglycol-terephthalic acid), the chemical structure of which is shown in Fig. 5.22. The precursors for the polymers in this figure are presented in Fig. 5.23. Those for PET are terephthalic acid (top) and 1,2-ethanediol (ethylene glycol) shown below it. The polymerization reaction (Eq. 5.22) is a sequence

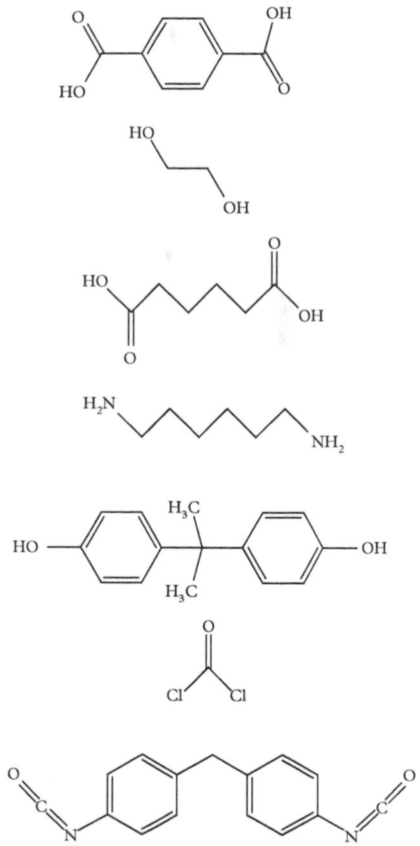

Figure 5.23 *Precursors for functional group polymers.*

of esterification, or condensation, reactions in which water is the co-product. It becomes a polymerization due to the presence of the functional groups at each end of the precursor. With an amine precursor, a condensation reaction with an acid will afford an amide, R–C(O)–NH–R′. Reaction between 1,6-hexanedioic (adipic) acid 1, 6-diaminohexane (hexamethylene diamine) (Fig. 5.23) affords the polyamide called Nylon 6,6.

$$ArC(O)OH + ROH \rightarrow ArC(O)OR + H_2O. \tag{5.22}$$

Polycarbonates (Fig. 5.22) are also manufactured on a 4–5-million-tonne scale annually as strong, often transparent thermoplastic structural materials which are good electrical insulators. The most commonly used diol is called Bisphenol A (Fig. 5.23, fourth down), which itself is synthesized from condensation of phenol and acetone. The most commonly used condensation reaction for polymerization is with carbonyl chloride (phosgene) formed from CO and Cl_2; the eliminating co-product is HCl (Eq. 5.23):

$$ArOH + COCl_2 \rightarrow ArC(O)OAr + HCl. \tag{5.23}$$

Removing the use of chlorine from this process is beneficial by avoiding atmospheric release. An alternative route has been developed that involves a transesterification using diphenylcarbonate (Fig. 5.24). In this case the side product is phenol. However, this can be recycled by the sequence of Eqs 5.24 and 5.25. The first of these provides a route to dimethylcarbonate, which can be formed from methanol, oxygen, and carbon monoxide using a catalyst such as CuCl or a supported copper site. Transesterification is then used to afford dimethylcarbonate (Eq. 5.25).

$$2\,CH_3OH + \frac{1}{2}O_2 + CO \rightarrow (CH_3O)_2C(O) + H_2O \tag{5.24}$$

$$(CH_3O)_2C(O) + 2\,PhOH \rightarrow (PhO)_2C(O) + 2\,CH_3OH. \tag{5.25}$$

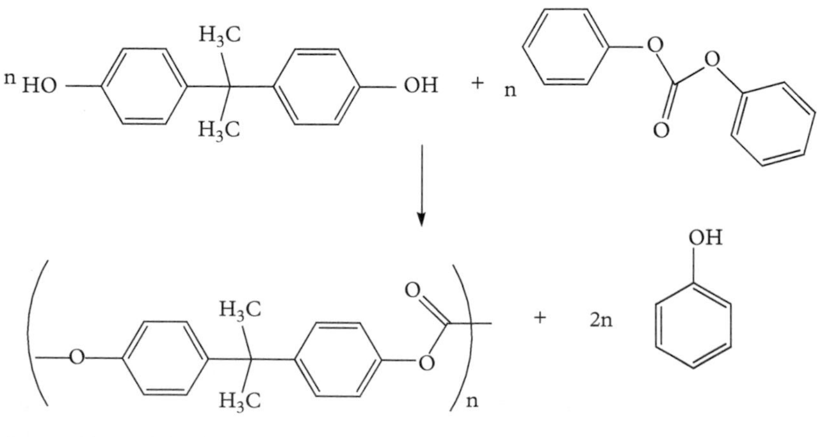

Figure 5.24 *Chlorine-free route to polycarbonate.*

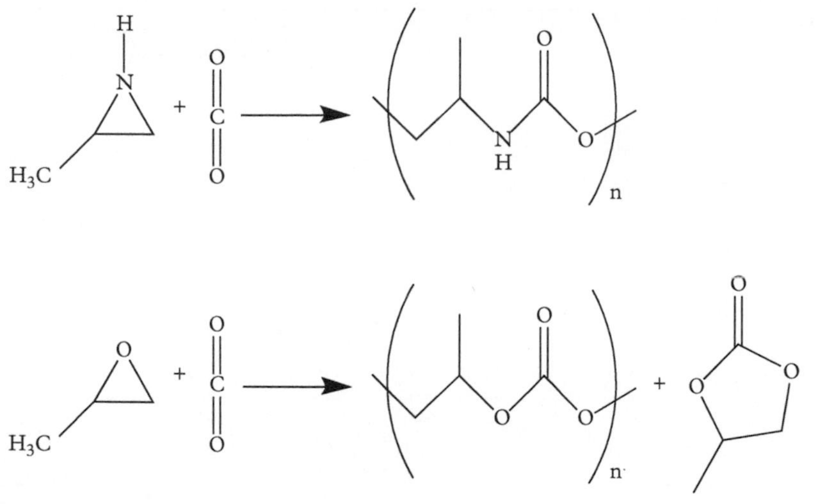

Figure 5.25 *Routes to polymers from CO_2.*

The general reaction for the formation of a carbamate, or urethane, from an isocyanate and an alcohol adds an alkoxide to the carbon and a proton to nitrogen (Eq. 5.26). This forms the basis of the production of polyurethanes which annually is in the region of 22 Mt. One of the widely used variants is that shown in Fig. 5.22 (bottom), which is a product of the reaction between of MDI (methylene diphenyl diisocyanate) (Fig. 5.23, bottom) and 1,2-ethanediol.

$$Ar - N = C = O + ROH \rightarrow Ar - N(H) - C(O)OR. \qquad (5.26)$$

Both polycarbonates and polyurethanes contain CO_2 units in their functional link. As a result there is considerable interest in the incorporation (fixing) of gaseous CO_2 into these materials.[32] Indeed, strained ring aziridines and epoxides can be opened up to afford polyurethanes[33] and polycarbonates,[34] respectively (Fig. 5.25). The reaction with the aziridine will occur in sc-CO_2 at 100 °C without a catalyst. The conversions of the epoxides do require catalysts. These can alter the direction of the reaction towards the cyclic carbonate, polycarbonates, and some polyethers. These catalysts also provide the opportunity for tacticity control as described for polypropylene. For these products manufactured on a scale of several Mt per annum, this fixing of CO_2 will make a contribution, albeit a minor one.

5.2.3 Polymer Waste

Key current issues are fate of used plastics and the release of polymers into the environment. The estimated production of plastics for different industrial sectors in 2015[35] is presented in Fig. 5.26. It is evident that for the largest sector, packaging, the annual waste generation is very similar to that of production. For other sectors such as building and construction and also industrial machinery the

[32] M. Aresta, A. Dibenedetto, and A. Angelini, 'Catalysis for the Valorization of Exhaust Carbon: from CO_2 to Chemicals, Materials and Fuels. Technological Use of CO_2', *Chem. Rev.*, 2014, **114**, 1709–42.

[33] Z.-Z. Yang, L.-N. He, J. Gao, A.-H. Liu, and B. Yu, 'Carbon dioxide utilization with C-N formation: carbon dioxide capture and subsequent conversion', *Energy Environ. Sci.*, 2012, **5**, 6602–39.

[34] F. D. Bobbink, A. P. van Muyden, and P. J. Dyson, 'En route to CO_2-containing renewable materials: catalytic synthesis of polycarbonates and non-isocyanate polyhydroxyurethanes derived from cyclic carbonates', *Chem. Commun.*, 2019, **55**, 1360–73.

[35] R. Geyer, J. R. Jambeck, and K Lavender Law, 'Production, use and fate of all plastics ever made', *Sci. Adv.*, 2017, **3**, e1700782.

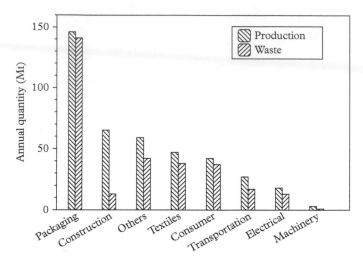

Figure 5.26 *Estimated world production and waste generation of plastics in 2015.*

36 V. Bisinella, P. F. Albizzati, T. F. Astrup, and A. Damgaard (eds) *Life Cycle Assessment of Grocery Carrier Bags*, Danish Environmental Protection Agency, Env. Proj. 1985, 2018.

37 V. Dufaud and J.-M. Basset, 'Catalytic Hydrogenolysis at Low Temperature and Pressure of Polyethylene and Polypropylene to Diesels and Lower Alkanes by a Zirconium Hydride Supported on Silica-Alumina: A Step Toward Polyolefin Degradation by the Microscopic Reverse of Ziegler-Natta Polymerization', *Angew. Chem. Int.*, Ed., 1998, **37**, 806–10.

38 S. Lambert and M. Wagner, 'Environmental performance of bio-based and biodegradable plastics: the road ahead', *Chem. Soc. Rev.*, 2017, **46**, 6855–71.

39 I. Taniguchi, S. Yoshida, K. Hiraga, K. Miyamoto, Y. Kimura, and K. Oda, 'Biodegradation of PET: Current Status and Application Aspects', *ACS Catalysis*, 2019, **9**, 4089–4105.

40 Y. Tokiwa, B. P. Calabia, C. U. Ugwu, and S. Aiba, 'Biodegradability of Plastics', *Int. J. Mol. Sci.*, 2009, **10**, 3722–3742.

majority of the production has a significant lifetime in the product. For consumer and institutional products and textiles, waste generation is a high proportion of production. A careful life-cycle assessment of grocery bags concludes that alternatives to single-use LDPE bags (which have been re-used once as a waste bin bag before incineration) must be used several times to have a comparable climate change effect, and multiple times if all environmental factors are taken into account.[36] For example, the factors were estimated to be 2 and 35 for recycled polyester PET and 52 and 7,100 for cotton! However, much waste does not track a completed orderly life cycle, resulting in a loss of energy and unintended environmental release.

Other chemical options could include conversion of petrochemical-based polymers into other chemical products rather than providing combustion fuel. One potential solution would be to carry out hydrogenolysis of a polyalkene converting it into light alkanes, as has been demonstrated using Zr–H catalyst sites on silica-alumina.[37] However, the hydrogenolysis of long-chain alkanes probably initially requires an interaction with the C–H bonds, and almost all realistic samples will contain impurities, with heteroatoms providing strongly competing interactions. The polymers with functional links (Fig. 5.22) appear to offer a wider spectrum of potential cleavage mechanisms.

As might be expected, PE is classed as a non-biodegradable polymer, since its hydrophobic nature limits interaction with enzymes.[38] Extremely low degradation is observed, depending upon the type of PE and conditions, but it can degrade only partially in decades without an oxidant, initiator, or UV irradiation. However, even with its ester linkages, potentially susceptible to hydrolytic cleavage, PET also has extremely low biodegradability.[39] In large part this can be due to the low accessibility. This can be related to the melting temperature (T_m) of a polymer[40] which is governed by the thermodynamic parameters according to Eq. 5.27:

$$T_m = \frac{\Delta_m H}{\Delta_m S}.$$
(5.27)

Typical T_m values are given in Table 5.3. The high melting temperatures of polyamides and polyurethanes can be ascribed to the large $\Delta_m H$ values due to the strong inter-chain hydrogen bonding between the N–H and carbonyl groups. The T_m of PET is similar (260 °C) and is atypical of the temperatures for aliphatic polymers in Table 5.3. This is ascribed to the very low values of $\Delta_m S$ due to the maintenance of the rigid polymer frame of the aromatic backbone. One approach has been to develop PET-degrading microorganisms with designed enzymes, PET*ase*, to cleave PET.[39] This incorporates the amino acid tyrosine with an aromatic side chain to orient the PET molecule by π–π stacking.

An alternative approach to achieving some degree of biodegradability of fossil-fuel-derived polymers would be to choose structures with lower T_m values, although clearly there will be applications for which this is inappropriate. The aliphatic polyesters and polycarbonates appear to be viable options. A polyester option is polycaprolactone (PCL) (Fig. 5.27). The precursor monomer, ϵ-caprolactone, is manufactured from oxidation of cyclohexane to cyclohexanone followed by reaction with per-acetic acid ($CH_3C(O)OOH$). Hydrolytic degradation of PCL occurs in soil, at a rate sensitive to soil type and the density of microbial communities. the timescales are much reduced from those for PE, to the order of a few years.

Polycarbonates from bisphenol A (Fig. 5.22) have a T_m of 147 °C and are not considered to be biodegradable.[41] Indeed, there is considerable interest in the study of CO_2-derived aliphatic polycarbonates which are susceptible to enzymic action.[42]

5.3 Chemicals from Biomass Alternatives

From biomass, extraction of starch and oils provides well-established routes to ethanol and diesel fuels.[43] The residue is called lignocellulose. Lignocellulose is the main structural plant material in cell walls. It has three main components: cellulose (40–50 %), based upon D-glucose units; hemicellulose (25–35 %), a polysaccharide containing D-xylose, L-arabinose, D-mannose, and D-galactose units; and lignin (15–20 %), a complex structure of aromatic rings and oxygen-containing side chains and links. Cellulose, $(C_6H_{10}O_5)_n$ (Fig. 5.28), is probably then the most prevalent organic compound on Earth. It is formed from a condensation polymerization of D-glucose and clearly has a much higher oxygen content than fossil-fuel-derived polymers. The glucose can be reformed by acid or enzymic catalytic hydrolysis. Glucose can be isomerized into fructose which itself can be dehydrated to the chemical precursor 5-hydroxymethylfurfural (HMF).[44] Fermentation of glucose can be used to form lactic acid, generally as the racemic form (Fig. 5.28). The major sugar in hemicellulose is xylose, which can be

Table 5.3 *Melting temperatures of aliphatic polymer types.*

Polymer	T_m (°C)
polyester	60
polycarbonate	65
polyurethane	180
polyamide	240–265

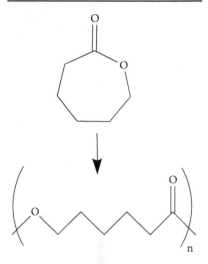

Figure 5.27 *Ring opening polymerization of ϵ-caprolactone to PCL.*

[41] T. Artham and M. Doble, 'Biodegradation of Aliphatic and Aromatic Polycarbonates', *Macromol. Biosci.,* 2008, **8**, 14–24.

[42] M. Scharfenberg, J. Hilf, and H. Frey, 'Functional Polycarbonates form Carbon Dioxide and Tailored Epoxide Monomers: Degradable Materials and Their Application Potential', *Adv. Funct. Mater.,* 2018, **28**, 201704302.

[43] D. M. Alonso, J. Q. Bond, and J. A. Dumesic, 'Catalytic conversion of biomass to biofuels', *Green Chem.,* 2010, **12**, 1493–1513.

[44] S. Chen, R. Wojcieszak, F. Dumeignil, E. Marceau, and S. Royer, 'How Catalysts and Experimental Conditions Determine the Selective Hydroconversion of Furfural and 5-Hydroxymethylfurfural', *Chem. Rev.,* 2018, **118**, 11023–117.

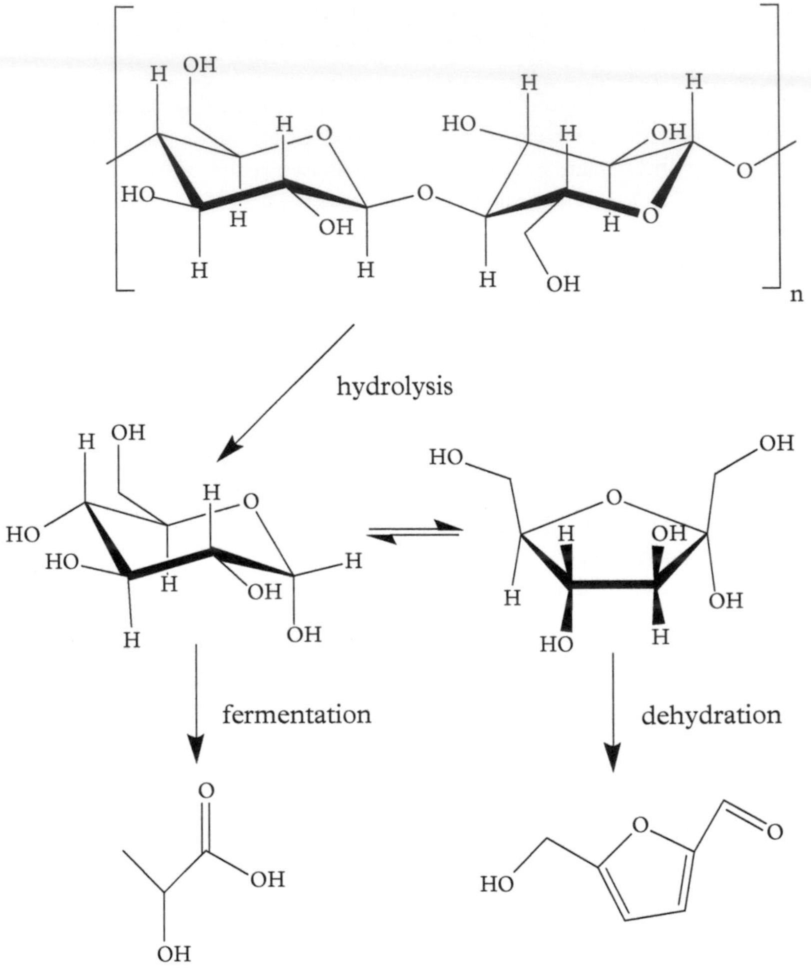

Figure 5.28 *Cellulose and some of its products.*

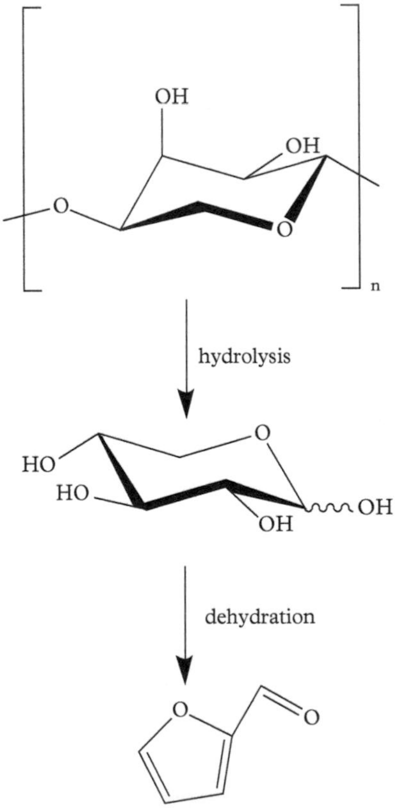

Figure 5.29 *A fructose unit from hemi-cellulose and furfural derived from it.*

extracted from the polymeric precursor. This in turn can be dehydrated with acid catalysis to form furfural (FFR) (Fig. 5.29).

Biodiesel is manufactured from the extracted vegetable oils, which are mostly triesters formed between glycerol and long-chain fatty acids (Fig. 5.30). These chains may be the same or different in the same molecule. The biodiesel is formed by a transesterification with a short-chain alcohol, depicted as ethanol in this figure. This is generally catalysed by a strong base such as NaOH or KOH. The mixture of the long-chain esters is the basis of the biodiesel and glycerol is a by-product. Fuels can also be generated from the products of the pentoses and hexoses derived from lignocellulose. For example, an aldol condensation between furfural and acetone affords an oxygenated C_8 centre, and a second such step

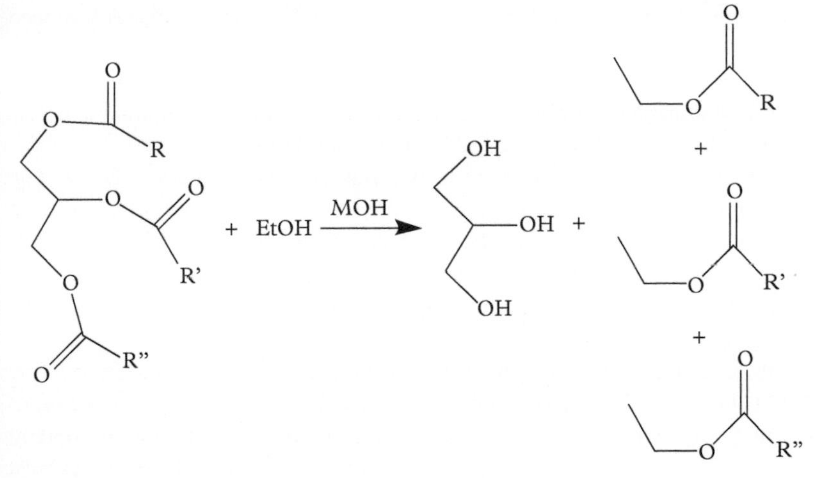

Figure 5.30 *Synthesis of biodiesel.*

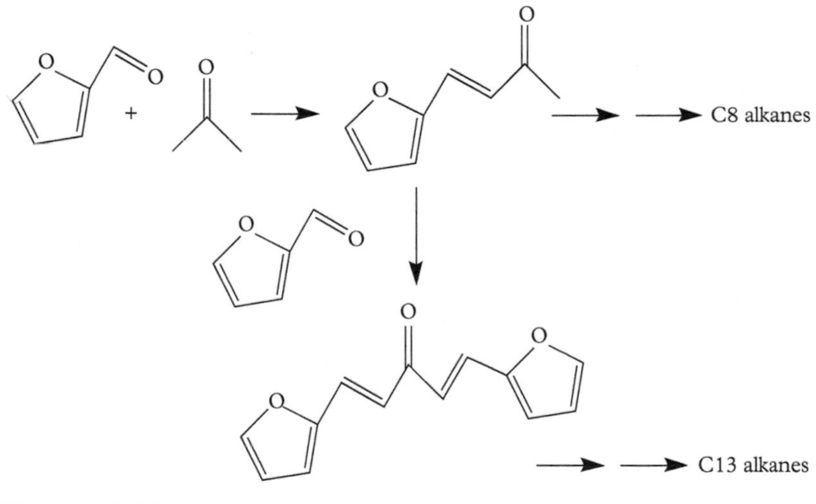

Figure 5.31 *Liquid fuels from furfural.*

leads to a C_{13} product (Fig. 5.31). Following metal (PGM) catalysed reductions, these can be converted into alkanes. Similarly, from the hexose-derived precursor HMF parallel reactions will afford C_9 and C_{15} alkanes.[45]

The other product highlighted in Fig. 5.28 is lactic acid. This has the components to form a polyester, and can do so as in Fig. 5.32. Dehydration can generate both cyclic and polymeric esters, with the cyclic ester (the lactide) being entropically favoured at high temperatures (200 °C). The resulting polymer is called polylactide or, erroneously, polylactic acid (PLA). Lactic acid has a chiral centre and exists as *D*- and *L*- forms or as a racemic (*rac*) mixture. The *rac*-PLA is an amorphous, glassy polymer with a T_m of 173–178 °C. The chiral form, generated by a ring-opening polymerization of the L,L′-lactide, is labelled PLLA.

[45] G. W. Huber and J. A. Dumesic, 'An overview of aqueous-phase catalytic processes for production of hydrogen and alkanes in a biorefinery', *Catal. Today*, 2006, **111**, 119–132.

[46] J. H. Song, R. J. Murphy, R. Narayan, and G. B. H. Davies, 'Biodegradable and compostable alternatives to conventional plastics', *Phil. Trans. R. Soc. B*, 2009, **364**, 2127–39.

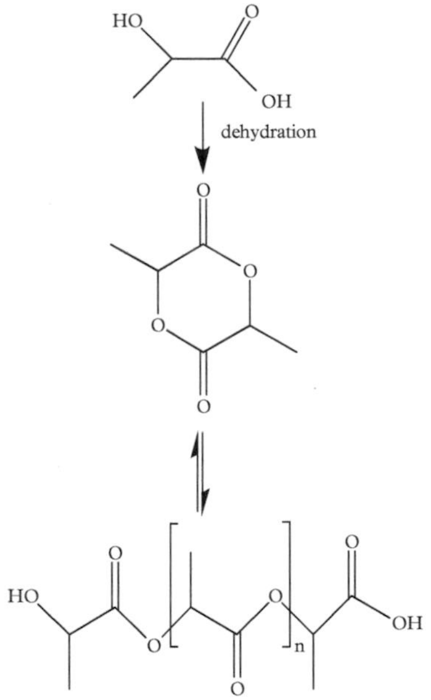

Figure 5.32 *Formation of polylactide (PLA).*

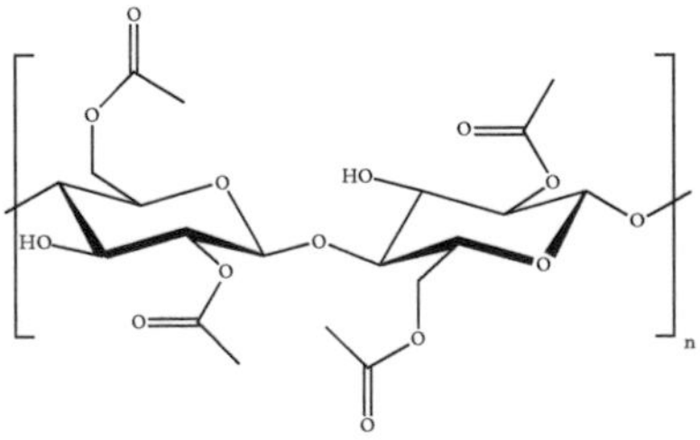

Figure 5.33 *Structure of cellulose acetate.*

It has increased crystallinity and an increased T_m by 40–50 °C. PLLA has been described as a biodegradable polymer from biorenewable resources. However, the biodegradability properties are very dependent upon particular conditions.[36] Testing by simulation of home composting and of degradation in aquatic media showed that decomposition is negligible after several months.[46]

Closer to the structure of the natural material is cellulose acetate polymer (Fig. 5.33). The production of this material, formed from the reaction of wood pulp with ethanoic acid and its anhydride, was $\approx$ 614 kT in 2016, and its use is largely in fibres, including cigarette filters. Left in the open, degradation is extremely slow, but it can be biodegraded in months in some active soils via the initial loss of the acetate groups.

Global starch production in 2015 was estimated to be 85 Mt, and production in the EU (2107) 11 Mt. Of this, *c*.40% was in non-food use, mainly in paper manufacture. Starch is a mixture of two molecular structures. The minor component (20–25 %), amylose, is a helical polymer consisting of linked glucose rings, whilst the majority structure type, amylopectin, is branched (Fig. 5.34). Animal starch contains glycogen, which is a more highly branched structure than amylopectin. Both motifs contain many hydroxyl groups and thus have high water affinity. In composting studies, potato-starch-based materials were shown to be rapidly degraded.[46] In a mixed material construction incorporating the caprolactone-derived polymer PCL as an overlayer, the starch did confer compostability. However, this was not observed for polypropylene compounded with starch granules. There is much scope for blending of materials for particular purposes.

5.4 Solar Energy Conversion

As described in section 1.4, the input of solar energy to Earth is in excess of 10,000 times the average power demand. It averages at 342 W m^{-2}. The solar spectrum peaks in the visible region with a wavelength of 500 nm and energy of 2.28 eV (Fig. 1.12). There is clearly a potential as an alternative energy source, which can be achieved by the **photovoltaic** (PV) effect. This conversion of light to electrical energy was first observed by Becquerel in 1839 using an electrode of AgCl. The first true photocell was provided by Fritts is 1876, using the semiconductor properties of selenium which has a band gap of 1.85 eV at 300 K (Section 3.2, Table 3.3, Fig. 3.3).

The first silicon-based PV cell was constructed in the Bell Laboratories in 1954 and had a conversion efficiency of 4%. The types of semiconductor are illustrated in Fig. 5.35. Silicon, which has a diamond structure (Fig. 3.4), is an example of an intrinsic semiconductor. Its band gap at 300 K is 1.14 eV, a wavelength of 1,090 nm in the near IR. At 0 K the conduction band of silicon is empty, but at 300 K there is some excitation to it from the valence band. However, there is only a small degree of excitation, since the value of $k_B T$ is 0.0259 eV at that temperature. The Fermi–Dirac distribution defines this probability, f, as (Eq. 5.28):

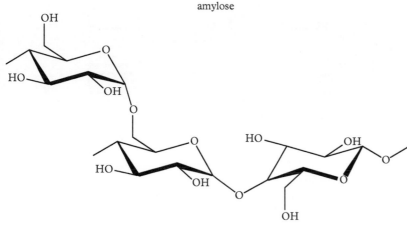

amylose

amylopectin

Figure 5.34 *Structural motifs in the components of starch.*

$$f = \frac{1}{1 + e^{\left(1 + \frac{E_c - E_F}{k_B T}\right)}}.$$ (5.28)

The energy of the band gap at 0 K, E_g, is the difference between the energies at the bottom of the conduction band (E_c) and those of the top of the valence band (E_v). The energy of the Fermi level (E_F) at 0 K for a semiconductor is the mean of E_c and E_v. At 300 K this small proportion of the valence electrons is transferred from the valence band, creating holes (shown as empty circles) balancing the charge of the of the electron (shown as filled circles) in the conduction band. The effect of adding isolated dopant atoms is also illustrated in Fig. 5.35. Doping of a phosphorus atom into the diamond structure provides an extra electron than is required for the filling of the valence band. At 0 K, because of the higher electronegativity of phosphorus, the energy of this donor dopant, E_d, resides below that of the conduction band, and the Fermi level is close to that energy. At 300 K, the extra electron can be delocalized into the conduction band in what is termed an *n*-type semiconductor, rendering the phosphorus sites cationic. Conversely, the

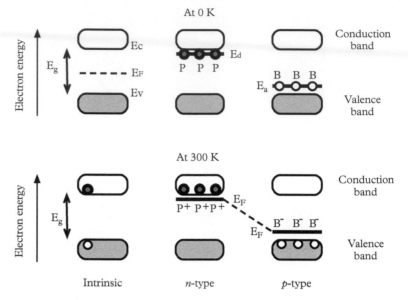

Figure 5.35 *Types of semiconductor at 0 K and ambient temperature.*

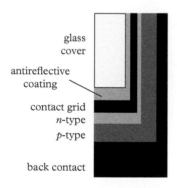

Figure 5.36 *Schematic of a PV cell.*

effect of boron doping is to add electron-acceptor sites with an energy just above that of the conduction band, at an energy E_a. At room temperature these electron holes are delocalized into the valence band forming a p-type semiconductor and locally the boron atoms acquire a negative charge. The Fermi level is just above the valence band, close to the energy E_a.

A schematic of a PV cell is shown in Fig. 5.36. Light passes through a glass window, an antireflective coating, and a contact grid to a thin p-type silicon and then across a p–n junction to the n-type silicon. Current is taken from the contact grid and the back contact plate. The p–n junction is not two independent pieces of silicon bonded to each other but has graded doping to create the diode. As shown in Fig. 5.34, the Fermi level will differ in the two separated materials. However, there is a common level in a diode without a biased voltage. This is achieved by a migration of electrons from the n-type region into the p-type. This creates a depletion zone where the separated electron-hole pairs are reduced in number. It also creates a bias voltage, and the result of this is that the energy levels of the bands are bent, as shown in Fig. 5.37. The effect of an incoming photon is also shown in this figure (drawn as impinging in the depletion layer). The photon excites a valence electron into the conduction band leaving a hole in the valence band. The electron follows the track of lowest potential into the n-type side of the diode whilst the hole migrates within the valence band into the n-type side. This creates a voltage difference at the collector, typically ≈ 0.5 V. The current will be dependent upon the flux of the incoming light. Arrays of cells provide a current at 12 V DC, and an inverter is used to convert this to main voltage AC. The efficiencies

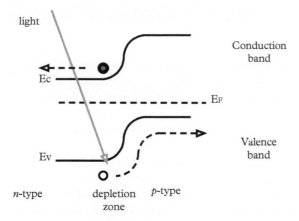

Figure 5.37 *Charge separation in a PV cell.*

of commercial crystalline silicon PV modules is $\approx$ 17–19%, with polycrystalline silicon operating at slightly lower efficiencies (15–16%).[47] The overall installed capacity in 2015 was *c*.230 GWp, but this rose to *c*.515 GWp through 2018.[48]

The highest efficiency for a single cell of monocrystalline silicon reported by 2108 was 26.7%, but the highest recorded for all types is 46%.[48] This latter figure was achieved by two additional features. First, optical devices are used to collect and concentrate the light onto a smaller area (by a factor of 500–1,000). Second, the detector itself is a multi-junction device incorporating semiconductors with different band gaps to give better absorption across the solar spectrum. A schematic diagram of such a cell is presented in Fig. 5.38. There are three different semiconductors arranged so that light impinges on the elements in order of decreasing band gap. The effect of this is presented in Fig. 5.39. The

[47] *World Energy Resources. Solar, 2016*, https://www.worldenergy.org, World Energy Council, London, 2016.

[48] *Photovoltaics Report*, © Fraunhofer ISE, 2018.

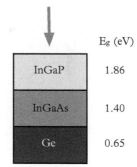

Figure 5.38 *Schematic of a multi-junction PV cell.*

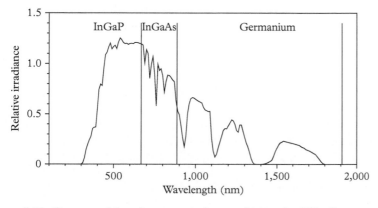

Figure 5.39 *Coverage of the solar spectrum by a multi-junction PV cell.*

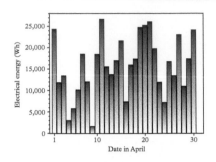

Figure 5.40 *Output of a monocrystalline silicon PV installation in central southern England in 2019. Annual insolation* $\approx$ *1,050 kWh m^{-2}.*

[49] M. Raugei, S. Sgourides, D. Murphy, V. Fthenakis, R. Fischknecht, C. Breyer, U. Bardi, C. Barnhart, A. Buckley, M. Carbajales-Dale, D. Csala, M. de Wild-Scholten, G. Heath, A. Jæger-Waldau, C. Jones, A. Keller, E. Leccisi, P. Mancarella, N. Pearsall, A. Siegel, W. Sinke, and P. Stolz, 'Energy Return on Energy Invested (ERoEI) for photovoltaic solar systems in regions of moderate insolation: A comprehensive response', *Energy Policy*, 2017, **102**, 377–84.

Table 5.4 *World production in 2018 of key elements in thin-film PV cells.*

Element	Production (tonnes)
cadmium	26,000
copper	2.1×10^7
gallium	410
indium	750
selenium	2,800
tellurium	440

largest band gap here is in the red part of the visible spectrum afforded a III-V semiconductor with a large spread of electronegativities. Light of higher energy than the absorption edge can be absorbed and generate a photocurrent; light of lower energy can pass through to the next element in the unit. By changing phosphorus to the less electronegative arsenic, the band gap is reduced and the absorption edge moves into the near IR. Thus the central section of the wavelength range can provide a second component of the overall photocurrent. Finally, the longest-wavelength light can be absorbed by germanium which has an absorption edge in the mid IR.

The energy return for a PV installation is also strongly dependent upon the location. Whilst latitude is the most substantial factor to govern the solar flux density, this is also dependent upon the particular site and prevailing conditions. As shown in Fig. 5.40 for an unshaded, south-facing location the yield can vary significantly from day to day with varying cloud conditions. Estimates of the energy return on energy invested (EROI, Table 1.1) vary considerably, but a detailed consideration of the situation in Switzerland (insolation c.1,000–1,400 kWh m^{-2} y^{-1}) estimated the lifetime EROI to be 8–9.[49] This factor is also expressed as the energy pay-back time (EBPT), which is generally highest for monocrystalline Si.[49] Even so, this is estimated to be $\approx$ 3.3 yr in Germany (insolation c.1,000 kWh m^{-2} y^{-1}), reducing to $\approx$ 1.75 yr in Sicily (insolation c.1,925 kWh m^{-2} y^{-1}).[47]

Water use comes only from PV manufacture and maintenance of modules; this has been estimated at 118 L MWh^{-1}. Emission levels have also been estimated: $\approx$ 57 kg CO$_2$e MWh^{-1}, which compares to 180 kg CO$_2$e MWh^{-1} from burning natural gas. SF$_6$ is employed in cleaning reactors used in silicon production and thus might be tightly contained due to its very high GWP (Section 1.9).

Thin film technologies using CdTe (E_g 1.5 eV at 300 K) were first produced in 2004 and their market share increased rapidly up to 2009. It has been noted that the demand for tellurium (Section 3.2.6) was 900% of the annual production in 2008.[30] Production of CdTe-based units did decline rapidly between 2009 and 2017. As can be seen in Table 5.4, the annual production of cadmium in 2018 was moderate (26,000 ton); production of tellurium (section 3.2.6) is much lower. A potential replacement is copper indium gallium selenide (CIGS). Copper (section 3.3.4) is the most abundant of all these elements, and its production is not a limit on the availability of CIGS materials. Selenium has a considerably higher production than tellurium and should not be the limiting factor. Thin film based devices were introduced in 2007 and are growing in application. Materials of the general formula CuIn$_x$Ga$_{1-x}$Se$_2$ can be prepared through from $x = 0$ to 1, with the band gap increasing from 1.0 eV for CuInSe$_2$ to 1.7 eV for CuGaSe$_2$. These can reduce the energy pay-back time, EBPT, and greenhouse gas emission by a factor of 2 or more. However, it may be that the availability of gallium and indium limit their application, but production of PV modules did increase by nearly a factor of 2 from 2014 to 2017.

5.5 Nuclear Energy Conversion

The processes of energy creation (ΔE) in the nuclear industry can be traced to the mass loss (Δm) associated with a nuclear fission process, according to the Einstein equation (Eq. 5.29), the basis of the conservation of energy-mass:

$$\Delta E = \Delta mc^2. \tag{5.29}$$

As shown in Fig. 3.56, typically a nuclear fuel charge, for example in a pressurised water reactor (PWR), would consist of 4% ^{235}U and 96% ^{238}U. In a fuel charge before reprocessing, *c.*75% of the ^{235}U would be converted into fission products and $\approx$ 3% of the ^{238}U would be converted into plutonium isotopes. Two of these are fissile, ^{239}Pu and ^{241}Pu. Overall the mass of the fission products in the used fuel is $\approx$ 5%, with 3% coming from the ^{235}U charge and 2% from the fission of plutonium derived from ^{238}U. Each of these sources provides a spectrum of processes. As described in Eq. 3.41, fission from ^{235}U occurs after capture of a thermal neutron to create the transient ^{236}U. ^{235}U, ^{239}Pu, and ^{233}U (derived from a thorium-fuelled reactor) all tend to split into a nucleus of mass around 95 and another of mass $\approx$ 135–140. The major fission products from ^{235}U are given in Table 5.5. These show something of the chemical complexity of fuel reprocessing, the opportunities provided by element synthesis (e.g. for technetium), and challenges for safe storage of waste.

The thermal yield from nuclear reactors comes from an array of primary processes. One example of such is Eq. 5.30, a fission channel from neutron activation of ^{235}U. The mass changes in amu are given in Eq. 5.31, from which a loss in the rest mass (Δm) can be calculated to be 0.1928. This corresponds to 2.8773×10^{-11} J, or 1.733×10^{10} kJ mol-^{235}U^{-1} undergoing this process.

$$^{235}_{92}U + ^{1}_{0}n \rightarrow ^{144}_{56}Ba + ^{90}_{36}Kr + 2^{1}_{0}n \tag{5.30}$$

$$235.0439301 + 1.0086652 \rightarrow 143.922953 + 89.919517 + 2.0173304. \tag{5.31}$$

A schematic diagram of a PWR is shown in Fig. 5.41. Here the uranium oxide fuel rods are held in a cladding, typically a Zircaloy. This is a grade of >95% zirconium and very low hafnium content to provide a low neutron cross-section. The neutron flux in the reactor is moderated by rods of relatively high neutron cross-section. These are often alloys of Ag (65–85%), In (2–20%) Cd (0–10%), Sn (0–5%), and Al (0–1.5%) encased in a stainless steel (SS) or an Inconel alloy. Inconel is one of the types of **superalloy** designed for high strength under extreme conditions of temperature and pressure. A passivating oxide layer forms on it for chemical protection. Nickel is the major component ($\approx$ 40–70%), with chromium as the second metal ($\approx$ 14–30%). The remaining components can include Fe, Mo, Nb, Ta, or Co. In addition to the control rods, Inconel can also be employed in the stream generator and cladding in the pressure vessel. As shown in Table 5.4 the mined production of cadmium in 2018 was of the order

Table 5.5 *Higher-yield fission products from neutron activation of* ^{235}U.

Yield (%)	Isotope	$t_{1/2}$
6.8	^{133}Cs-^{134}Cs	2.07 y
6.3	^{135}I-^{135}Xe	6.57 h
6.3	^{93}Zr	1.53 My
6.1	^{99}Mo	65.9 h
6.1	^{137}Cs	30.2 y
6.1	^{99}Tc	211 ky
5.8	^{90}Sr	28.9 y

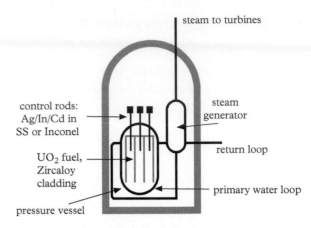

Figure 5.41 *Schematic of a pressurised water reactor.*

of 26,000 tonnes, and that of silver was similar. The supply of indium, which is principally used in the application of indium tin oxide (ITO) coatings of display screens, is a key factor in the development of nuclear power stations,[30] as well as for the development of CIGS-based solar cells (Table 5.4).

Typically a 1 GW output power station will require a thermal input from the steam generator of 3 GW due to the intrinsic energy efficiency of the heat engine. The majority of the nuclear energy is within the kinetic energy of the particles formed, together with secondary nuclear reactions and other contributions from components within the reactor. The nuclear reactor required to deliver 3 GW of thermal energy would contain $\approx$ 100 t of enriched uranium (113 t of enriched UO_2). The annual consumption of fissile material is *c.*1 t, with a mass consumption of 1 kg converted into energy. The lifetime of a fuel charge is typically 4 years, and so amounts to an annual consumption of the charge of 25 t. Replenishment by new natural uranium would require an annual supply rate of 250 t.[50] The supply of both fuel and indium are the potential limiters on expansion.

[50] https://www.nuclear-power.net.

5.6   Hydrogen as a Fuel

In Section 4.1.2, the major method for production of hydrogen gas, namely steam methane reforming, was described. The co-product is CO_2. In Section 4.2.2 the capture and storage of CO_2 (CCS) from post- and pre-combustion gas streams was also described as differing means of effectively utilizing hydrocarbons as low-carbon-emitting sources of reduction by H_2. This is termed blue hydrogen; the non-CCS methods are 'coloured' according to their carbon source: black (from coal), grey (from natural gas), and brown (from lignite). Green hydrogen is the term ascribed to hydrogen production from non-fossil-fuel sources.[51] Potentially

[51] *The Future of Hydrogen. Seizing Today's Opportunities*, International Energy Agency report to the G20, 2019, http://www.iea.org.

generation of H_2 gas by the water-splitting reaction (Eq. 5.32) could provide the gas without the processing required for the pre-combustion removal of CO_2. It is highly endothermic ($\Delta_r H°$ 286 kJ mol^{-1}, 2.96 eV or 419 nm) and endoergic ($\Delta_r G°$ 237 kJ mol^{-1}) and will require energy input either photo- or electro-chemically.

$$H_2O_{(l)} \rightarrow H_{2(g)} + \frac{1}{2}O_{2(g)}. \tag{5.32}$$

5.6.1 Water Splitting in Nature

In nature, the closest approach to this is in photosystem II which precedes photosystem I in the electron-transfer chain. It is present in green plants and cyanobacteria, in which different parts of the reaction sequence within Eq. 5.33 are carried out by specific components.

$$2\,H_2O_{(l)} \rightarrow 4\,H^+ + O_{2(g)} + 4\,e. \tag{5.33}$$

The electron flow starts with the binding of two water molecules to a catalytic centre with Mn and Ca ions, with a formula of the type $Mn_4CaO_xCl_{1-2}(HCO_3)_y$. This water-oxidizing site has been investigated extensively by X-ray diffraction of a light-activated dark state and the core unit is believed to have the structure shown in Fig. 5.42.[52] There are four different Mn centres in a core. Three of the manganese and the calcium ions are in a distorted cubane with four bridging oxo ligands. The extra-cluster manganese and the calcium are both found to have two coordinated water molecules. Binding by carboxylate-containing amino acid residues completes octahedral coordination at all manganese sites and seven-coordination at the calcium.

The synthesis of this cluster is photoactivated, with a low quantum efficiency ($\approx 1\%$).[53] Manganese in aqueous conditions is generally as Mn(II). There is a high-affinity binding site for Mn(II) and an efficient photooxidation to Mn(III). A second Mn(II) can then bind, and a second photon is required to oxidize this to a second Mn(III). A suite of these processes effects the formation of the water oxidising cluster (WOC) in Fig. 5.42.

Light absorption is through chlorophyll a at a magnesium-containing chlorin ring (Fig. 5.43). The absorption at 680 nm (1.82 eV) causes charge separation in the photosystem, creating a radical cation at the chlorin complex, labelled P680$^+$. The quantum efficiency of the photochemical cycle is high (> 90%). This radical cation is neutralized by the electrons coming from the manganese cluster. Four photons are required to complete the cycle and re-oxidize the cluster for the next pair of water molecules. Four electrons are delivered to the electron-transfer chain, to supply those passed by photosystem I to effect the reduction of CO_2.

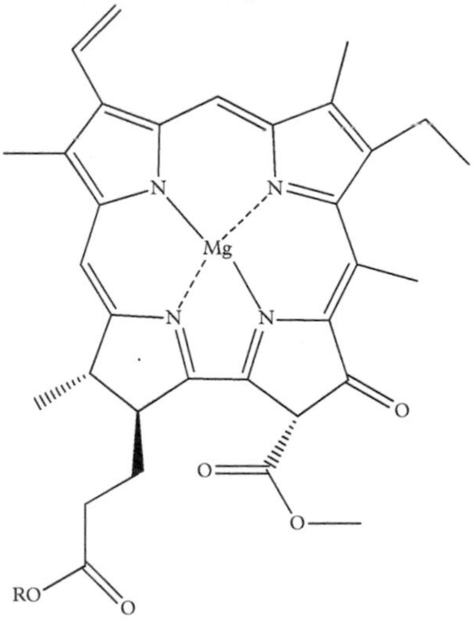

The changes in the manganese–calcium cluster have been represented by the scheme in Fig. 5.44. There are five species. Starting from the aquated complex, S_0, there are four photon-induced removals of electrons leading to the end of the cycle at S_4. This is a short-lived transient. In a thermal (dark) reaction, O_2

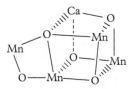

Figure 5.42 *Structure of water oxidizing Mn_4CaO_5 cluster.*

[52] Y. Umena, K. Kawakami, J.-R. Shen, and N. Kamiya, 'Crystal structure of oxygen-evolving photosystem II at a resolution of 1.9 Å', *Nature*, 2011, **473**, 55–60.

[53] H. Bao and R. L. Burnap, 'Photoactivation: The Light-Driven Assembly of the Water Oxidation Complex of Photosystem II', *Front. Plant Sci.*, 2016, **7**, 578.

Figure 5.43 *Structure of chlorophyll a.*

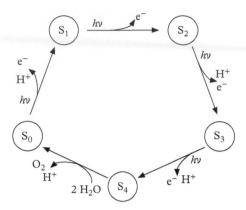

Figure 5.44 *Outline of the steps in water oxidation in photosystem II.*

[54] H. Chen, D. A. Case, and G. C. Dismukes, 'Reconciling Structural and Spectroscopic Fingerprints of the Oxygen-Evolving Complex of Photosystem II: A Computational Study of the S_2 State', *J. Phys. Chem. B*, 2018, **122**, 11868–82.

[55] M. Suga, F. Akita, M. Sugahara, M. Kubo, Y. Nakajima, T. Nakane, K. Yamashita, Y. Umena, M. Nakabayashi, T. Yamane, T. Kakano, M. Suzuki, T. Masuda, S. Inoue, T. Kimura, T. Nomura, S. Yonekura, L.-J. Yu, T. Sakamoto, T. Motomura, J.-H. Chen, Y. Kato, T. Noguchi, K. Tono, Y. Joit, T. Kameshima, T. Hatsui, E. Nango, R. Tanaka, H. Naitow, Y. Matsuura, A. Yamashita, M. Yamamoto, O. Nureki, M. Yabashi, T. Ishikawa, S. Iwata, and J.-R. Shen, 'Light-induced structural changes and the site of O=O bond formation in PSII caught by XFEL', *Nature*, 2017, **543**, 131–35.

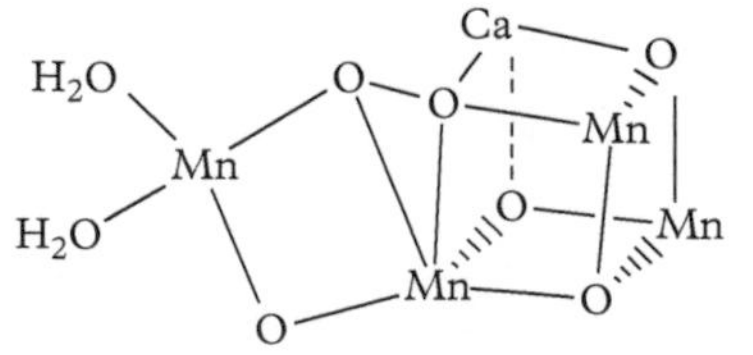

is evolved from the S_4 state of the cluster and two new H_2O ligands bound. The two states S_0 and S_1 are stable in the dark, and have been the subject of an X-ray diffraction study.[52] S_2 and S_3 are metastable with lifetimes of seconds and without light will revert to S_1. The definition of all the oxidation states is still the subject of debate. In the light-activated S_2 state, the three sites are considered to be Mn(III) and another Mn(IV).[54] Devices for generating X-rays with pulse lengths of the order of tens of fs are called **free electron lasers** (XFELs). X-ray diffraction then is observed with a time slice shorter than that of metal–ligand vibrations. The crystal can be destroyed by the process, but the experiment is carried out with a large population of microcrystals and each crystal is only studied once. This gives an observation which is radiation-damage free and can be timed in a sequence of chemical excitations by laser pulses in the visible region. In this way, the structure of state S_3 with a lifetime of 220 μs was investigated by a large team.[55] The results indicated that the O–O bond is created in this step (Fig. 5.45). One of the oxygen atoms is thought to come from the fourfold site on Fig. 5.42 bridging the calcium to three manganese ions. The water molecules on the dangling manganese are still in place.

This illustrates that effecting water splitting photochemically will require several steps to be achieved with high step efficiency. As in photosystem II the oxidation of water and release of four electrons will require four sequenced photochemical steps. The sites to achieve this may require photoactivation. The light-absorption site will need to achieve charge separation with high quantum efficiency and have an electron-accepting excited state in a location with a fast electron-transfer rate from the water-oxidizing site. The electrons released from the water-oxidizing complex may either be channelled into closely located chemical reduction complex complexes (e.g. photosystem I which reduces CO_2) or be transferred to an electrode in a photo-electrochemical device.

Figure 5.45 *PS II cluster in the S_3 state.*

5.6.2 Hydrogen from Electrolysis

The electrochemical production of H_2 has already been described in Section 4.1.2, essentially as a side-product of the manufacture of Cl_2 by the chlor-alkali process; this involves the electrolysis of brine, concentrated aqueous NaCl. Replacing the electrolyte with an alkaline solution (NaOH or, generally, KOH) provides the reagents for an **alkaline electrochemical cell** (AEC). This type of cell has similarities to the chlor-alkali cell. O_2 is generated by Eq. 5.32 carried out at the anode ($E°$ 1.23 V), and the reduction of protons to H_2 occurs at the cathode. Temperatures are in the 60–80°C range and pressures up to 30 bar. The separator between the two half-cells is electrically insulating and a hydroxyl ion conducting diaphragm. Typically it is 50–500 μm thick and constructed of zirconia, ZrO_2, on a polysulphone polymer (e.g. Fig. 5.46). These are rigid polymers with high strength and temperature stability. The energy required to achieve the electrolysis at a minimum is the enthalpy rather than the Gibbs free energy, and this corresponds to a voltage of 1.48 V. In reality perfect efficiency cannot be achieved, and this extra barrier is termed an **overpotential**. Typically the cell voltage is in the range 1.8–2.4 V. To achieve this, the electrodes bear catalysts to reduce the energy barriers for the formation and desorption of the gases. These catalysts are generally based on nickel and its alloys and provide a voltage efficiency of 50–70%.

An alternative approach in the proton exchange membrane electrolysis cell (PEMEC) (Fig. 5.47). Here water, rather than strong alkali, is employed. It is introduced to the anode emitting O_2 from the same side of the device, with temperatures at 50–80 °C and pressures in the 30–80 bar range. The protons generated are transferred across a membrane and reduced to H_2 at the cathode. Generally the proton-conducting polymer membrane is a fluorinated sulphonic acid polymer (See Fig. 4.8), such as *Nafion* ©. Voltage efficiencies are similar to those of AECs. These devices are presently employed less as they require higher installation costs. Current catalysts are platinum-group metals: Pt or Pt–Pd for the cathode and RuO_2 or IrO_2 for the anode.

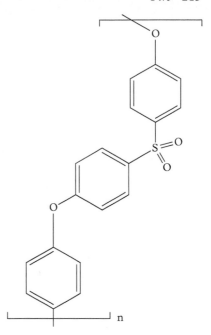

Figure 5.46 *Example of a polysulphone polymer.*

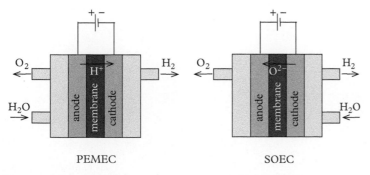

Figure 5.47 *Schematics of a proton exchange membrane electrolysis cell (PEMEC) and a solid oxide electrolysis cell (SOEC).*

The third option is the solid oxide electrolysis cell (SOEC), and this is presently less well developed. Here the water as steam is introduced to the cathode, and H_2 is emitted from this side. Charge balance is maintained by migration of oxide anions across this membrane which are oxidized in the anodic side of the device. Operating temperatures are in the region of 650–1,000 °C and pressures are under 25 bar. High temperatures are needed for rapid oxide ion transfer, which is effected using yttria (Y_2O_3) stabilized zirconia (YSZ). For every two Y(III) ions replacing Zr(IV) ions, an O^{2-} vacancy is generated thus promoting oxide conductivity. The maximum conductivity is attained at 8–9% of yttria. This substitution also causes the structure of the ZrO_2 lattice to change from the complex one that is stable at room temperature (monoclinic with seven-coordinate zirconium) to the cubic structure of CaF_2. This can be explained by the higher ionic radius of Y(III) over Zr(IV) expanding the structure and favouring eight-coordination. Platinum-group metals are avoided in the electrodes in these devices. Both electrodes have YSZ as the base material, with nickel incorporated into the cathode and a lanthanum strontium manganate perovskite, $La_{0.8}Sr_{0.2}MnO_3$, in the anode. A further attractive feature is that the electrical efficiency is higher, currently 74–81%. These devices can also be used in reverse as fuel cells, consuming hydrogen and oxygen and generating electricity, and thus could be grid-balancing devices. The electrolysis cells are generating H_2, and O_2, from water, but the 'greenness' of the gas is dependent upon the electricity generation supplied to the cells.

It is estimated that in 2018 the production of hydrogen was mainly dedicated (69 Mt) but a significant proportion was as a by-product (48 Mt) (Fig. 5.48),[51] with the chlor-alkali process an important component of this second stream. The great majority of the production of hydrogen was from fossil fuels (natural gas 196 Mtoe, coal 75 Mtoe, and oil 2 Mtoe). Of the dedicated stream <0.4 Mt of H_2 is being produced as blue hydrogen by by CCS, and <0.1 Mt as green H_2 from renewables. Further green H_2 is produced as a by-product but this is estimated to be <0.3 Mt. Presently the green hydrogen economy is some way off.

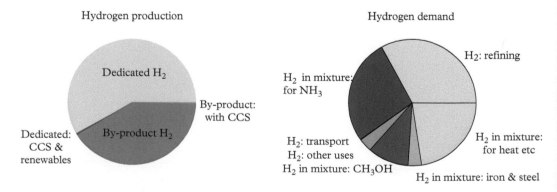

Figure 5.48 *Estimated production methods and demand for H_2 in 2018.*

The demand for hydrogen is dominated by chemical production (Fig. 5.48). Pure H_2 is used mainly for the oil refining (38 Mt) and ammonia production (31 Mt), and, mixed with other gases for methanol (12 Mt) and iron and steel (4 Mt) production. These are all fixed installations and currently <0.01 Mt is used in transport. Overall the annual energy demand for H_2 is $\approx$ 330 Mtoe and results in annual emission of $\approx$ 830 $MtCO_2$, about 2% of the total. The potential contributions to lowering the overall greenhouse gas emissions are broadly in two ways: reduction in the CO_2 emissions in its manufacture, and as a replacement for fossil fuels. Further installations of pre-combustion CCS units and reduction in the costs of non-fossil-fuel electricity supplies combined with improved efficiency in electrolysis cells are seen as the key factors for expanding blue and green H_2, respectively.

5.6.3   Water Splitting Using Sunlight

As compared to photosynthesis, the electrolysis processes replace light energy with electrical energy. Considering the water-oxidizing system in photosystem II (Fig. 5.44), the four photons needed to oxidize water are generated at the anodes (Fig. 5.47) whilst the chemical reduction, as performed by four more photons in photosystem I, is achieved in a separate space at the cathode. An alternative approach, not yet contributing significantly to the production of H_2, is a photochemical device effecting artificial photosynthesis. Such a device is illustrated in Fig. 5.49.[56] This is based on colloidal titania (TiO_2) supporting two different surface materials in aqueous solution. The platinum is acting as the cathode in an electrolysis cell and the chemical reduction engine of photosystem I, whilst the RuO_2 is the anode in the electrolysis cell and water-oxidizing cluster on photosystem II. The energy is provided photochemically rather than electrochemically. A sensitizer molecule (shown as S) absorbs light to provide the excited state S^*. The excited state is potentially reducing, like chlorophyll, and

[56] M. Grätzel, 'Artifiical Photosynthesis: Water Cleavage into Hydrogen and Oxygen by Visible Light', *Acc. Chem. Res.*, 1981, **14**, 376–84.

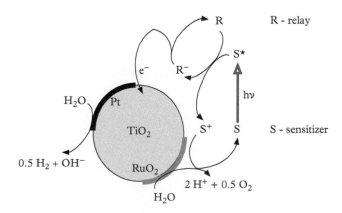

Figure 5.49 *Schematic of a bifunctional redox catalyst for water splitting.*

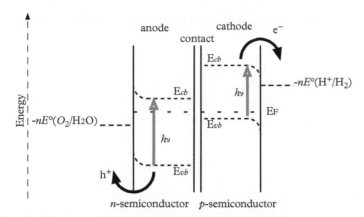

Figure 5.50 *2,2′-bipyridyl (2,2′-bipy).*

Figure 5.51 *Methylviologen (MV^{2+}).*

an electron is transferred to a further additive termed a relay molecule, *R*. The oxidized form of the sensitizer, S^+, effects the oxidation of water on the RuO_2 catalyst. The reduced relay molecule, R^-, transfers its electron to the conduction band of the titania which allows it to migrate to the platinum catalyst to reduce adsorbed water.

The sensitizers investigated included $[Ru(2,2'\text{-bipy})_3]^{2+}$ and its derviatives, where the ligand is a bidentate di-imine (Fig. 5.50). This Ru(II) complex (of a *4d* element) will be low spin and has the electronic configuration $t_{2g}{}^6e_g{}^\circ$. The excited states generated by absorption in the visible or near UV are metal-to-ligand charge transfer (MLCT) in nature and so in effect they can be represented as $[Ru^{III}(2,2'\text{-bipy})_2(2,2'\text{-bipy})^-]^{2+}$. The electron in the ligand π^* orbitals can be transferred to a relay molecule. The one employed is the doubly quaternized 4,4′-bipy ion (Fig. 5.51) labelled MV^{2+} and also known as the herbicide *paraquat*. The radical cation MV^+ so formed is deep blue in colour and can transfer the extra electron to the titania particle. Under the optimum conditions the efficiency (quantum yield) for H_2 production was 6%; the O_2 was retained in solution adsorbed on the TiO_2.

An alternative photo-electrochemical device is to have membrane-separated electrodes as in an electrolysis cell, for example with a Pt cathode and TiO_2 anode, similar to the PEMEC in Fig. 5.47. But the electrical supply is provided by the illumination of the semiconductor anode to effect water oxidation, giving photo-generated current. This has the attractive possibility of removing the need for the sensitizer and relay molecules but will increase the requirements on the materials forming the electrodes.

Research has aimed at improving the quantum efficiency, avoiding the use of platinum group metals, and creating alternatives to linking an electrolysis cell to solar-derived electricity.[57] Colloidal approaches, as in Fig. 5.49, conventional and nanostructured (to provide higher surface areas) electrodes are all under investigation as alternatives. The earth-abundance of an element is seen as an important factor.[58] Photoactive semiconductors are a means of providing oxidation and reduction sites from the electron-hole charge separation created on light absorption. The band-gap range is bounded by that needed by the energy required by the electrolysis of water (> 1.23 eV) and the solar spectrum (< 3.0 eV).[59]

These approaches include a tandem photocathode/photoanode cell (Fig. 5.52). This comprises two semiconductor materials in electrical contact. Absorption of a photon by the *n*-semiconductor photocathode can promote an electron from the valence band to the conduction band thus creating a hole in the valence band. In semiconductor parlance, the cationic hole oxidizes the water at the electrode surface to O_2; in a more chemical parlance, the electrons released by the oxidation of water (Eq. 5.33) refill the conduction band. At the cathode, light absorption at the *p*-semiconductor transfers an electron to the conduction band which then effects the reduction to H_2 at the cathode surface. The electrical contact between the two semiconductors completes the circuit as electrons can flow from the conduction band of the anode to the valence band of the cathode.

[57] F. E. Osterloh, 'Inorganic nanostructures for photoelectrochemical and photocatalytic water splitting', *Chem. Soc. Rev.,* 2013, **42**, 2294–2320.

[58] X. Zou and Y. Zhang, 'Noble metal-free hydrogen evolution catalysts for water splitting', *Chem. Soc. Rev.,* 2015, **44**, 5148–80.

[59] X. Chen, S. Shen, L. Guo, and S. S. Mao, 'Semiconductor-based Photocatalytic Hydrogen Generation', *Chem. Rev.,* 2010, **110**, 6503–70.

Figure 5.52 *Schematic of a tandem photocathode/photoanode cell for water splitting.*

Note that the energy levels within the semiconductor are shown as being curved near the surfaces, which is the effect of the differing potentials across the device. An efficiency of 12% can be achieved using a GaAs photoanode and a GaInP$_2$ photocathode, and this can be increased to 18% with a multijunction shell. These values are approximately half of the theoretical efficiencies of such devices. However, the photoelectrodes undergo corrosion. So although the efficiency is attractive, the stability is not. An approach to this problem has been described as the artificial leaf (Fig. 5.53).[60] Less is demanded of the electrodes than in a tandem cell. As in photosystem II, the light gathering is in a separate region of the device, and can be effected by an amorphous silicon photocell. Additionally, a chemical barrier was incorporated between that and the anode to enhance stability.

Using a light source with the properties of sunlight at AM1.5, the coverage of the solar spectrum could be approximately matched with a series of three semiconductors, akin to the approach in photovoltaics described in section 5.4 (Fig. 5.39), but using a principal direction of illumination along rather than across as shown in Fig. 5.53. The aim was to utilize earth-abundant elements as much as possible. The photocell system was protected by anodic oxidation by an indium tin oxide (ITO) layer (50 nm thick). The top photoactive layer was formed of amorphous silicon, giving an optimum quantum efficiency near 400 nm. The two other layers were two different a–Si–Ge compositions providing maximum quantum efficiencies at ≈ 600 and 750 nm. This was bounded by a ZnO layer. The anode is deposited onto the outside of the ITO film as a thin layer of a material called Co-OES (OES denotes oxygen evolving complex). This is formed from aqueous Co(II) with phosphate and self-generates cubane-type clusters similar to those in the Mn$_4$Ca cluster in PS II. The cathode is on the other side of the sandwich and was a NiMoZn alloy. The overall solar-to-fuel efficiency (SFE) is given by Eq. 5.34, where ϕ(PV) is the quantum efficiency of the photovoltaic cell

[60] D. G. Nocera, 'The Artificial Leaf', *Acc. Chem. Res*, 2012, **45**, 767–76.

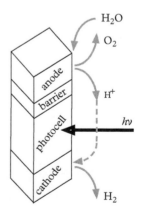

Figure 5.53 *Schematic of a synthetic leaf.*

and ϕ(WS) that of the water-splitting photoreaction. Overall, the *SFE* was up to 4.7% emanating from a ϕ(WS) of $\approx$ 60% and ϕ(PV) of *c*.7.7%.

$$SFE(\%) = \phi(PV).\phi(WS). \tag{5.34}$$

Development projects are in progress to enhance the efficiency, output, and sustainability of photo-electrochemical water splitting, for example by optimizing a tandem cell with haematite (Fe_2O_3) photoelectrodes.[61] Another approach has been to utilize a Fresnel focusing lens to concentrate sunlight onto a multijunction solar cell, to achieve relatively hugh current density. The solar cell, consisting of III–IV semiconductors sited at the anode of a PEMEC (Fig. 5.47).[62] Outdoor tests demonstrated *SFE* of *c*.20%. With a triple-junction III–V solar cell (InGaP/GaAs/GaInNAsSb) driving two different PEMEC systems, *SFE* values of *c*.30% have been demonstrated.[63] It can be expected that, either by integral devices or by coupled PV-electrolyser systems, new options for the manufacture of green H_2 are to be envisaged in future.

5.6.4   Fuel Cells

A fuel cell is a device in which the energy released by combustion is converted into electrical energy. It is the reverse of an electrolysis process (Section 5.6.2). The options for the combustion are legion, but the emphasis has been on oxidation of potentially renewable fuels such as methanol, ethanol, and hydrogen. One interest in deriving energy from H_2 is to avoid the CO_x and particulate by-products; another is that by electrical conversion rather than direct combustion NO_x formation is avoided. These last two benefits arise whatever the source of H_2.

Transport accounts for *c*.18% of annual greenhouse gas emissions worldwide (a total of *c*.50 Gt) with nearly 50% of that being due to the use of cars and motorbikes.[64] In contrast, the consumption of H_2 for transport in 2018 is estimated to be 10,000 t (Section 5.6.2, Fig. 5.48). It is estimated that the worldwide fleet of vehicles in 2018 was 5,200 cars (mostly in Japan and the USA), 115 buses (mostly in China and the USA), and over 11,600 forklift trucks (mostly in the USA).

The type of fuel cell employed in vehicles is the polymer electrolyte membrane fuel cell (PEMFC) (Fig. 5.54) which is related to the PEMEC in Fig. 5.47. The two half-cell equations are as in Eqs 5.35 and 5.36.[65] The membrane, like that in the PEMEC, is generally a sulphonated PTFE which has high proton conductivity, chemical stability, and tensile strength. The oxygen is provided by an air supply and the H_2 gas is of high purity. Typically a single cell produces electricity at 0.6–0.7 V. These are stacked to generate the voltage needed, that being $\approx$ 650 V in a Toyota system.

$$Anode: H_{2(g)} \rightarrow 2H^+_{(aq)} = 2e^-, \tag{5.35}$$

$$Cathode: \frac{1}{2}O_{2(g)} + 2H^+_{(aq)} + 2e^- \rightarrow H_2O_{(l)}. \tag{5.36}$$

[61] A. Vilanova, T. Lopes, C. Spenke, M. Wullenkord, and A. Mendes, 'Optimized photoelectrochemical tandem cell for solar water splitting', *Energy Storage Mater.*, 2018, **13**, 175–88.

[62] A. Fallisch, L. Schellhase, J. Fresko, M. Zedda, J. Ohlmann, M. Steiner, A. Bösch, L. Zielke, S. Tiele, F. Dimroth, and T. Smolinka, 'Hydrogen concentrator demonstrator module with 19.8% solar-to-hydrogen conversion efficiency according to the higher heating value', *Int. J. Hydrog. Energy*, 2017, **42**, 26804–15.

[63] J. Jia, L. C. Seitz, J. D. Benck, Y. Huo, Y. Chen, J. W. D. Ng, T. Bilir, J. S. Harris, and T. F. Jaramillo, 'Solar water splitting by photovoltaic-electrolysis with a solar-to-hydrogen efficiency of over 30%', *Nat. Commun.*, 2016, **7**, 13237.

[64] I. Staffell, D. Scamman, A. V. Abad, P. Balcombe, P. E. Dodds, P. Ekins, N. Shah, and K. R. Ward, 'The role of hydrogen and fuel cells in the global energy system', *Energy Environ. Sci.*, 2019, **12**, 463–491.

[65] T. Wilberforce, Z. El-Hassan, F. N. Khatib, A. Al Makky, A. Baroutaji, J. G. Garton, and A. G. Olabi, 'Developments of electric cars and fuel cell hydrogen electric cars', *Int. J. Hydrogen Energy*, 2017, **42**, 25695–734.

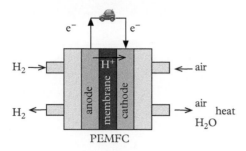

Figure 5.54 *Schematic of a polymer electrolyte membrane fuel cell (PEMFC).*

As is evident from Fig. 5.54, some energy emanates as heat rather than electrical energy. Low-temperature proton-exchange membrane fuel cells (LT-PEMFCs) operate at 60–80 °C with an electrical efficiency in the range of 40–60%.[66] There are significant energy barriers to overcome at the electrodes, which are reduced by catalysts. The material of choice for both of the electrodes is polycrystalline platinum on carbon, a conducting support. The more difficult process to catalyse is the oxygen reduction reaction (ORR). The basis for choosing platinum is illustrated with a modification of the volcano plot in Fig. 5.3.[67] As depicted in Fig. 5.55, platinum displays the highest activity of all available metals. Those with higher group numbers appear to have a lower binding energy for oxygen than is optimum, although Ag is close. The earlier members of the transition series bind oxygen too strongly and will in effect form a surface oxide barrier. Palladium alone approaches the performance of platinum.

The ORR has been the subject of much study and disagreement, but the importance of the species HOO has emerged.[68] The scheme in Fig. 5.56 summarizes the steps, utilizing adsorption modes indicated by density functional theory.[69] Adsorption of O_2 is proposed to occur in a bridging fashion involving three Pt atoms, and this reacts with net H (i. e. H^+ from solution and e^- from the electrode) to form a stable hydroperoxo species. A second addition of 'H' affords bound H_2O_2; with some metals this stage can result in desorption of hydrogen peroxide. This can then cleave to form two hydroxyl groups and two further additions of 'H' effect the elimination of two molecules of water and regenerate the initial site.[67]

The active sites are very susceptible to poisoning by impurities in the gas streams. Ammonia is inimical to these catalysts and low limits for CO (< 10 ppm) and H_2S (< 0.1 ppm) are required. These impurities are associated with H_2 production from fossil fuels rather than electrochemical processes. Incorporation of ruthenium into the catalyst raises the threshold for CO poisoning to 30 ppm. From the volcano plots (Figs 5.3, 5.55), it would be anticipated that the chemisorption energy of CO would be higher for Ru than Pt. This can be ascribed to the trends in electronegativity across the *d*-block (Sections 2.5 and 3.3). The energy of the valence *d*-orbitals for the less electronegative group-8 element will

[66] T. Wilberforce, A. Alaswad, A. Palumbo, M. Dassisti, and A. G. Olabi, 'Advances in stationary and portable fuel cell applications', *Int. J. Hydrogen Energy*, 2016, **41**, 16509–22.

[67] Z. W. Seh, J. Kibsgaard, C. F. Dickens, I. Chorkendorff, J. K. Nørskov, and T. F. Jaramillo, 'Combining theory and experiment in electrocatalysis: Insights into materials design', *Science*, 2017, **355**, eaad4998.

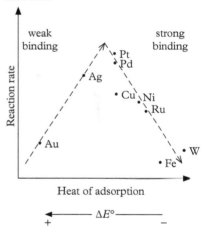

Figure 5.55 *Volcano plot for the oxygen reduction reaction(ORR).*

[68] A. Gómez-Marín, J. Feliu, and T. Edson, 'Reaction Mechanism for Oxygen Reduction on Platinum: Existence of a Fast Initial Chemical Step and a Soluble Species Different from H_2O_2', *ACS Catalysis*, 2018, **8**, 7931–43.

[69] H. Sun, J. Li, S. Almheiri, and J. Xiao, 'Oxygen reduction on a Pt(111) catalyst in a HT-PEM fuel cell by density functional theory', *AIP Advances*, 2017, **7**, 085020.

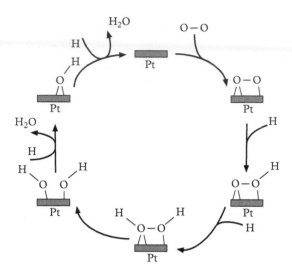

Figure 5.56 *Catalytic cycle for the platinum-catalysed oxygen-reduction reaction (ORR).*

be higher and thus closer to that of the CO π^* orbitals giving enhanced back-bonding. However, the enhanced threshold for CO tolerance cannot be due to that alone as the Ru sites will be quickly blocked whilst on-stream.

Platinum and ruthenium adopt different close-packing structures (cubic and hexagonal, respectively), and so can readily be distinguished for particles large enough to provide X-ray diffraction patterns. A PtRu/C electrode catalyst displayed a broadened *fcc* pattern as compared to Pt/C shifted to longer scattering angles.[70] The shift in the diffraction peaks indicates a decrease in the unit cell dimension, *a*, as expected by the rearranged Bragg equation Eq. 5.37, where λ is the X-ray wavelength, *d* the interlayer distance between the scattering planes, and 2θ the scattering angle. This is consistent with the majority of the metal particles being an alloy with the smaller ruthenium atom located within a platinum lattice.

$$d = \frac{n\lambda}{2\sin\theta},$$
(5.37)

The line width is indicative of the length, L (Å), of an array of ordered crystallographic planes perpendicular to the scattering plane, according to the Debye–Scherrer equation Eq. 5.38. The linewidth, β, is corrected for any instrumentation line width (typically measured by the linewidth of a powdered sample of NaCl) and is expressed in radians. The constant, K, is a shape factor, and typically is close to unity (values of 0.9 are usually employed for layered clay materials). This mosaic length L is often described as the mean particle size, but in reality the particle size can be larger by virtue of disordered components or conglomerates of microcrystallites. Nevertheless, in this case the line broadening indicates a reduction in the size of the ordered microcrystallites.

[70] P. P. Lopes and E. A. Ticianelli, 'The CO tolerance pathways on the Pt-Ru electrocatalytic system', *J. Electroanal. Chem.,* 2010, **644**, 110–16.

$$L = \frac{K\lambda}{\beta cos\theta} \qquad (5.38)$$

However, high-temperature (HT) treated catalysts show evidence of particle growth and phase separation. Thus, compared to the Pt/C catalyst, there are broadly two types of additional components that can be present: PtRu/C and Ru/C. The former displays evidence of there being an electronic effect on the platinum, which enhances CO tolerance, and the latter will provide an independent mechanism and can lead to CO methanation at the anode. Also, the high-temperature treatment will reduce the surface/bulk atomic ratio of platinum.

All of these effects provide drivers and opportunities for catalyst development. As described in Section 3.3.5, catalysts in all areas form the major demand (102 t) for platinum (Table 3.17). It is estimated that the platinum required for the stack of fuels cells required for a medium-sized vehicle is 30 g, about ten times more than in a diesel autocatalyst.[64] Production of platinum fell from 2017 to 2018 to *c*.160 t. The annual production of ruthenium is much less than that of platinum and so the price is very sensitive to changes in demand; the price rose 212% from 2017 to 2018. Accordingly, means of reducing the platinum content by beneficial alloying, particle size control, core–shell structures (with a more sustainable element internal to an active shell), and alternative materials are seen as essential for sustainable development.[67,71]

There are other options available to PEMFCs which may be appropriate for other applications, including phosphoric acid fuel cells (PAFCs), molten carbonate fuel cells (MCFCs), and solid oxide fuel cells (SOFCs). The last of these is like the solid oxide electrolysis cells (Fig. 5.47) used in reverse (Section 5.6.2). These cells can be poisoned by sulphur (< 2 ppm) and NH_3 ($< 0.5\%$) but can use CO as a fuel. The PAFC is diagramatically similar to the PEMFC (Fig. 5.54), but the conducting medium is liquid phosphoric acid, H_3PO_4, within a silicon carbide matrix at an operating temperature in the 150–200 °C range. These devices also employ Pt/C as the electrode catalysts. However, they are more tolerant of sulphur (< 50 ppm) and CO ($< 1\%$). A MCFC has different electrode reactions (Eqs 5.39, 5.40):

$$Anode: H_{2(g)} + CO_3^{2-} \rightarrow H_2O + CO_2 + 2e^-, \qquad (5.39)$$

$$Cathode: \frac{1}{2}O_{2(g)} + CO_2 + 2e^- \rightarrow CO_3^{2-}. \qquad (5.40)$$

The basis of this type of cell is shown in Fig. 5.57. In this case, again O_2 is reduced at the cathode. With CO_2 present, carbonate is created and it is CO_3^{2-} which is the charge carrier through the electrolyte. Water is now formed at the anode from the oxidation of H_2 to protons with the carbonate providing the oxide. The evolved CO_2 can be recycled to the cathodic side of the cell. The electrolyte is a mixture of group-1 carbonates (Li, Na, K) and is molten at the operating temperatures (600–700 °C). It is supported by a matrix formed by a

[71] I. E. L. Stephens, J. Rossmeisl, and I. Chorkendorff, 'Toward sustainable fuel cells', *Science*, 2016, **354**, 1378–79.

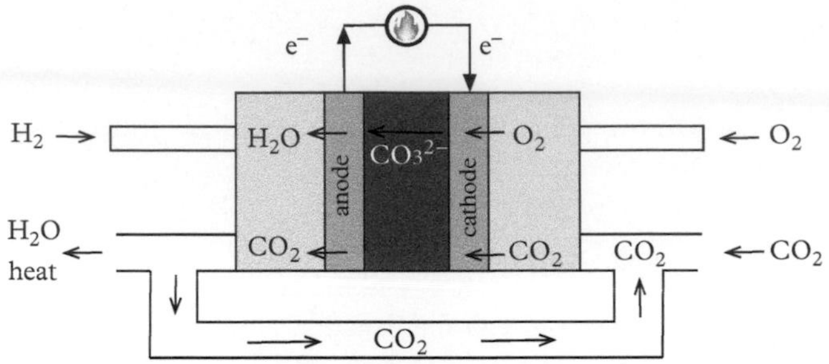

Figure 5.57 *Schematic of a molten carbonate fuel cell (MCFC).*

ceramic such as $LiAlO_2$. Nickel is the active element utilized in the electrodes. The anode is generally formed from a nickel alloy with some aluminium or chromium; these elements form $LiMO_2$ which prevents sintering of the nickel particles. The cathode is under oxidizing conditions and the nickel is as NiO, generally partially lithiated. There is a tendency for the NiO to dissolve in the molten carbonate and so stabilizers like MgO can be incorporated. They are more tolerant of sulphur (< 10ppm) and ammonia (< 1%) than PEMFC and SOFC devices. Like SOFCs they can use CO as a fuel.

The overall efficiencies for converting fuel gas energy to power and to heat vary with type.[72] As shown in Table 5.6, presently the electrical efficiency of PEMCs is lower than that of the other types. It also demonstrates that in a combined heat and power (CHP) mode of use, efficiencies of $\approx$ 90% are available.

[72] I. Stafell, 'Zero carbon infinite COP heat from fuel cell CHP', *Appl. Energy*, 2015, **147**, 373–85.

Table 5.6 *Fuel cell efficiencies (%).*

Type	Electrical	Thermal
PEMFC	35–39	55
SOFC	45–60	30–45
PAFC	42	48
MCFC	47	43

[73] M. Winter and R. J. Brodd, 'What are Batteries, Fuel Cells, and Supercapacitors?', *Chem. Rev.*, 2004, **104**, 4245–70.

5.7 Batteries and Supercapacitors

These are two related classes of electrochemical energy storage devices (ESDs).[73] Batteries are the more familiar items, consisting of one or more electrochemical cells. Two important classes of battery are primary and secondary. A primary battery is delivered in a charged state and delivers the energy stored in the electrodes once only before recycling (or not). Secondary batteries can be recharged and thus must be able to cycle reversibly through charge–discharge processes.

The electrode reactions of the most-used primary batteries, the alkaline cell, are as in Eqs 5.41 and 5.42. The reduction potentials of these two processes are -1.26 V and -0.15 V, respectively, and the overall standard potential for the cell, given by Eq. 5.43, will be -1.41 V. The cell potential will differ from this due to the non-standard concentrations, and these cells are nominally assigned a voltage of 1.5 V.

$$Anode: Zn(s) + 2OH^-_{aq} \rightarrow ZnO_{(s)} + H_2O_{(l)} + 2e^- : E^\circ - 1.26\,V, \quad (5.41)$$

$$Cathode: 2MnO_{2(s)} + H_2O_{(l)} + 2e^- \rightarrow Mn_2O_{3(s)} + 2OH^-_{(aq)} : E^\circ - 0.15\,V. \quad (5.42)$$

$$E^{\circ}_{cell} = E^{\circ}_{cathode} - E^{\circ}_{anode}. \tag{5.43}$$

The stainless steel can of the battery acts as the cathodic connection and is in contact with a paste of MnO_2 with carbon. The electrode reaction here is the reduction of Mn(IV) to Mn(III). Zinc powder, which provides the majority of the chemical energy, is dispersed in a gel containing the electrolyte, aqueous KOH. This cell cannot be recharged readily and for more than a few cycles hence it is a single-use device. As can be seen in Table 5.7, both manganese and zinc are produced in large scale and so this is a viable option for the storage of electrical power; however it is a more effective use of materials to be able to recharge over many cycles.

A long-established rechargeable battery is the lead acid battery, which has an anode of lead and a cathode with PbO_2. This was the first rechargeable battery and was demonstrated by Gaston Planté in 1859. The electrode reactions are given in Eqs 5.44 and 5.45. Since the standard reduction potentials for the two half-cells are 1.685 and -0.356 V, respectively, the cell voltage is ≈ 2.04 V, which, by Eq. 2.82, equates to 394 kJ mol^{-1}. The overall reaction (Eq. 5.46) is a conproportionation reaction from Pb(o) and Pb(IV) to 2 Pb(II). This cell is shown in schematic form in Fig. 5.58. In operation metallic lead is oxidized to Pb(II), releasing electrons and creating the negative plate. On interaction with sulphate it precipitates on the electrode surface as $PbSO_4$. Electrons entering the cell at the second electrode (the positive plate) reduce PbO_2 also to Pb(II), which likewise precipitates as $PbSO_4$ at the electrode surface with water also being formed. In the recharging process the flow of material is reversed. Thus $PbSO_4$ is reduced at the cathode reforming lead, and oxidized at the anode reforming PbO_2.

$$Cathode: PbO_{2(s)} + HSO^-_{4(aq)} + 3H^+_{(aq)} + 2e^- \rightarrow PbSO_{4(s)} + 2H_2O_{(l)}, \tag{5.44}$$

$$Anode: Pb(s) + HSO^-_{4(aq)} \rightarrow PbSO_{4(s)} + H^+_{(aq)} + 2e^-, \tag{5.45}$$

$$Pb_{(s)} + PbO_{2(s)} + 2H_2SO_{4(aq)} \rightarrow 2PbSO_{4(s)} + 2H_2O_{(l)}. \tag{5.46}$$

Typically the majority of the mass of the lead acid battery is lead based (25% Pb and 35% PbO$_x$), with the electrolyte the bulk of the remainder ($\approx 26\%$).[74] There is a variety of cell constructions. A common form of automobile battery is a pack of six cells in series to deliver a 12 V output. These can use aqueous acid electrolytes (H_2SO_4/H_2O as 35:65) with electrodes comprised of lead alloys. The alloys are based upon PbSb or PbCa alloys which are less soft than lead itself. Other additives are incorporated to provide long-term stability (Se, Cd, Sn, As). Spongy lead and a PbO_2 paste can be inset into lead alloy frames to form the electrodes. The paste is a mixture of carbon black as conductor, $BaSO_4$ as seed for the formation of $PbSO_4$ crystals, and sulphonated lignin as a proton-conducting binder. In so-called sealed devices, the electrolyte is supported by a

Table 5.7 *Mine production (t) in 2018 of elements involved in battery manufacture.*

Element	Production(t)
zinc	1.3×10^7
manganese	1.8×10^7
lithium	8.5×10^4
lead	4.4×10^6
nickel	2.3×10^7
cadmium	2.6×10^4
cobalt	1.4×10^5
lanthanides	1.7×10^5

[74] X. Zhang, L. Li, E. Fan, Q. Xue, Y. Bian, F. Wu, and R. Chen, 'Toward sustainable and systematic recycling of spent rechargeable batteries', *Chem. Soc. Rev*, 2018, **47**, 7239–7302.

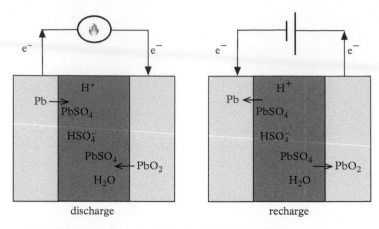

Figure 5.58 *Schematic of a cell of a lead acid battery.*

gel, typically silica, which also converts the electrolyte into a paste. The charge–discharge efficiency of automobile batteries may be $\approx 50\%$. Deep-cycle (to 95%) versions are also available, but this is achieved at a cost of a reduction in peak current. Recycling of lead acid batteries is well established, and required to avoid environmental pollution; no level of lead is considered to be safe by the World Health Organization (WHO).[75]

One key parameter for a battery system is the energy density in terms of the mass of a device and also its volume (Fig. 5.59). Given the density of lead it is unsurprising that the mass-related energy density is relatively low. Also the volume-related energy density is low, for example, as compared to the alkaline battery. Many other cell designs have been developed with the aim of improving the energy density of rechargeable batteries. Some progress was achieved with the nickel-cadmium (NiCd) invented by Walter Junger in 1899. In the 1980s these were essentially the only type of small secondary cells, and were the major use of cadmium. In the early 1990s nickel–metal hydride (NiMH) and, shortly after, lithium-ion devices, each of which has higher energy density, resulted in a reduction in the production of NiCd cells from 1995. The dominant driver for the phasing out of NiCd batteries is the toxicity of cadmium. Cadmium compounds are soluble and mobile and so contamination can spread easily, rendering landfill an unacceptable option for spent batteries. The primary affect is kidney damage, with a string of secondary affects. An EU Directive on batteries has restricted cadmium levels to the low value of 0.002%.

The nickel–metal hydride battery is a much more modern invention (1967), and has the advantage over lead acid batteries of being based upon lower Z and hence lighter elements. The half-cell reactions are shown in Eqs 5.47 and 5.48. Electrons are released by the reduction of the hydroxyl anion by a metal hydride and at the positive plate the Ni(III) material β-NiO(OH) is reduced to β-Ni^{II}(OH)$_2$, which has the *brucite*, Mg(OH)$_2$, structure (Fig. 5.60).[76] In that the oxide ions form a hexagonal-type ABAB stacking lattice. The Ni(II) sites

[75] *Lead poisoning and health*, 2018, https://www.who.int/news-room/fact-sheets/detail/lead-poisoning-and-health.

[76] M. Casa-Cabanas, M. D. Radin, J. Kim., C. P. Grey, A. Van den Ven, and M. R. Palacín, 'The nickel battery positive electrode revisited: stability and structure of the β-NiO(OH) phase', *J. Mater. Chem. A*, 2018, **6**, 19256–65.

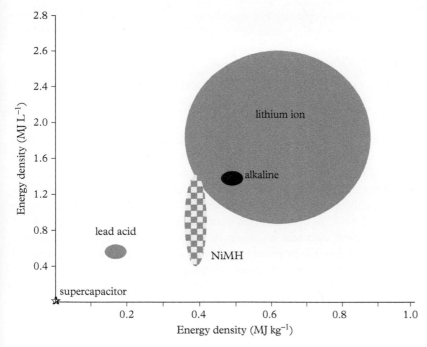

Figure 5.59 *Energy density by mass and volume for some batteries and supercapacitors.*

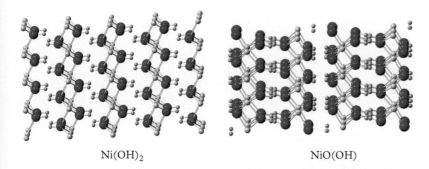

Figure 5.60 *Structures of Ni(OH)₂ and NiO(OH); red = O, grey = Ni, pink = H.*

are edge-sharing octahedra bonded to triply bridging hydroxyl groups. There are alternating nickel-centred and hydroxyl protons through the structure, with the latter providing a hydrogen bonding opportunity for migrating hydroxyl anions and water. A sequence of structural changes takes place up to the formation of β-NiO(OH). In this oxidized form the oxide ion arrangement switches to a *ccp* type of ABC packing. The nickel centres are now high-spin d^7 in configuration and thus may undergo a Jahn–Teller distortion (Section 3.3). This affords a series of Ni–O distances from 1.91 to 2.12 Å. There are two types of hydrogen sites

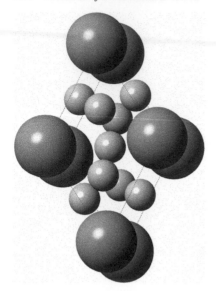

Figure 5.61 *Unit cell of LaNi₅; large = La, small = Ni.*

which are partially occupied. The overall chemical reaction is given in Eq. 5.49. The standard electrode potential for the cathode reaction is favourable at 0.49 V.

$$Cathode: NiO(OH)_{(s)} + H_2O_{(l)} + e^- \rightarrow Ni(OH)_{2(s)} + OH^-_{(aq)}, \quad (5.47)$$

$$Anode: OH^-_{(aq)} + MH_{(s)} \rightarrow M_{(s)} + H_2O_{(l)} + e^-, \quad (5.48)$$

$$MH_{(s)} + NiO(OH)_{(s)} \rightarrow M_{(s)} + Ni(OH)_{2(s)}. \quad (5.49)$$

The anionic terminal contains the metal hydride (MH) and the one utilized will be one of a myriad of materials, many of which are based upon the intermetallic compound LaNi₅ (Fig. 5.61). This is quite distinct from close-packed structures. The smaller nickel atoms are in two different locations, one with five nickels as nearest neighbours and the other with six (Ni 2.48, 2.52 Å). The shortest La–Ni distance is 2.91 Å. In hydrogen-rich materials very significant expansion of the metal interatomic distances are observed (≈ 0.1 Å) consistent with interstitial hydrides being formed. In practice the lanthanum is replaced by an extracted mixture (mischmetal) of lanthanides which requires less refining (typically Ce 50–55%), La (18–28%), Nd (12–18%), Pr(4–6%). The metal hydride can be considered to be in equilibrium with H₂ and the LaNi₅ phase (Eq. 5.50). In that circumstance the electrode reaction can be rewritten as Eq. 5.51, releasing an electron. The standard reduction potential for this half cell is -0.828. Thus, from Eq. 5.43, $E°_{cell}$ should be 0.49 + 0.83 = 1.32 V. The nominal voltage of a NiMH cell is given as 1.2 V.

$$x\frac{1}{2}H_2 + LaNi_{5(s)} = LaH_xNi_{5(s)} \quad (5.50)$$

$$OH^-_{(aq)} + \frac{1}{2}H_2 \rightarrow H_2O_{(l)} + e^-. \quad (5.51)$$

A schematic of the structure of a NiMH cell is presented in Fig. 5.62. This can be viewed either as a cross-section through a flat cell, or through a cylindrical

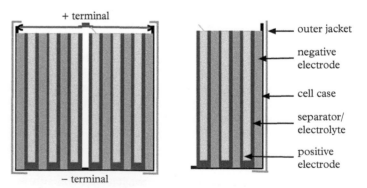

Figure 5.62 *Schematic of a NiMH or a lithium-ion battery cell and its components.*

cell which is shaped like a Swiss roll. The metal hydride negative electrode is in contact with the cell casing which provides the negative terminal. Next to this is the electrolyte-carrying separator housing the positive electrode. A connector links this to the stub positive terminal. That also includes a pressure-release valve. The electrodes comprise layers of powdered active materials. For the negative terminal the metal hydride typically absorbs up to 1 or 2% hydrogen stoichiometrically, and the design makes the quantity of the positive electrode material, NiO(OH) set in nickel powder, the limiter on the charge of the cell. The electrolyte is aqueous KOH electrolyte with a polyalkene or polyamide material acting as the separator. As is evident from Fig. 5.59, NiMH cells can have higher energy density than lead acid batteries, and indeed more than the NiCd devices they have replaced. Charge/discharge efficiencies are higher than those of most lead acid batteries, being in the range of 66–92%; however, self-discharge rates are also higher.

NiMH batteries are in widespread use for secondary, rechargeable energy storage. This encompasses household batteries and electric vehicles.[77] In terms of mass, the major component of a rechargeable battery in the steel case (44%) with nickel and nickel hydroxide the second-largest component (24–30%).[74] Nickel is produced in substantial quantities (Table 5.7) and does not present such high toxicity as cadmium or lead. The metal hydride of the negative electrode is present to a component of 8–13%, and it is estimated that lanthanides comprise $\approx$ 7% of NiHM batteries; this equates to about 1 g in an AAA battery, 60 g in a power tool and 2 kg in a hybrid electric vehicle.[78] Consumption of rare-earth oxides in 2015 for use in batteries was estimated to be $\approx$ 8%; the demand for NiMH batteries is expected to grow.[79]

The remaining battery type in Fig. 5.59 is the lithium-ion battery. The structure of a cell can also be envisaged from Fig. 5.62. Here lithium ions are within the electrolyte and separator, generally as $LiPF_6$ or similar salts. The chemically inert, large anion allows the salt to be soluble in polar organic solvents such as carbonate esters and ethers. In other structures the electrolyte may be a polar polymer such as polyethylene oxide (PEO), polyacrylonitrile, or polymethylmethacrylate (PMMA). A separator consisting of a porous polymer (e.g. PE or PP) film is also utilized. The positive plate (often copper) bears, generally, a high-oxidation-state metal oxide, such as one derived from $Co^{IV}O_2$. In a polymer battery, these materials are mixed with a polymer binder.

The structure of CoO_2 could not be refined but it is considered to consist of two close-packed layers of oxides with cobalt(IV) ions occupying octahedral holes between them (Fig. 5.63), with a uniform Co–O bondlength of 1.836 Å.[80] Each Co(IV) is $3d^5$, and a low spin state would have one unpaired electron giving a magnetic moment of 1.73 μ_B; the observed value, however, is 0.18 μ_B and it may be that the oxide displays metallic properties. Each of these sandwich layers is separated by another without cobalt ions. The extreme of the electrode reaction (Eq. 5.52) results in the reduction of cobalt to low-spin Co(III), which is evidenced partly by an increase in the Co–O bondlength to 1.916 Å. The lithium ions intercalate into the layer vacant in CoO_2 adopting octahedral sites. In

[77] Including my electric bike!

[78] M. Petranikova, I. Herdzik-Koniecko, B.-M. Steenari, and C. Ekberg, 'Hydrometallurgical processes for recovery of valuable and critical metals from spent car NiMH batteries optimized in a pilot plant scale', *Hydrometallurgy*, 2017, **171**, 128–141.

[79] B. Zhou, Z. Li, and C. Chen, 'Global Potential of Rare Earth Resources and Rare Earth Demand for Clean Technologies', *Minerals*, 2017, **7**, 203.

[80] T.Motohashi, Y. Katsumata, T. Ono, R. Kanno, M. Karppinen, and Y. Yamauchi, 'Synthesis and Properties of Co_2, the x=0 End Member of the Li_xCoO_2 and Na_xCoO_2 Systems', *Chem. Mater.*, 2007, **16**, 5063–66.

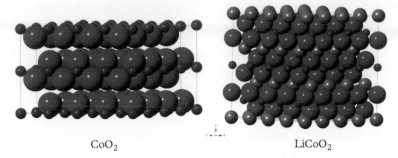

CoO$_2$ LiCoO$_2$

Figure 5.63 *Structure of CoO$_2$ and LiCoO$_2$; red = O, blue = Co, grey = Li.*

[81] J. B. Goodenough, 'Evolution of Strategies for Modern Rechargeable Batteries', *Acc. Chem. Res.*, 2013, **46**, 1053–61.

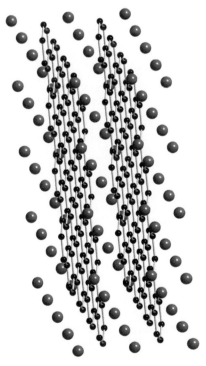

Figure 5.64 *Structure of LiC$_6$; black = C, grey = Li.*

[82] *The Batteries Report 2018*, RECHARGE, Brussels, Belgium, http://www.rechargebatteries.org.

practice, starting from LiCoO$_2$, charging can only be carried out reversibly from Li$_{1-x}$CoO$_2$ up to $x = 0.55$.[81] The standard electrode potential of the cathode is in the region of 1.0 V.

The negatively charged electrode is based upon graphite (Eq. 5.53). In the reduced form, charging has caused the migration of lithium ions from the cathode and this lithium is intercalated between the graphite layers shown in Fig 3.4. An example of such a material is LiC$_6$ (Fig. 5.64). The overall cell equation is given in Eq. 5.54, involving the transfer of Li$^+$ from an intercalation site from a cobalt oxide lattice to graphite. The small size of the cation makes the structural changes relatively small so helping with the discharge–recharge cycle and also providing stabilization by virtue of higher lattice energy. The standard reduction potential, $E°$, of the anode is -2.84 V, slightly less negative than that of the Li$^+$/Li couple (-3.04 V). So overall $E°_{cell}$ is $\approx$ 3.8 V.

$$Cathode: CoO_{2(s)} + Li^+_{(solv)} + e^- \rightarrow LiCoO_{2(s)}, \tag{5.52}$$

$$Anode: Li(C)_{(s)} \rightarrow Li^+_{(solv)} + C_{(s)} + e^-, \tag{5.53}$$

$$CoO_{2(s)} + Li(C)_{(s)} \rightarrow LiCoO_{2(s)} + C_{(s)}. \tag{5.54}$$

The relatively high energy density of lithium-ion batteries, and durability and low self-discharge, have been major factors in the increase in their production and adoption. This has extended out from electronic devices to transport and industrial applications, with sales expected to rise from *c.*60 MWh in 2015 to *c.*140 MWh in 2020.[82] The overall composition of a lithium-ion battery, like a NiMH one, has the outer steel casing as the major component in terms of mass (40%), with, in order, the cathode (15–27%), anode (10–18%), and electrolyte (10–16%) comprising most of the active materials.[74] About 2 wt% of a lithium-ion battery is lithium. The lithium content of the battery of an electric vehicle contains 6 to10 kg, with that of a hybrid vehicle about 10% of that.

The key elements also include the cathode components, which include cobalt, the least abundant *3d* element in the upper continental crust of the Earth. A plot

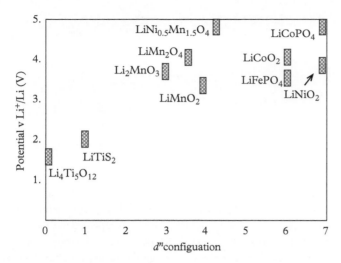

Figure 5.65 *Potentials of some lithium-ion battery electrode materials.*

showing the types of materials developed for lithium-ion batteries is shown in Fig. 5.65. These include layered materials like $LiCoO_2$, but other structure types, like *spinels* such as $LiMn_2O_4$ and *olivines* such as $LiFePO_4$.[83] Higher potentials are generally derived from materials containing the more electronegative elements of the higher group numbers for which oxidation requires higher energy. There are also structural dependencies, with the layered materials tending to have lower potentials. This indicates that, by avoiding this type of structure, other more abundant but less electronegative elements such as manganese and iron may be used as alternatives to cobalt and still maintain a high energy density. Indeed, $LiMn_2O_4$ and $LiFePO_4$ have been commercialized.

A related type of device is a supercapacitor. Unlike a battery, in which the energy is stored by chemical change, the supercapacitor stores electrical energy. The energy density is based on the conversion of the electrical units into energy, E, by Eq. 5.55 (C capacitance in F,[84] and V the voltage in V). As indicated in Fig. 5.59 they have a very low energy density, but they have the property of being able to discharge at high rates (high power, P), as given in Eq. 5.56, where R is the equivalent series resistance (in Ω).[85]

$$E = \frac{1}{2}CV^2 \ (Wh) \qquad (5.55)$$

$$P = \frac{V^2}{4R} \ (W). \qquad (5.56)$$

A conventional capacitor (Fig. 5.66, left) consists of a sandwich of two electrodes and a thin insulating material (a dielectric).[86] At a negatively charged terminal, positively charged volumes within the insulator will be attracted and form a layer close to the electrode; similarly, negatively charged localities will be

[83] N. Nitta, F. Wu, J. T. Lee, and G. Yushin, 'Li-ion battery materials: present and future', *Mater. Today*, 2015,**18**, 252–264.

[84] A Farad (F) is a capacitance in which 1 coulomb is stored with a potential difference of 1 volt.

[85] Y. Ma, H. Chang, M. Zhang, and Y. Chen, 'Graphene-Based Materials for Lithium-Ion Hybrid Supercapacitors', *Adv. Mater.*, 2015, **27**, 5296–5308.

[86] A. Noori, M. F. El-Kady, M. S. Rahmanifar, R. B. Kaner, and M. F. Mousavi, 'Towards establishing standard performance metrics for batteries, supercapacitors and beyond', *Chem. Soc., Rev.*, 2019, **48**, 1272–1341.

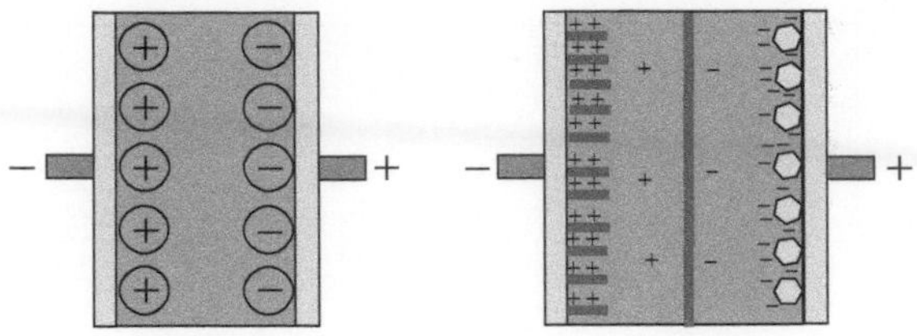

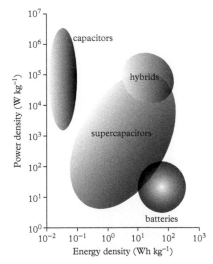

Figure 5.66 *Comparison of a conventional and a lithium-ion hybrid supercapacitor (LIHS).*

attracted to a positively charged region. Typically, conventional capacitors have capacitances of the order of μF with potential differences of tens or hundreds of volts. In a supercapacitor the insulating dielectric is replaced with a material with a high internal surface area to house a much larger array of charges at each electrode. In an electrochemical device there is an intrinsic capacitance. An electrolyte solution is polarized by the electrode charges so that, for example, positive ions will concentrate near the negative electrode and create partitioning of the solution. This surface layer near the electrode is called the Helmholtz double layer and this effect forms the basis of an electric double-layer capacitor. If activated carbon (AC) is used for these electrodes, with $N(CH_3)_4BF_4$ as electrolyte in acetonitrile or propylene carbonate as the electrolyte medium, then the maximum voltage is 2.7 V. A development from this is the lithium-ion hybrid supercapacitor (LIHS), also called a lithium-ion capacitor (LIC). This is shown schematically in Fig. 5.66 (right) in a charged state. At the negatively charged plate lithium ions are largely associated with graphite as in the anode in the lithium-ion battery. At the positively charged plate, on the other side of an ion-permeable membrane separator, the anions, such as BF_4^- or PF_6^-, are collected on a high-area activated carbon (AC) or graphene-derived electrode. The voltage is limited by the redox boundaries of the solvent, and with ionic liquids affording a 4 V range and organic electrolytes 2.5–3.0 V. As shown in Fig. 5.67, these devices can deliver higher energy and power density than supercapacitors. Although presently they do not provide as high an energy storage density as batteries, they are complementary in nature since they can provide orders of magnitude higher power. It can be envisaged that there will be considerable development with electrochemical devices of these kinds.

Figure 5.67 *Power and energy density of devices.*

5.8 Questions

1. The iridium-containing salt $(Ph_4As)[IrI_2(CO)_2]$ **A** can be synthesized from $IrCl_3.xH_2O$, NaI, and aqueous 2-methyoxyethanol.
    (i) What are the coordination number, oxidation state, and d^n configuration of the iridium? Propose a structure for the anion in **A**. How many valence electrons are in the coordination sphere of the iridium?

(ii) As described in section 5.1.2, ethanoic acid is manufactured by carbonylation of methanol. Compound **A** will catalyse this reaction, in the presence of CH_3I. This process is poisoned by $(NBu_4)I$ but promoted by $[RuI_2(CO)_3]_2$. Propose a catalytic cycle for the iridium-catalysed component of methanol carbonylation.

2. (i) Using the Anderson–Schulz–Flory distribution of the weight fraction, W_n, of a hydrocarbon of n carbon atoms, calculate the fraction of mass in the gasoline fraction ($n = 7$ to 9) that would be created by an iron-based Fischer–Tropsch catalyst with an α value 0.7.

$$W_n = n(1 - \alpha)^2\alpha^{(n-1)}. \qquad (5.57)$$

(ii) What would be the effect of a potassium co-promoter?

(iii) What hydrocarbon products would be favoured with that promoter?

3. In a study to control the blending of HDPE and LLDPE by polymerization of ethene alone, a mixture of catalyst precursors of the types $[NiBr_2(diimine)]$ **D** and $[Zrcp_2Cl_2]$ **E** were employed with an MAO promoter. Using complex **D**, chain branches were observed at a frequency of 26 per 1,000 CH_2 groups, whilst branches were negligible using complex **E**. The branching frequency was 10 per 1,000 using a mixture of the two catalysts. Also, using complex **E**, carrying out the reaction under H_2 reduced the mean molecular weight of the polymer by a factor of 12, but the effect for complex **D** was small.

(i) Propose steps for the activation of complex **D** by MAO and hence draw a catalytic cycle for the onset of polymerization.

(ii) Account for the differences in branching frequency between the two catalysts.

(iii) Account for the effect of H_2 on the polymer molecular weight using complex **E**.

4. The most common source of thorium is the mineral *monazite*, from which thorium phosphate, $Th_4(PO_4)_4(P_2O_7)$, can be extracted at *c.*7 wt%. One isotope of thorium predominates, ^{232}Th, 99.98%. Nuclear fuel is formed by neutron capture and two sequential β^- decay processes. The energy available to a thermal nuclear reactor is reported to be 200.1 MeV per atom.

[1 eV = 96.485 kJ mol^{-1}]

(i) Identify the isotope acting as the nuclear fuel.

(ii) Estimate the nuclear energy that could be available from the thorium in 1 kg of the monazite ore.

5. A multijunction photovoltaic cell has been constructed from sandwiches of the following materials (with band gaps in eV): indium tin oxide (Sn doped In_2O_3, ≈ 4.0), ZnO (3.3), CdS (2.42), $CuIn_{(1-x)}Ga_xSe_2$ (1.0–1.7), and molybdenum (metallic).

    (i) Account for the variation in band gap between $CuInSe_2$ and $CuGaSe_2$.

    (ii) Propose a structure for the multijunction cell indicating the roles of these materials.

6. Solid oxide fuel cells (SOFC) have a structure like that of the SOEC in which the electrolyte and electrodes are solid-state oxide conductors (Fig. 5.47).

    (i) What are the electrode reactions for a SOEC cell using (a) H_2 and (b) CH_4 as fuel?

        Typically, SOFCs operate at 700–800 °C using a nickel catalyst incorporated at the cathode.

    (ii) What might be the drawbacks of such a device? An alternative solid-state device, called a protonic ceramic fuel cell (PCFC), is being developed which uitlizes a proton conductor solid electrolyte and electrodes, with the cathode also being a proton and electron conductor.

    (iii) What are the electrode reactions for a PCFC cell using (a) H_2 and (b) CH_4 as fuel?

        One set of PCFC devices has used $BaZr_{0.8}Y_{0.2}O_{(3-\delta)}$, a perovskite material, as the electrolyte and basis of the anode, which also contained nickel or copper. The cathode materials also contained cerium, as in for example $BaCe_{0.6}Zr_{0.3}Y_{0.1}O_{(3-\delta)}$. Such devices operated at 500 °C.

    (iv) What advantages may this type of cell have over SOFCs?

7. Lithium-ion batteries can be made with a variety of cathodes including Li_xCoO_2 and $Li_xM(PO_4)$.

    (i) Draw an energy-level diagram, to show the electron flow during the discharge of a lithium metal/Li_xCoO_2 battery. How will it differ if the anode is changed to Li/C?

    (ii) The cell voltage for $Li_xM(PO_4)$ is 4.1 V for M = Mn and 3.4 V when M = Fe. Account for this trend.

Water

6.1 Properties of Water

The properties of water can be introduced using a phase diagram (Fig. 6.1) drawn similarly to that for CO_2 in section 4.2.2 (Fig. 4.22). As we all know, under ambient conditions water exists as an equilibrium between liquid and vapour and hence, unlike for CO_2, the triple point lies much below atmospheric pressure at any point of land (273.16 K, 611.657 Pa). The critical point of water occurs at much higher temperature and pressure than that of CO_2 (647.1 K, 22.053 MPa) and so we rarely experience sc-H_2O ($\rho = 0.322$ g cm^{-3}). Salination increases the parameters of the critical point and the critical point of seawater at the pressure required (298.5 bar) is attained at a depth of $\approx$ 2,960 m. The fluids emanating from hydrothermal vents at such depths are generally of lower salinity than seawater, and as a result the fluids escaping from these vents are supercritical.

Clearly too the solid–liquid curve is complex in nature. This is due to the fact that different solid phases are involved at different temperature/pressure regimes. Around ambient pressures and 273 K, ice forms the phase I$_h$ (hexagonal), and this is the structure observed by the large majority of ice in the biosphere. A small proportion ($\approx$ 1%) is of a second phase I$_c$ (cubic) which is stable below $-100\,^\circ$C. Three further phases III, V, and VI are involved with the cusp region as the pressure increases, and ice VII is in equilibrium with liquid water at high temperatures and pressures. Yet more phases can be formed away from the solid–liquid curve. The structures of two different forms of ice are presented in Fig. 6.2. The near-ambient form, ice I$_h$, can be seen to form a hexagonal network of oxygen atoms stacked on top of each other. Each oxygen is shown as coordinated to four hydrogen atoms. This will be a mixture of O–H bonds (0.985 Å) and hydrogen bonds. In ice VII, the structure can be seen to be much more closely packed and indeed the density is $\approx$ 1.65 g cm^{-3}. The mixture of the two types of links: O–H..O gives additional complexity to the networks that can form; and in ice VII this results in two interpenetrating lattices. The O–H bond length remains roughly constant in these structural changes but the hydrogen-bonded O...O distances decrease markedly with increasing pressure: from 2.98 Å at ambient pressure to 2.54 Å at 300 kbar.[1] The different phases can be identified by vibrational spectroscopy, and these include different amorphous phases formed under different temperatures and pressures. Low-density amorphous (LDA) ice forms from rapid cooling of liquid water before the ordered structure of I$_h$ can

[1] Y. Yoshimura, S. T. Stewart, M. Somayazulu, H. K. Mao, and R. J. Hemley, 'Convergent Raman Features in High Density Amorphous Ice, Ice VII and Ice VIII under pressure', *J. Phys. Chem. B.,* 2011, **115**, 3756–60.

Elements of a Sustainable World. John Evans, Oxford University Press (2020). © John Evans.
DOI: 10.1093/oso/9780198827832.003.0006

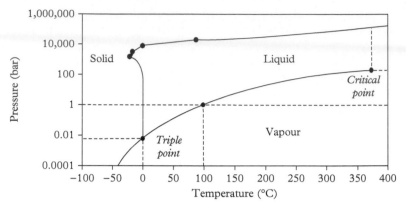

Figure 6.1 *Phase diagram of water.*

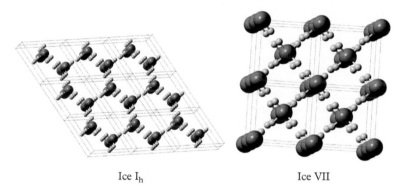

Ice I$_h$ Ice VII

Figure 6.2 *Structure of ambient (left) and high-pressure (right) forms of ice; large = O, small = H.*

assemble. Applying pressure to ice I$_h$ affords high-density amorphous (HDA) ice, and a very-high-density amorphous (VHDA) phase. It might be expected that aspects of the structures of I$_h$ ice and ice VII might be present in the LDA and HDA phases, but without X-ray or neutron diffraction there is an element of speculation about these amorphous phases.

Speculation about the structure of liquid water is still evident, but experimental evidence favours it being one uniform phase.[2] The local structure of this phase varies continuously from a low-density to a high-density form with increasingly pressure.[3] In the low-density (ambient) form, at an instant, the highest proportion of water molecules have four nearest neighbours subtending tetrahedral angles (Fig. 6.3). This domination of the tetrahedral angles continues at least to the fourth shell. There is also a very significant proportion of molecules with some of these sites vacant, providing three or two nearest neighbours. In the higher-density form the preponderance shifts to higher numbers of nearest

[2] A. K. Soper, 'Is water one liquid or two?', *J. Chem. Phys.*, 2019, **150**, 234503.

[3] A. K. Soper and M. A. Ricci, 'Structures of High-Density and Low-Density Water', *Phys. Rev. Lett.*, 2000, **84**, 2881–4.

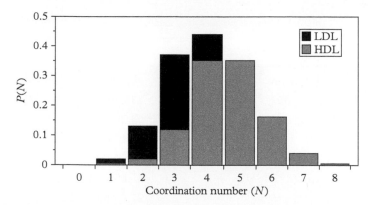

Figure 6.3 *Distributions of coordination numbers of water molecules in low (LDL) and high (HDL) density liquid forms.*

neighbour molecules (five to eight) with much-reduced under-coordination. Accordingly the bond angle distributions show a high degree of bimodal character, with a peak in probability developing near 70–80°. The structures are dynamic, and the indications are of water molecules having shells of neighbours in nested tetrahedra linked by hydrogen binding. Under ambient conditions, there are vacancies in these structures, but at higher pressures the density increases as more molecules are packed into the volume, replacing gaps with higher packing densities.

A spectrum of the physical properties of water is presented in Table 6.1. It shows that water has very significant values for the enthalpy of fusion and vaporization, these stages requiring partial and then complete disruption of the hydrogen-bonded network. It quantifies, too, the increase in density on melting of ice I_h, giving the iceberg effect. Water displays another anomaly in that the density increases further on warming to 4 °C, as the balance between liquid structures changes. This provides the ice-on-pond effect which allows aquatic life to continue in the more dense liquid underneath the frozen surface. The still higher density of sea water enhances both of these effects.

There is a strong moderating effect on atmospheric temperature provided by the equilibria between the phases of water, as discussed in Section 1.10. The basis for this can be estimated by comparing the values of $\Delta_{vap}H°$ and C_p at 25°C (Eq. 6.1). The energy absorbed by a given mass of water evaporating is similar to that of 600 times that quantity being raised in temperature by 1°C. The residence time of water in the atmosphere is in the order of days and evaporation/condensation are relatively rapid processes which moderate and quickly respond to other the effects of longer-term atmospheric changes.

$$\frac{\Delta H_{vap}}{C_p} = \frac{44,000}{75} = 587. \tag{6.1}$$

Table 6.1 *Physical properties of water.*

Property	Value
melting point	273.15 K
boiling point	373.13 K
heat capacity (C_p) (25 °C)	75.28 J mol^{-1} K^{-1}
$\Delta_{fus}H$	6.01 kJ mol^{-1}
$\Delta_{vap}H°$	44.0 kJ mol^{-1}
$\Delta_{vap}H$ (373.15 °C)	40.68 kJ mol^{-1}
$\Delta_{fus}S$	22.0 J mol^{-1} K^{-1}
$\Delta_{vap}S°$	118.89 J mol^{-1} K^{-1}
$\Delta_{vap}S$ (373.15 °C)	109.02 J mol^{-1} K^{-1}
density (25 °C)	0.9970474 g mL^{-1}
density (4 °C)	0.999972 g mL^{-1}
density (0 °C)	0.9998396 g mL^{-1}
density ice (0 °C)	0.9167 g mL^{-1}
density sea water (average, 25 °C)	1.0236 g mL^{-1}
viscosity	0.8937 mPa s^{-1}
surface tension (25 °C)	71.97 dyn cm^{-1}
pK_a	13.995
pH seawater	8.2
electrical conductivity	0.05501 μS cm^{-1}
thermal conductivity	0.6065 W m^{-1} K^{-1}
dielectric constant (25 °C)	78.54
vapour pressure (0 °C)	0.6113 kPa
vapour pressure (25 °C)	3.1690 kPa
vapour pressure (100 °C)	101.32 kPa

The effect of melting ice is rather less, but still very significant (Eq. 6.2). The energy absorbed by a given mass of ice melting is similar to that of 80 times that quantity being raised in temperature by 1 °C.[4] This adds a second effect to the removal of polar and continental ice in addition to that of reducing the surface reflectivity to solar irradiation. Ice melting/freezing provides a means of moderating seasonal changes in atmospheric temperature and has also absorbed much energy which otherwise would have added to the increase in global atmospheric temperature; the scope for this is decreasing.

[4] Another way of phrasing this is that it takes approximately the same amount of energy to melt 80 g of ice (a volume of a drinking glass or mug) as it takes to heat that quantity of water from room temperature to boiling point.

$$\frac{\Delta H_{fus}}{C_p} = \frac{6,010}{75} = 80. \tag{6.2}$$

The vapour pressure of water at $0\,°C$ can be seen to be low ($c.0.006$ bar) compared that at boiling point, which, by definition, is atmospheric pressure. There is a strong curvature to the increase with temperature, which is shown in terms of the saturation vapour density (kg m^3) in Fig. 6.4. From $15\,°C$ to $60\,°C$, this increases by a factor of ten. Hence climate excursions now can create regions of very warm air—even $55\,°C$—which can absorb substantially more water vapour than is feasible at lower temperatures. As a result a drop in temperature will result in higher levels of precipitation.

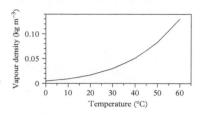

Figure 6.4 *Saturation vapour density of water.*

6.2 Distribution of Water

The broad distribution of water on the surface of Earth is shown in Table 6.2.[5] As is evident, oceans are the dominant location and comprise 99% of saline water; the bulk of the remainder of saline water is in groundwater (0.95%). The main location of freshwater is in ice and snow (69.6% of freshwater), the vast majority of which is the Antarctic ice sheet (61.7% of freshwater). The major remaining source of freshwater is groundwater (30.1% of freshwater). Approximately half of the volume in lakes is saline and as a result lakes house just 0.26% of freshwater. Of the global total the contributions held by the atmosphere, swamps, rivers, and life are relatively small. It is estimated that the water content in the mantle of the Earth considerably exceeds that in the oceans, some dissolved in minerals as such or as OH$^-$ anions.

[5] https://www.usgs.gov/special-topic/water-science-school.

Table 6.2 *Occurrence of water.*

Region	Quantity (km^3)	% total
oceans, seas, bays	1.338×10^9	96.54
icecaps, glaciers, snow	2.064×10^7	1.74
groundwater	2.34×10^7	1.69
soil moisture	1.65×10^4	0.001
ground ice, permafrost	3.0×10^5	0.022
lakes	1.764×10^5	0.013
atmosphere	1.29×10^4	0.00093
swamps	1.147×10^4	0.00083
rivers	2.12×10^3	0.00015
biology	1.12×10^3	0.000081

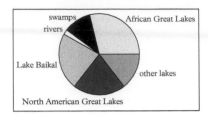

Figure 6.5 *Distribution of surface fresh-water.*

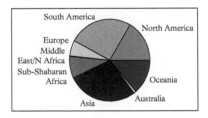

Figure 6.6 *Distribution of river flow.*

⁶ *Leaving No One Behind*, the UN World Water Development Report 2019, UNESCO, Paris, 2019.

In terms of utility, it is the surface freshwater which provides the potential reservoir of $\approx$ 104,600 km³ (Fig. 6.5). The majority of this occurs in three regions of freshwater lakes: the African Great Lakes, Lake Baikal, and the North American Great Lakes. All other lakes contain $\approx$ 15,200 km³, 14.5% of the total freshwater. This is augmented by 10,530,000 km³ of fresh groundwater (including aquifers) which is accessible through springs and wells. However, depleting these surface reservoirs has considerable environmental consequences affecting life systems, food supplies, and local climate. A more sustainable supply is provided by precipitation from the atmosphere, estimated to be 505,000 km³ annually. Of this $\approx$ 398,000 km³ is estimated to fall over oceans, leaving annual precipitation of $\approx$ 107,000 km³ over land, comparable to the instantaneous freshwater reservoir provided by lakes, rivers, and swamps. Over 40% of this results in river flow ($\approx$ 47,000 km³). It is difficult to specify an appropriate maximum for water extraction from river systems. One estimate is 9,000 km³, about 20% of the river flow. In 2014 the withdrawal of water globally was 4,000 km³, with the expectation of this rising to near 4,500 km³ in 2040.⁶ It is worth noting though that the net consumption of water from local sources was $\approx$ 1,400 km³, with the majority returning to source via drainage. However, the potential supply base from river flow is distributed unevenly across the regions of land (Fig. 6.6), with South America and Asia containing over half of this. Of this, over half emanates in two basins: the Amazon and Orinoco (6,500 km³), and the southern Asian basin which includes the Brahmaputra, Irrawaddy, and Mekong (8,000 km³). River flow in the Middle East and North Africa and also in Australia together represent only 1.3% of the global total, indicating that water stress will be in the form of a patchwork.

Based on the world population in 2014 of 7.73 billion, in 2014 the annual withdrawal and consumption of water per capita was $\approx$ 517,000 and 181,000 L, respectively. That corresponds to a daily withdrawal and consumption per capita of 1,416 and 496 L. This can be compared to the mean daily supply delivered

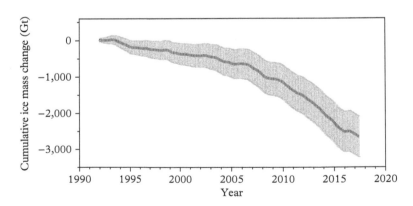

Figure 6.7 *Cumulative change in the ice mass of the Antarctic from 1992 to 2017 with error estimates.*

to UK households of 150 L per capita (in 2018); this excludes all other calls on water.[7]

This analysis is based on longstanding measurements, but there is a question about how much recent changes in ice mass have affected it. The cumulative changes in the ice mass of Antarctica from 1992 to 2017 have been reported,[8] and the changes overall are shown in Fig. 6.7. The results show a turn in the cumulative ice mass which has been decreasing since 1993, with evidence of an increase in slope since 2007. The overall reduction is $\approx$ 2,500 Gt, which equates to $\approx$ 2,500 km^3 of water volume. This is clearly very substantial but the broad, global estimations in Table 6.2 are little affected (so far!).

6.3 Distribution of Elements in Water

The typical elemental composition of the major solutes in seawater is shown in Table 6.3. This example is for a salinity of 3.5%. In the oceans, salinity varies between regions and depths, but the compositional ratios remain very similar. Salinity is lowest near regions of melting ice (*c*.3.2%) and highest in largely enclosed seas and ocean regions near land deserts. There are thirteen elements

Table 6.3 *Solute elements present in seawater above 1 ppm, with the composition of two freshwater samples and the drinking water limits (UK).*

Element	Seawater mg L^{-1}	Rhine leaving Alps mg L^{-1}	Stream, igneous rocks mg L^{-1}	Drinking water limits mg L^{-1}
Cl	19,400	1.1	0.06	
Na	10,800	1.4	0.16	200
Mg	1,290	7.2	0.24	
S	904	36 (SO_4^{2-})	1.3 (SO_4^{2-})	
Ca	411	40.7	1.68	
K	392	1.2	0.31	
Br	67.3			0.01 (BrO_3^-)
C	28.0	114 (HCO_3^-)	5.4 (HCO_3^-)	
N	15.5			50 (NO_3^-)
				0.5 (NO_2^-)
F	13			1.5
Sr	8.1			
B	4.45			1
Si	2.9	3.7	0.7	

[7] The International Water Association, https://iwa-network.org.

[8] The IMBIE team, 'Mass balance of the Antarctic ice sheet from 1992 to 2017', *Nature*, 2018, **558**, 219–22.

Table 6.4 *Elements in seawater between 1 ppb and 1 ppm.*

Element	Concentration (mg L^{-1})
Ar	0.450
Li	0.170
Rb	0.120
P	0.088
I	0.064
Ba	0.021
Mo	0.01
Ni	0.0066
Zn	0.005
Fe	0.0034
U	0.0033
As	0.0026
V	0.0019
Al	0.001
Ti	0.001

Table 6.5 *Elements in seawater between 1 ppt and 1 ppb.*

Element	Concentration (mg L^{-1})
Cu	0.0009
Se	0.0009
Sn	0.00081
Mn	0.0004
Co	0.00039
Sb	0.00033
Cs	0.0003

[9] S. Fendorf, H. A. Michael, A. van Geen, 'Spatial and Temporal Variations in Groundwater Arsenic in South and Southeast Asia', *Science*, 2010, **328**, 1123–27.

present at a level of over 1 ppm (mg L^{-1}). Two examples of freshwater sources are also shown. One is of the River Rhine as it leaves the Alpine region. The solutes reflect the nature of the rocks in the Alps, showing extraction of calcium and magnesium carbonates and sulphates. The sample from igneous rocks (the Cascades in Washington State in the USA) show much reduced ionic content after flowing for a relatively short distance over impervious rocks. As can be seen from this table, both freshwater samples would pass the hurdles shown for drinking water, although there are many additional ones.

There are a further fifteen elements which present between ppb and ppm by weight (Table 6.4). These include two group-1 elements and a group 2 from the *s*-block, and from the *p*-block Al, P, As, I, and Ar. There are five row-4 elements of the *d*-block (Ti, V, Fe, Ni, Zn) and one from row 5 (Mo). Finally there is one row-7 element of the *f*-block, namely uranium; at this concentration there should be 1 mg of uranium in each 300 L. Two elements are present in a higher abundance than is indicated by their concentration in seawater: Al and Ti. This can be ascribed to the high stability of their oxides giving very low solubility. The same is also true for iron, which is present under oxidizing conditions as Fe(III). There are key elements for metallo-enzymes in this set, making them available to aquatic life.

Finally, there are a large number of elements present at concentrations between 1 ppt (par per trillion) and 1 ppb (Table 6.5), and most others are present at even lower concentrations. Those in Table 6.5 also include elements important for life, such as Cu, Se, and Mn, and the remaining group-18 gases with stable isotopes. The list includes gold, with a mg of gold spread over 91,000 L!

As is evident from Table 6.3, the elemental content on freshwater is very dependent upon the nature of the rocks encountered, and also the location and residence time. An important example of this is highlighted by the occurrence of arsenic in drinking water. This can result in the chronic form of arsenic poisoning, the disease arsenicosis, substantially caused by arsenates substituting for phosphates in biochemical processes. Arsenicosis can have a wide range of consequences including skin cancer and liver disease.

Arsenic is mostly bound in sulphide ores, and the most common mineral is *arsenopyrite*, FeAsS. Under weathering, the arsenic can be oxidized to arsenate, which can also associate with iron as $Fe^{III}As^{V}O_4$. The solubility of ferric arsenate is very low under neutral conditions; this is increased in acidic and basic media. Hence, arsenic can be more readily transported under some industrial waste conditions. However, even under unpolluted environments, arsenates can be washed into groundwater.[9] Underground biological oxygen demand can exceed supply, and *reductase* enzymes can effect microbial reduction of arsenic and iron (Eq. 6.3). This increases the mobility of the arsenic, as the weak acid, $As(OH)_3$, and its salts and complexes. Under even more anoxic conditions, SO_4^{2-} can also provide the oxidant in microbial biochemical processes, being reduced to sulphide. In turn this can sequester the arsenic again.

$$FeAsS \rightarrow Fe^{III}As^V \rightarrow Fe^{II}As^{III} \rightarrow MAsS_x \quad (6.3)$$

As a result of this sequence there are regions in aquifers with higher levels of naturally occurring arsenic in solution than the WHO guideline limit of 10 μg L^{-1}. These occur to different depths in different countries taking water from the Himalayas. Two different situations are evident, in Cambodia and West Bengal in India. Wells close to the surface ($\approx$ 10 m) in the Mekong valley in Cambodia have a relatively low risk ($\approx$ 30%) of exceeding the WHO limit. But for a depth band below this to 60 m the risk rises to $\approx$ 50% and As concentration can reach 1 mg L^{-1}. At similar depth in the Ganges Valley in West Bengal, the probability of exceeding the WHO limit exceeds 70% and remains at 40% even down to a depth of 400 m, thus necessitating appropriate purification procedures.

6.4 Water in Processing

Just as there are estimates of the embodied energy (EE) and embodied carbon (EC) of a material, there is almost always an associated burden on water resources for the train of processes from extraction to production. The usual title for this is virtual water (VW) (Section 2.9). Estimates of VW are less evident even than those for EE and EC, and so the picture is rather patchy. There must be a caveat that VW estimates will vary widely between production sites, but nevertheless these do provide some idea of scale. As shown in Chapter 7, Fig. 7.15, water consumption is dominated by agriculture. The VW value of a material will substantially arise from industrial processing (overall this was $\approx$ 130 km^3 in 2014).[10] However, there will be other inputs from energy production ($\approx$ 77 km^3 in 2014) and from agriculture ($\approx$ 1140 km^3 in 2014) for biomass materials. Estimates for the virtual water per kg are presented in Table 6.6. This shows three classes of materials: metals, polymers, and a variety of materials in common use (CFRP is carbon fibre reinforced polymer).[11,12] The table compares materials of different types of use and density on a mass-related basis. The diverse range of materials show very widely differing water requirements. Dense inorganic materials like brick, concrete, and glass have relatively low values. Polymers are typically an order of magnitude or so higher, with wood-based materials a further decade larger. In this table, the natural fibres have by far the highest mass-based values. Between them there are very large differences, with flax and jute much lower than cotton, and all far lower than wool.

The list of metals also displays considerable variation. The VW value tends to be lower for elements derived from ores with high concentration. For the very electropositive elements it increases with the energy required for reduction. The value for tellurium is relatively low, perhaps because it is generally extracted from flue ashes and the estimation may consider this as an aside in which the major constituents are ascribed the whole of the water consumption. Many life-cycle

Table 6.5 *Continued*

Element	Concentration (mg L^{-1})
Ag	0.00028
Kr	0.00021
Cr	0.0002
Hg	0.00015
Ne	0.00012
Cd	0.00011
Ge	0.00006
Xe	0.000047
Ga	0.00003
Pb	0.00003
Zr	0.000026
Bi	0.00002
Nb	0.000015
Y	0.000013
Au	0.000011
Re	0.0000084
Hf	< 0.000008
He	0.0000072
Sc	< 0.000004
La	0.0000029
Nd	0.0000028
Ta	< 0.0000025
Ce	0.0000016
Eu	0.0000013
W	< 0.000001

[10] *Leaving No One Behind*, the UN World Water Development Report 2019, UNESCO, Paris, 2019.

[11] M. F. Ashby, *Materials and the Environment*, Butterworth-Heinemann, Elsevier, Oxford, 2nd. ed., 2018.

[12] M. F. Ashby, *Materials and Sustainable Development*, Butterworth-Heinemann, Elsevier, Oxford, 2016.

Table 6.6 *Virtual water (L kg^{-1}) for materials of different types.*

Metals	VW	Polymers	VW	Materials	VW
Al	1,200	butyl rubber	110	alumina	57
Cr	470	polyamides	180	bamboo	700
Co	580	polycarbonate	270	brick	5.5
Cu	310	polyester	200	cement	37
Dy	55	polyethylene	58	concrete	5.1
Fe	50	PET	130	cotton	8,000
Li	470	polylactide	69	CFRP	1,400
Mg	980	PMMA	76	Flax	3,100
Mo	360	polypropylene	39	glass fibre	95
Nd	55	polystyrene	140	jute	3,200
Ni	230	PTFE	460	paper/card	1,000
Te	26	polyurethane	98	plywood	750
Ti	110	PVC	210	Pyrex	32
W	150	silicones	330	soda–lime glass	18
U	3,500	starch plastics	170	wool	170,000

[13] B. Sprecher, Y. Xiao, A. Walton, J. Speight, R. Harris, R. Kleijn, G. Visser, G. J. Kramer, 'Life Cycle Inventory of the Production of Rare Earths and the Subsequent Production of NdFeB Rare Earth Permanent Magnets', *Environ. Sci. Technol.*, 2014, **48**, 3951–58.

[14] R. Pell, F. Wall, X. Yan, J. Li, X. Zeng, 'Temporally explicit life cycle assessment as an environmental performance decision making tool in rare earth project development', *Miner. Eng.*, 2019, **135**, 64–73.

[15] B. J. Glaister, G. M. Mudd, 'The environmental costs of platinum-PGM mining and sustainability: Is the glass half-full or half-empty?', *Miner. Eng.*, 2010, **23**, 438–50.

analyses of rare-earth production and their conversion into magnets do not include water consumption, concentrating on waste-water purity.[13] The values in Table 6.6 for the *4f*-block metals appear to be rather low, but recent analyses are broadly consistent with this.[14]

In contrast, the use of water in the extraction of platinum-group metals is considerable, substantially due to the employment of grinding and flotation methods to produce concentrates.[15] The mean consumption is $\approx$ 390 m^3 kg 4E PGM^{-1} (390,000 L kg 4E PGM^{-1}) with the range being a factor of 8. Since much of the production of PGMs takes place in the relatively dry area of southern Africa, water consumption may be a critical factor.

6.5 Desalination of Water

The overall demands for water and the disparate distribution can result in local and regional shortages. As a result, desalination to increase water supply is envisaged as an important component of inclusive development,[10] with Clean Water and Sanitation forming number 6 of the UN Strategic Development Goals. For many locations with coast-lines there is the potential for increasing the supply by separation of water from its solutes, predominantly those listed in Table 6.3.

From 2008 to 2016 global water production by desalination was projected to have doubled to 38 km^3,[16] and the reported production in August 2019 was at a rate of 45 km^3.[17] That is $\approx$ 3% of the consumption of processed water.

The initial desalination plants were based upon distillation. As indicated by the parameters in Table 6.1, the high values of the heat capacity, C_p, and, particularly, ΔH_{vap} make this an energy-intensive process. This can be reduced by vacuum distillation and multi-step methods (multi-effect distillation, MED) and provide an electrical energy equivalent consumption of 6.5–11 kWh m^{-3}. As a result, substantial low-carbon energy supplies are required nearby to avoid significant elevation of CO_2e levels. The majority of the large-scale plants constructed in recent years ($\approx$ 65% in 2015)[18] and planned for the future are based on a different physical process called **seawater reverse osmosis** (SWRO). These have a much lower electrical energy demand, in the range 3–5.5 kWh m^{-3}.

Osmosis is an example of a **colligative** property, as are the elevation of boiling point and depression of freezing point by a solute. For an ideal solution, the effect is due to the number of particles of the solute present. In the case of NaCl in water, there are two particles per formula: Na$^+$ and Cl$^-$. In the situation shown in Fig. 6.8, the solvent and solute are separated by a semipermeable membrane. In terms of the liquid phases this is a non-equilibrium condition. The equilibrium would be achieved if the solute and solvent thoroughly mixed to form a more dilute solution. Thus the tendency is for solvent molecules to flow into the solution. This semipermeable membrane allows only solvent molecules to pass through it, and not the solute. The **osmotic pressure**, Π, is that required to prevent this flow from occurring, as is depicted in the diagram by the difference between the solution column's height and that of the solvent. This is described by the **van't Hoff equation** (Eq. 6.4), which is closely related to the ideal gas equation. Here V is the volume of the solution containing n moles of solute particles.

$$\Pi V = nRT. \tag{6.4}$$

This tendency is the reverse of what is required for desalination, which is to retain the solute molecules on one side of the membrane and allow the water molecules to permeate through. In **reverse osmosis**, Fig. 6.9, a pressure is required to overcome the osmotic pressure and drive the flow to separate out the solvent ($\approx$ 40 bar). Energy is required for the pressurized flow through the reverse osmosis cell. The solvent permeates through the membrane and so concentrates the saline solution. The minimum energy required for a desalination step is dependent upon the concentration of the seawater and the percentage of the water recovery. Taking a standard seawater of 3.5% salt and a recovery of 50%, the minimum energy required is 1.06 kWh m^{-3}.[16] Outside the osmosis steps, there is additional energy input required for seawater supply, with pretreatment (of raw seawater) and post-treatment (reduction of borates and chloride); this can require a further 1 kWh m^{-3}.

[16] M. Elimelech, W. A. Phillip, 'The Future of Seawater Desalination: Energy, Technology, and the Environment', *Science*, 2011, **333**, 712–7.

[17] International Desalination Association, Topsfield, Massachusetts, USA, https://idadesal.org.

[18] S. Marbach, L. Bocquet, 'Osmosis, from molecular insights to large-scale applications', *Chem. Soc. Rev.*, 2019, **48**, 3102–44.

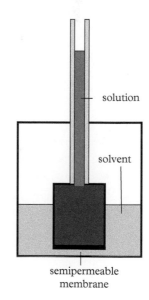

solution

solvent

semipermeable membrane

Figure 6.8 *Formation of osmotic pressure.*

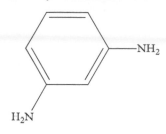

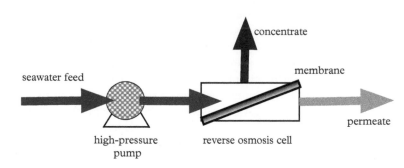

Figure 6.10 *Precursors to the aromatic polyamide in reverse osmosis membranes.*

[19] C. Klaysom, T. Y. Cath, T. Depuydt, I. F. J. Vankelecom, 'Forward and pressure retarded osmosis: potential solutions for global challenges in energy and water supply', *Chem. Soc. Rev.*, 2013, **42**, 6959–89.

The efficiency of the osmosis steps is influenced by the process design and membrane. A highly permeable membrane can reduce the overpressure required to drive the permeate through the membrane. However, this must be held in tension with the separation factor. Too high a transport efficiency can come with a loss of selectivity. The membranes generally employed are composites with a polyester web to act as the support (120–150 μm). A microporous polysulphone (see Fig. 5.46) (c.40 μm) provides a polar component, and the barrier (0.2 μ) is generally an aromatic polyamide which is chemically robust. The commonly used precursors are the diamine and tri acid-chloride shown in Fig. 6.10. Many other materials, such as carbon nanotubes, are being investigated for their utility.

Coupling two units as in Fig. 6.9 can improve the efficiency and provide a practical minimum energy consumption for reverse osmosis of 1.56 kWh m^3 for 50% recovery from 3.5% seawater. There have been many proposals for composite processes to increase efficiency and lower environmental impact. One such includes forward as well as reverse osmosis steps (Fig. 6.11).[19] In this example, recycled or waste water is used to dilute the incoming seawater feed by forward osmosis, which is thermodynamically favourable. This diluted seawater

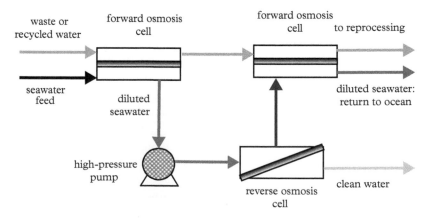

Figure 6.9 *Schematic of a reverse osmosis unit for desalination.*

Figure 6.11 *Coupling forward and reverse osmosis for desalination.*

enters the reverse osmosis cell which can be operated at a lower pressure due to this dilution. Clean water is then collected from the permeate. The seawater will now be more concentrated before a second forward osmosis unit dilutes it again to a state that can be returned to the ocean. The incoming water stream into the forward osmosis cell can be from waste water, providing the osmotic pressure is lower than that of the seawater sides of the membrane. It will become more concentrated in salts from the two forward osmosis steps.

An interesting prospect is to utilize the osmotic pressure streams of sea- and freshwater to generate electricity.[18] This so-called **blue energy** can be generated in a variety of ways. One is by pressure-retarded osmosis. In this the flow of freshwater through the semipermeable membrane drives a turbine of an electrical generator. This has been demonstrated to afford 3 W m^{-2} of membrane area.

An alternative method is reverse electro-dialysis (RED), depicted in Fig. 6.12. Forward osmosis occurs from seawater into freshwater through two different ion-selective membranes. One is anion-selective and transport to the anode effects a flow of electrons. This is continued to the cathode covered by freshwater that is augmented by the ion flow from the seawater through a cation-selective membrane. A higher power density of 6–7 M m^{-2} has been reported for this process,[20] a value considered to be economically viable. Indeed, a pilot plant was opened in 2014.[21] This is located at the Afsluitdijk barrier in the Netherlands at which there is a freshwater supply from the Ijsselmeer in addition to the seawater on the other side of the barrier. Initial data from a pilot plant in western Sicily was first reported in 2016.[22] This used brackish water and brine from a saltworks. An average power output of 40 W was achieved with a power density of 1.6 W m^{-2}. An assessment of the potential of this approach to electricity generation points out that locations will require there to be colocation of water supplies of different salinities.[23] The available flow rates and salinity differences are important aspects of the location. For the RED cell itself, the properties (permeability and cost) of the membranes can be deciding factors on the economic viability of an installation. With high-performance membranes, an energy yield efficiency of 30–40% could be attainable. The most economically viable locations would

[20] A. Daniilidis, D. A. Vermaas, R. Herber, K. Nijmeijer, 'Experimentally obtainable energy from mixing river water, seawater or brines with reverse electrodialysis', *Renew. Energy.*, 2014, **64**, 123–31.

[21] REDstack BV, Sneek, Netherlands, http://www.redstack.nl.

[22] M. Tedesco, C. Scalici, D. Vaccari, A. Cipollina, A. Tamburini, G. Micale, 'Performance of the first reverse electrodialysis pilot plant for power production from saline waters and concentrated brines', *J. Membrane. Sci.*, 2016, **500**, 33–45.

[23] F. Giacalone, M. Papapetrou, G. Kosmadakis, A. Taburini, G. Micale, A. Cipollina, 'Application of reverse electrodialysis to site specific types of saline solutions: A techno-economic assessment', *Energy*, 2019, **181**, 532–47.

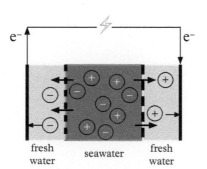

Figure 6.12 *Reverse electro-dialysis (RED) for electricity generation.*

be those with both brine and freshwater; this derives from the high levels of concentration differences providing more favourable ΔG of mixing. However, there is more potential energy generation, and CO_2 reduction, from river/seawater locations due to the higher flow rates.

6.6 Seawater as a Source of Elements

A comparison of the abundances of the elements in seawater and the Earth's crust is presented in Fig. 6.13. The dashed line indicates equality between the two environments. Relatively few elements lie above that line, in addition to the expected hydrogen and oxygen. For those with abundances of over 1 ppm these are chlorine, bromine, and sulphur. The dissolved inert gases krypton and xenon are also slightly more concentrated in seawater, but at a sub-ppb level. Excluding hydrogen and oxygen, following an extreme desalination, the concentration of the elements in the residue should increase by a factor of *c.*30. The solid line presents a threshold of seawater concentration at 1% of the crustal value. The additional elements within this band include the four most abundant metallic elements: sodium (10,800 ppm), magnesium (1,290 ppm), calcium (411 ppm), and potassium (392 ppm). Extending down to ppb concentrations, there is a second batch of elements in this band: lithium (170 ppb), strontium (8.1 ppm), boron (4.5 ppm), fluorine (13 ppm), iodine (6.4 ppb), and molybdenum

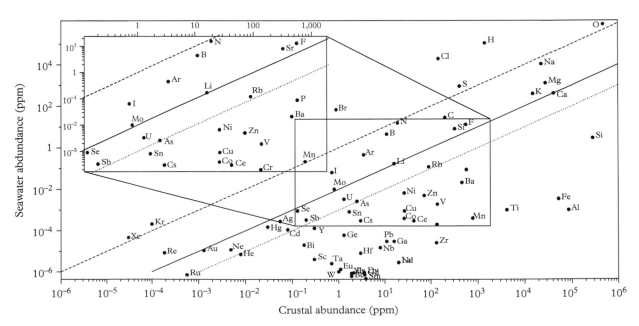

Figure 6.13 *Comparison of the abundances of elements in seawater and the Earth's crust.*

(10 ppb). Since seawater is much more homogeneous than the crust, it could be envisaged that an extraction process could be commonly applied to extraction from it, and as a result it may be viable to extract to a lower abundance level. Accordingly the dotted line indicates a band down to seawater having a 0.1% abundance level of the crust. For those elements with a concentration in seawater of over 1 ppb, this extends the choice of elements to include rubidium (120 ppb), arsenic (2.6 ppb), and uranium (3.3 ppb).

The mass of the dissolved salts in ocean waters is estimated to be 5×10^{16} tonnes, much in excess of the mass of minerals extracted annually (1×10^{11} tonnes).[24] The comparison though between the extraction of minerals from land and sea is complicated by the heterogeneity of the crust. The minerals that are extracted are generally from spaces in which the mineral in question is in considerably higher concentration than the mean value. The processing of large volumes of water has an energy cost, and in some cases the volumes are so large that it is very hard to envisage conversion from land to sea mining. For example, to match the annual production (2007 figures) of lithium, uranium, and gold would require the processing of 1.4×10^{11}, 2×10^{13}, and 2.3×10^{14} tonnes of water, respectively.

The most long-standing process is the extraction of salt by **solar evaporation** from seawater, and this has been extended to incorporate brines from SWRO, for example at Eilat.[25] From **lime softening** processing, commonly associated with water treatment works, addition of lime water, containing $Ca(OH)_2$, or soda ash (Na_2CO_3) affords a pH of 10–11, and $Mg(OH)_2$ can be precipitated (Eq. 6.5):

$$Mg^{2+}_{aq} + 2OH^- \rightarrow Mg(OH)_{2(s)} \downarrow. \tag{6.5}$$

Three alternative methods of separation are **electrodialysis** (ED), **membrane distillation/membrane crystallization** (MD/MDC) and adsorption/desorption/crystallization. Electrodialysis is illustrated in the schematic in Fig. 6.14.[25,26] It consists of a pair of electrodes providing the energy driver. These are within a layer of an electrode solution, which may be aqueous Na_2SO_4 in this case. The electrode reactions are the reduction of water at the cathode and its oxidation at the anode. Between these layers are many pairs of ion-selective membranes. Typically these contain either cationic functions (e.g. $-NR_3^+$), for anion-selective electrodes, or anionic functions (e.g. $-SO_3^-$), for cationic-selective electrodes. There are bonded on polymer side chains which allow singly charged ions to pass more readily than doubly charged ones. Thus in the case of an SWRO outlet the feed here would be a brine with a low concentration of doubly charged anions. With some membrane types these tend to be confined to the input chain and pass through. The other (concentrate) output channel will be enriched predominantly in NaCl, with some potassium and bromide ions also present.

A laboratory-scale scheme has been demonstrated taking a seawater reverse osmosis output and following that with electrodialysis to separate mono- and

[24] U. Bardi, 'Extracting Minerals from Seawater: An Energy Analysis', *Sustainability*, 2010, **2**, 980–92.

[25] P. Loganathan, G. Naidu, S. Vigneswaran, 'Mining valuable minerals from seawater: a critical review', *Environ. Sci.: Water Res. Technol.*, 2017, **3**, 37–53.

[26] S. E. Kentish, E. Kloester, G. W. Stevens, C. A. Scholes, L. F. Dumée, 'Electrodialysis in Aqueous-Organic Mixtures', *Sep. Purif. Rev.*, 2015, **44**, 269–82.

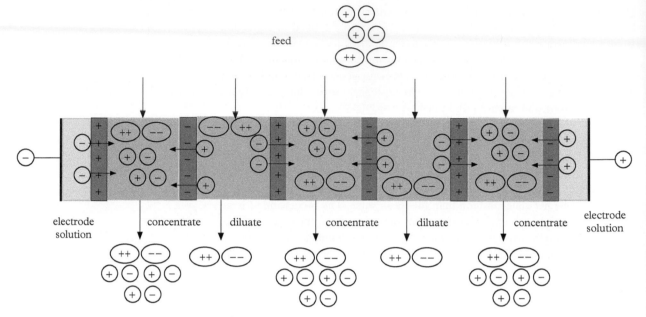

feed

electrode
solution

concentrate diluate concentrate diluate concentrate

electrode
solution

Figure 6.14 *Separation of salts by electrodialysis.*

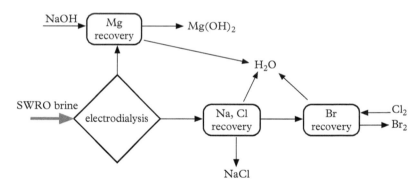

Figure 6.15 *Major element extraction from SWRO brine by ED.*

di-valent ions (Fig. 6.15). By addition of NaOH in a lime-softening step, $Mg(OH)_2$ could be crystallized. From the monovalent ion stream, evaporation could effect crystallization of NaCl. The residual solution would contain the more soluble NaBr, which in turn can be oxidized by Cl_2 to Br_2. It is worth noting that a nearby chloralkali process would provide the NaOH and Cl_2 required. Also, all of the processes liberate pure water. There has been little development in potassium isolation by such methods, although some potash materials are obtained as by-products from solar salt production. It is likely that the increase in demand for K-containing fertilizers will favour further developments.[27]

[27] A. Shahmansouri, J. Min, L. Jin, C. Bellona, 'Feasibility of extracting valuable minerals from desalination concentrate: a comprehensive literature review', *J. Clean. Prod.*, 2015, **100**, 4–16.

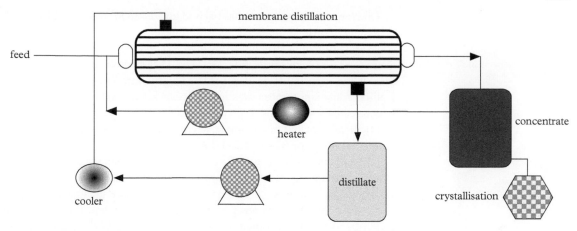

membrane distillation

feed

concentrate

heater

distillate

crystallisation

cooler

Figure 6.17 *Molecular distillation/crystallisation (MDC) schematic.*

An alternative use of membranes in the separation of seawater components is in MDC. One step in this is membrane distillation (MD) (Fig. 6.16).[25,28] The membrane material must be porous to vapour but not the liquid. As a result highly hydrophobic materials are required such as polypropylene and PTFE. Thus the water vapour is transported across the membrane from the hot feed to the cool permeate. In this case the permeate can be a water of high purity and the feed will concentrate in the salt content. This process can be coupled with crystallization equipment to effect MDC. The polymer membrane can be in the form of batches of hollow fibres. Thus the permeate can be collected from the fluid expelled from the tubes and the concentrated solution passed on.[29] A schematic presentation of such a process is given in Fig. 6.17. The hollow membrane tubes accept the feed and also the recycling loop which is both heated and pumped. As the concentration builds up in this loop a sequence of crystallization steps can be performed. The permeate from the MD unit can be collected in the distillate. This too can be recycled as a cooling stream to maintain the temperature gradient across the membranes.

In an estimation of energy costs for such a process for desalination, including nanofiltration (NF) units to reduce deposition, the MDC system was assessed as requiring twice the energy unit volume of water.[30] However, the water recovery was increased from 50% to 100% and brine disposal was not necessary.

A further technique considered to be more appropriate to more dilute minerals is concentration through adsorption/desorption processes. Adsorption will need to be specific, highly efficient, and reversible. Of the elements shown in the expanded box in Fig. 6.13, three group-1 metals, Li, K, and Rb, are potential targets for seawater mining, given their economic value.[25] In addition uranium may also be considered, being very different chemically and so will be considered separately.

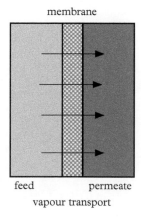

membrane

feed   permeate

vapour transport

Figure 6.16 *Membrane distillation (MD).*

[28] B. P. Pramanik, K. Thanavadivel. L. Shu, V. Jegatheesan, 'A critical review of membrane crystallization for the purification of water and recovery of minerals', *Rev. Environ. Sci. Biotechnol.*, 2016, **15**, 411–39.

[29] E. Curcio, G. Di Profio, E. Drioli, 'Recovery of fumaric acid by membrane crystallization in the production of L-malic acid.', *Sep. Purif. Technol.*, 2003, **33**, 63–73.

[30] E. Drioli, A. Criscuoli, E. Curcio, 'Integrated membrane operations for seawater desalination', *Desalination*, 2002, **147**, 77–81.

[31] G. Naidu, S. Jeong, Y. Choi, M. H. Song, U. Oyunchuluun, S. Vigneswaran, 'Valuable rubidium extraction for potassium reduced seawater brine', *J. Clean. Prod.*, 2018, **174**, 1079–88.

Figure 6.18 *A heulandite unit cell; blue = Al/Si tetrahedra, red = O.*

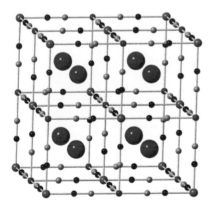

Figure 6.19 *A unit cell of $K_2CoFe(CN)_6$; large = K, brown = Fe, dark blue = Co, black = C, light blue = N.*

[32] G. Naidu, T. Nur, P. Loganathan, J. Kandasamy, S. Vigneswaran, 'Selective sorption of rubidium by potassium cobalt hexcyanoferrate', *Sep. Purif. Technol.*, 2016, **163**, 238–46.

One sequence investigated has been to utilize zeolite adsorption to extract the potassium and rubidium from SWRO brine.[31] A natural zeolite, *clinoptilolite*, is considered to have a structure similar to *heulandite*(Fig. 6.18). The aluminosilicate tetrahedra here create [6]-, [8]-, and [12]-rings (half of the [8]-rings can seen at the corners of the unit cell). The structure affords many options for high-coordination-number sites for the *s*-block ions to adopt. In one particular structure, the sodium ions display eight oxygen neighbours (Na–O 2.4–2.72 Å), and both calcium and potassium have eleven neighbours with wide variations in interatomic distance (Ca–O 2.2–3.2 Å; K–O 2.9–4.0 Å).

The binding constant to the adsorbent can be expressed as the Langmuir affinity constant, K_L (L mg^{-1}), and the maximum absorbed per unit mass of absorbent, Q_{max} (mg g^{-1}), by Eq. 6.6. Here the experimental parameters are: q_e, the mass of adsorbate per unit mass of absorbent; and C_e, the equilibrium concentration of the adsorbate in solution. The value of Q_{max} for a solution of potassium at the concentration of the brine sample (790 mg-K L^{-1}) was 209 mg g^{-1}, but this was reduced to 52 mg g^{-1} using the brine. The presence of Rb^+ and Sr^{2+} had little effect on potassium absorption; the smaller ions did show strong competition effects ($Ca^{2+} > Mg^{2+} > Na^+$). Nevertheless, potassium was concentrated by the absorption process, and the resulting material, derived from natural zeolites, was proposed as a fertilizer.

$$q_e = \frac{Q_{max}K_L C_e}{1 + K_L C_e}. \tag{6.6}$$

In this study, there was a stack of microfilter membranes comprising polysulphones and poly-vinylidenefluoride, $(-CH_2-CF_2-)_n$. The permeate showed that little rubidium had been retained by the adsorption/microfiltration sequence. Evidently concentrating rubidium will require a different absorbent, with a larger cavity size. One such is $K_2CoFe(CN)_6$, the structure of which is shown in Fig. 6.19. The core unit is the $[Fe^{II}(CN)_6]^{4-}$ anion. This has a low-spin $3d^6$ iron with strong-field ligands and this maximizes the LFSE (Section 3.3). The cobalt(II) ion ($3d^7$) is coordinated by the nitrogen atoms of six cyano ligands. It displays a magnetic moment of 1.0 μ_B, indicating the cobalt is also low spin and will also have a high LFSE. The result is a stable lattice with significant cavities for the potassium ions. Each is in the cavity created by twelve cyanide groups with the K–C/N distances 3.62 Å. This is also a commercial material (CsTreat) for caesium absorption. The values of Q_{max} for rubidium were found to be even higher than that of caesium (*c.*100 as opposed to 75 mg g^{-1}).[32] The smaller ions, Li^+, Na^+, and Ca^{2+}, displayed little propensity for absorption, as Eq. 6.7 will be unfavourable due to the higher heat of hydration of the smaller ions.

$$M_{aq}^+ + K_2^+ CoFe(CN)_6 = K_{aq}^+ + K^+ M^+ CoFe(CN)_6. \tag{6.7}$$

In the study on the effects of potassium reduction on rubidium binding, a similar active absorbent, $K_2CuFe(CN)_6$, on polyacrylonitrile, was shown to be over 80% efficient at absorption of rubidium from the permeate. This was an improvement by a factor of four from direct absorption from SWRO output; the rubidium could be liberated by concentrated aqueous KCl making the absorbent recyclable. Taken together, these studies show that a sequence of absorption–filtration steps could be employed to concentrate and extract the group-1 metals potassium and rubidium.

Since lithium is present at a slightly higher concentration in seawater than rubidium, this suggests that its extraction might also be feasible. All the alkali metal ions have very similar properties, differing mainly in their ionic radii. As the smallest of the group-1 elements (Fig. 2.16), lithium has the highest hydration enthalpy (Table 2.11), and so an absorbent must be well size-matched to Li^+ to overcome that. Mg^{2+} has a smaller radius and higher charge and hence may compete strongly for the lithium sites. As a result the Li^+/Mg^{2+} ratio is an important component in the design of a lithium-extraction process.

At one extreme is an example of industrial wastewater with lithium and magnesium concentrations of 1,269 and 0.0087 mg L^{-1}. The lithium concentration (1,823 mM) is *c.*70,000 higher than that in seawater (0.025 mM).[33] A schematic of this process is given in Fig. 6.20. The wastewater was first passed through an ion-exchange column to remove the group-2 ions and thus reduce precipitation of their salts (sulphate, carbonate) downstream. There followed three stages of concentration; one reverse osmosis and two electrodialysis. This could provide an increase in lithium concentration by a factor of *c.*12. From this concentrated solution ($\approx$ 2.2 M), addition of sodium carbonate can afford the precipitation of lithium carbonate.

An intermediate example is the waters from a saline lake, Great Salt Lake. Here the water sample contained 20 mg L^{-1} of Li^+, with the the four major *s*-block components in seawater also in large excess here (Na^+ 54.8, Mg^{2+} 4.3, K^+ 2.5, and Ca^{2+} 0.45 g L^{-1}).[34] The absorbent process employed is related to the process in lithium-ion batteries, for which manganese may be a substitute for

[33] Y. Qiu, H. Ruan, C. Tang, Y. Lao, J. Shen, A. Sotto, 'Study on Recovering High-Concentration Lithium Salt from Lithium using a Hybrid Reverse Osmosis (RO)-Electrodialysis (ED) Process', *ACS Sustainable Chem. Eng.*, 2019, 7, 13481–90.

[34] R. Marthi, Y. R. Smith, 'Selective recovery of lithium from the Great Salt Lake using lithium manganese oxide-diatomaceous earth composite', *Hydrometallurgy*, 2019, **186**, 115–25.

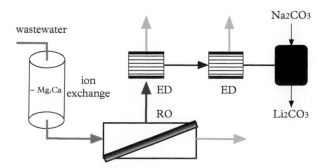

Figure 6.20 *Extraction of lithium from wastewater.*

Figure 6.21 *Structure of Li₂MnO₄;*
pink = MnO₆ octahedra, grey = LiO₄
tetrahedra.

[35] D. Su, H.-J. Ahn, G. Wang, 'Hydrothermal synthesis of α-MnO$_2$ and β-
MnO$_2$ nanorods as high capacity cathode
materials of sodium ion batteries', *J. Mater.*
Chem. A, 2013, **1**, 4845–50.

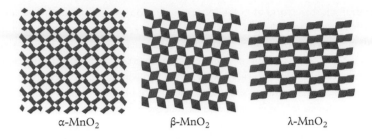

α-MnO₂   β-MnO₂   λ-MnO₂

Figure 6.22 *Structures of three forms of MnO₂; pink MnO₆ octahedra.*

cobalt. In the structure of LiMn$_2$O$_4$ (Fig. 6.21) the lattice structure is defined by
MnO$_6$ octahedra and the lithium ions occupy tetrahedral holes. That can provide
selectivity for lithium over sodium.

However, the selectivity is strongly related to the morphology of the MnO$_2$
used. Three forms of MnO$_2$ are shown in Fig. 6.22. As expected the structure
comprises linked octahedra. In α-MnO$_2$, there are two types of tunnel. The
smaller ones described as a (1×1) channel has one MnO$_6$ unit along each
edge, and these units share vertices. The O..O distance across the tunnel is 3.3
Å, and this is sufficiently large to allow intercalation of Na$^+$.[35] The second type
of channel in the α form is a (2×2) channel, in which the side of the channel
comprises two edge-sharing octahedra. The square of oxygen atoms on the inside
of the channel has edge and diagonal lengths of 3.54 and 5.00 Å, respectively.
This proves to be too large a cavity for stable intercalation of Na$^+$, and its binding
is stronger to β-MnO$_2$, which only has the (1×1) type of channel.

The third morphology shown in Fig. 6.22 is that of the λ polymorph. From the
angle of view, this looks to be a (2×1) channel, but in this case the octahedra are at
two different elevations. The resultant channel (Fig. 6.23) comprises a puckered
ring of six oxides, similar to a chair conformation with nearest O..O distance
of 2.72–2.75 Å, and a geometry nearer to tetrahedral. Using this polymorph
composited with diatomaceous earth (an amorphous silica), high selectivity could
be achieved for lithium extraction over sodium and potassium. However, leaching
of manganese from the absorbent did occur. This could be reduced by the
oxidizing acid H$_2$S$_2$O$_8$, implying that the leaching mechanism involves reduction
to Mn(II).[34]

Granulated λ-MnO$_2$ has been utilized in an absorption column as the pre-
liminary concentration step in recovering lithium from seawater. The metals
collected on the absorption column were eluted by aqueous HCl and then passed
through a cation-exchange resin containing sulphonic acid (Ar-SO$_3$H) groups on
a framework of polystyrene cross-linked by divinylbenzene. Such strongly acidic
sites will be relatively weak Lewis bases and thus interact weakly with monovalent
cations. The stronger Lewis acid sites of the group-2 cations from seawater, and
Mn^{2+} emanating from the absorbent column, will have longer residence times
on the extraction columns. Stronger ligands are required for separation of the

Figure 6.23 *The channel structure of*
λ-MnO₂; red = O, pink = Mn.

singly charged ions, and this was effected by the extractants in Fig. 6.24 on a resin column and a high pH (12). Thus the long-chain diketone can provide β-diketonate ligands, and the phosphine oxide can act as a monodentate neutral ligand. The affinity for such ligands would be expected to decrease with the ionic radii of the cations. Thus lithium is the last of the ions to elute. Taking these steps together the concentration of lithium was by a factor close to 7,000 (Fig. 6.25). This in turn provided the basis for further concentration using vacuum distillation allowing precipitation of lithium as its carbonate.[36] Although this was a laboratory experiment, samples were obtained by extraction from the sea. Optimization of the material for initial sorption material can be envisaged in the future, but a pathway to isolation of a useful lithium salt has been demonstrated.

The concentration of uranium presents a still larger challenge. In terms of concentration lithium is over 50 times more concentrated than uranium, but in terms of mol L^{-1} the ratio is $\approx$ 1,800, with uranium being present at a level of 14 nM. Nevertheless, the potential for nuclear energy from the dissolved uranium has stimulated much research.[37]

The potential chemistry of uranium in seawater is considerably wider than that of the alkali metals. A Frost diagram, which shows the relative stability of oxidation states (Section 2.7), indicates that U(IV) would be the most favoured oxidation state under standard conditions, with U(VI) also viable; U(V) would tend to disproportionate to a mixture of its neighbours. However, seawater does not present standard conditions: it is considerably more alkaline (pH 8.2), more dilute in uranium, and contains a range of counter ions.[38] Under these more basic conditions, the concentrations of anions of weak acids increase, which will favour coordination to the metal cations. These can be strong donor ligands that stabilize high oxidation states by charge stabilization and also increased electron donation. For example, moving from H_2O to OH^- to O^{2-} increases the σ donation propensity of the oxygen and also adds π donation. As a result, apart from anoxic conditions, U(VI) becomes the dominant oxidation state.

In acidic conditions the aquated uranyl anion is the dominant uranium species. This has a linear O–U–O unit. With a weakly coordinating anion like perchlorate (ClO_4^-), a structure in the crystal shows five aquo ligands and a pentagonal bipyramidal seven-coordinate structure (Fig. 6.26). Carbonate is present at a concentration of 2.2 mM in seawater and under less acidic conditions HCO_3^- will become readily available. At pH values higher than $c.5$, the predominant species

Figure 6.24 *Extracting agents for separation of Li$^+$.*

[36] S. Nishihama, K. Onishi, K. Yoshizuka, 'Selective Recovery Process of Lithium from Seawater Using Integrated Ion Exchange Methods', *Solvent Extr. Ion Exch.*, 2011, **29**, 421–31.

[37] C. W. Abney, R. T. Mayes, T. Saito, S. Dai, 'Materials for the Recovery of Uranium from Seawater', *Chem. Rev.*, 2017, **117**, 13935–14013.

[38] F. Endrizzi, C. J. Leggett, L. Rao, 'Scientific Basis for Efficient Extraction of Uranium from Seawater.1: Understanding the Chemical Speciation of Uranium under Seawater Conditions', *Ind. Eng. Chem. Res.*, 2016, **55**, 4249–56.

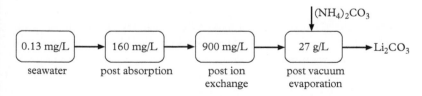

Figure 6.25 *Concentrations of Li$^+$ through extraction from seawater.*

Figure 6.26 *Partial structures of* $[UO_2(H_2O)_5](ClO_4)_2$, *showing the location of the closest anions to the uranium (left). The structure of* $Ca_2(UO_2)(CO_3)_3(H_2O)_{11}$ *(right) shows the coordination modes of calcium to the uranium-coordinated carbonate groups. Red = O, green = Cl, black = C, pink = H, small grey = U, large grey = Ca.*

all contain the tris-carbonato complex, also shown in Fig. 6.26. The uranium coordination site again has a *trans* O–U–O unit, here with three bidentate carbonate ligands giving a hexagonal bipyramidal geometry. Throughout the pH range of 5–9 the distributions of species are similar: $Ca_2(UO_2)(CO_3)_3(aq)$ (50–60%), $Mg(UO_2)(CO_3)_3{}^{2-}$ (20–25%), $Ca(UO_2)(CO_3)_3{}^{2-}$ (15–20%), and $(UO_2)(CO_3)_3{}^{4-}$ (5–10%). Some extraction systems have been targeting the anion $(UO_2)(CO_3)_3{}^{4-}$. However, it appears that the alkali metal ions in combination with $HCO_3{}^{-}$ can compete with binding agents with the majority of the uranium sequestered as an aquated neutral $UO_2{}^{2+}/CO_3{}^{2-}/Ca^{2+}$ complex.

A wide variety of adsorbents have been developed for this process. In terms of uranium uptake, anatase (TiO_2) has proven to be one of the most effective inorganic adsorbents in environmental seawater. However, in terms of the mass ratio polymer-based materials have provided an increase of a factor of 4 on this (to 4–6 mg U g^{-1} adsorbent). Given the nature of the environment, the properties of the polymer would have to be close to an inversion of those desired for natural decay of plastics (Section 5.2.3). So polyethylene (PE) is an appropriate chain for this purpose with side chains to bear coordinating ligands which can compete with $CO_3{}^{2-}$. A grafting technique called radiation-induced graft polymerization (RIGP) has been employed to create copolymers, for example of PE and polyacrylonitrile (PAN) (Fig. 6.27). Radiolysis generates radicals and

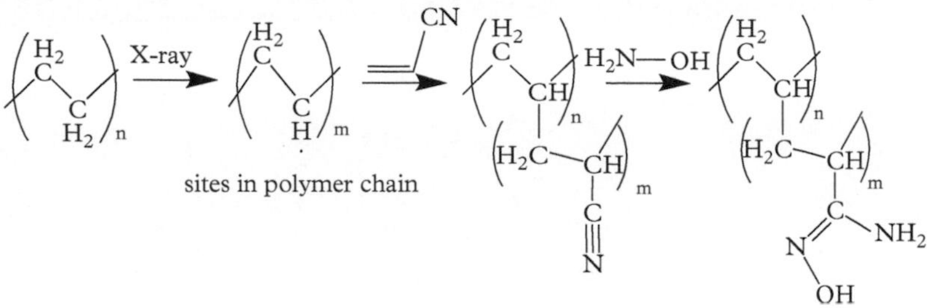

Figure 6.27 contains a chemical reaction scheme at the top of the page.

Figure 6.27 *Formation of PE–PAN copolymer by radiation-induced grafted polymerization (RIGP).*

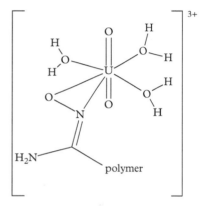

these sites can be the initiation points for a radical polymerization. In this way a polyacrylonitrile side chain can be grafted onto the polyethylene backbone. On reaction with hydroxylamine (NH_2OH) the polymer can provide an amidoxime ligand for uranium. This ligand has a variety of possible coordination options involving deprotonation of the hydroxyl group to provide a mono-anionic ligand. This could be a chelating mode forming a five-membered ring. However, known structures favour an η^2 coordination by the N–O bond, as shown in a possible adsorbed uranium site presented in Fig. 6.28. This will create a positively charged polymer surface. Whilst this may attract the anion $(UO_2)(CO_3)_3{}^{4-}$, it is less clear that this is the case for the major species in seawater, the neutral cluster $Ca_2(UO_2)(CO_3)_3$.

Major difficulties remain before a technological solution can be envisaged that will provide a durable absorbent that is effective for 14 nM uranium in a seawater mix. Many very-high-surface-area materials have been investigated such as ligand-functionalized mesoporous silicas and metal–organic framework materials (MOFs). Practically, these powdered materials must be presented in a way that can be sustained in environmental seawater. There is also strong competition for the ligand binding sites afforded by *d*-block elements in seawater, notably nickel, iron, and vanadium. The vanadyl unit, $[O=V^{IV}]^{2+}$ has some similarities to the uranyl group and thus is a challenge for selectivity.

One route investigated to overcome this problem is by protein engineering to provide a binding site appropriate for a pentagonal bipyramidal coordination site, as adopted by the uranyl aquo complex (Fig. 6.26).[39] Starting from a promising protein derived from a stable methane-producing bacterium, *Methanobacterium thermoautotrophicum*, mutations were introduced to closely match the target coordination site affording super uranyl-binding protein (SUP). The site developed is shown in Fig. 6.29. The two carboxylate ligands in the equatorial belt are from glutamate and aspartate residues in different helices of the protein, with the the uranyl group in the cleft between them. A third residue, an arginate, provides a hydrogen-bonding interaction with one of the oxygens of the uranyl group. The selectivity for uranium is very high compared to other metals in seawater, in the range of 10^6 to 10^7 for the most part. Even for Cu^{2+}, which generally provides

Figure 6.28 *Possible structure of a uranium site adsorbed onto an aldoximine-functionalized polymer.*

[39] L. Zhou, M. Bosscher, C. Zhang, S. Özçubukçu, L. Zhang, W. Zhang, C. J. Li, J. Liu, M. P. Jensen, L. Lai, C. He, 'A protein engineered to bind uranyl selectively and with femtomolar affinity', *Nat. Chem.*, 2014, **6**, 236–41.

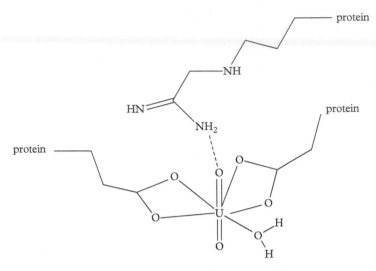

Figure 6.29 *Binding site in super uranyl-binding protein (SUP).*

the strongest binding of the $3d$-divalent cations according to the Irving–Williams series (Section 3.3), the uranyl group is favoured by over 10^3. Binding of uranyl compared to vanadyl, which could also be envisaged to locate in such a site, is higher by $\approx 10^4$.

Immobilized versions of the SUP using a maltose-binding protein and an amylose resin were very effective at removal of uranium from a dilute solution of uranyl salts (666 nM), and the resin could be recycled by washing with carbonate solutions. The SUP material and dissolved carbonate do compete; on one hand this provides a mechanism for uranium retrieval from the adsorbent, but on the other it reduced the capture by the protein. Nevertheless, a significant proportion of a realistic concentration of uranium (13 nM) in a synthetic seawater could be extracted by the immobilized SUP. This illustrates a potential approach which could perhaps be augmented by further protein engineering and immobilization methods. As yet, though, a technology to mine uranium from seawater remains an aspiration.[37]

6.7   Water Remediation from Heavy Metals and Organics

Remediation of water requires a wide range of approaches, depending upon the nature of contaminants and the required destination. In this section the purity guidelines employed are those for water useable for irrigation. For heavy metals, the legal limits for this use of water are about a factor of 10 higher than those for drinking water. Table 6.7 collects some observed levels of contamination for the

Table 6.7 *World average concentrations and limits of elements in water and soil (dw = dry weight).*

Metal	Upper crust mg kg^{-1}	River water μg L^{-1}	EU soil limits mg kg^{-1} dw	Irrigation water μg L^{-1}
As	1.8	0.13–2.71		100/100
Cd	0.1	0.06–0.61	1	10/5
Cr	35	0.29–11.46	75	100/8 Cr^{VI} 5 Cr^{III}
Cu	14	0.23–2.59	50	200/200
Pb	15	0.007–308	70	5,000/200
Hg	0.007		0.5	
Ni	19	0.35–5.06	50	200/500
Zn	52	0.27–27	150	2,000/1,000

Note: irrigation limits given as those of FAO/Canada.

upper crust of soil and river water and the guidelines or limits. For soil, the EU limits are those proposed for the EU; the EU does not have limits for irrigation water so those given are those of the FAO[40] and Canada.[41] There is also no limit set for arsenic in agricultural soil; limits on sediments vary from 5.9–33 mg kg^{-1} dw in Canada and the USA. The background soil contents are within the EU limits, and the irrigation water limits of the FAO cover the environmental values too. The Canadian guidelines for irrigation water for chromium and lead do bite into the observed range of background values.

There are a variety of parameters used to map the degree of contamination in an area. One is the **contamination factor** (*CF*), which is the ratio of an element compared to the background or reference concentration (Eq. 6.8). Each element then can be given a value of *CF*, and an overall **Pollution Load Index** (*PLI*) can be assigned to a location according to Eq. 6.9.

$$CFn = \frac{[element\ n]}{[reference\ of\ element\ n]} \tag{6.8}$$

$$PLI = \prod_1^n (CF_n)^{\frac{1}{n}}. \tag{6.9}$$

An alternative indexing parameter is called the **geoaccumulation index** (I_{geo}). This is estimated according to Eq 6.10. The reference used could be the background value, or also the legal value for the limit of that element in its type of environment. This will provide a scaling of contamination against health-assessed norms for contaminated soils and sediments. Values of I_{geo} correspond to different levels of pollution, as presented in Table 6.8.

[40] Food and Agriculture Organization of the United Nations.

[41] J. P. Vareda, A. J. M. Valente, L. Durães, 'Assessment of heavy metal pollution from anthropogenic activities and remediation strategies: A review', *J. Environ. Manage. 2019*, **246**, 101–18.

Table 6.8 *Bands of I_{geo} for pollution levels.*

I_{geo}	Level
≤ 0	unpolluted
0–1	unpolluted to slight
1–2	moderate
2–3	moderate to heavy
3–4	heavily to extremely
> 4	extremely high

$$I_{geo} = log_2\left(\frac{[element\ n]}{1.5 \times [reference\ of\ element\ n]}\right). \tag{6.10}$$

These can be used to assess the levels of pollution in soils from sites affected by the mining industry, as in Table 6.9. In the second column the example is for soils taken near a smelting works in Northern France. The samples exhibited a distribution of pollution levels of zinc, cadmium, and lead including areas where these values are extremely high. Analyses taken from the soil of cultivated land in the vicinity of the smelter showed that the maximum concentrations of these elements reduced with distance, but there were locations where they were still extremely high. Pollution by arsenic and mercury was also heavy in places. The next example is an old, disused mine in the UK. Even though no longer in use, contamination of the soil was extremely high in places in zinc, cadmium, and lead. Finally, there are soils examined from the Copperbelt Province in Zambia. Again there is a wide distribution of contaminations between sampling locations. Unsurprisingly, copper contamination can be extreme in places, and this was also the case for arsenic and lead; moderate contamination was evident for the group-12 elements.

The transport of the heavy metals can involve dust and watercourses, these being mostly untreated. Accordingly, particular industrial sites can develop their own fingerprint of heavy metals in the watercourses emanating from them. It is worth noting that the units are in mg L^{-1} in Table 6.10, not μg L^{-1} as in Table 6.7. Some elements are present to levels which are orders of magnitude greater than the recommended levels for irrigation. These can become problematic without appropriate treatment. It is apparent that in the tabulated example, the recommended concentrations for cadmium, copper, lead, nickel, and zinc were exceeded at times.

Table 6.9 *Geoaccumulation indexes (I_{geo}) for soils affected by mines.*

Metal	Smelter, France	Cultivated land, France	Old mine, UK	Copperbelt, Zambia
As	–	1.20–3.43	–	−1.67–4.98
Cd	−1.74–10.99	3.63–7.15	< 11.18	−2.91–3.06
Cr	–	−0.59–0.40	–	–
Cu	–	−0.44–1.14	0.12–1.88	−2.86–10.81
Pb	−1.53–7.91	2.17–5.65	3.95–10.89	−2.93–4.22
Hg	–	0.19–2.63	–	−4.25–1.89
Ni	–	−1.57–0.47	–	–
Zn	−1.38–8.64	0.94–4.80	2.04–9.42	−2.72–2.82

Table 6.10 *Heavy metal concentrations in wastewaters from selected industries (mg L^{-1}).*

Metal	Electroplating	Engineering plant	Chlor alkali plant	Used in irrigation
As	–	–	–	–
Cd	⋆	–	0.6	⋆–0.06
Cr	28.1	1,780	–	⋆–0.81
Cu	6.4	79.8	–	0.07–6.30
Pb	⋆	0.4	3.2	⋆–7.50
Hg	–	–	9.9	–
Ni	9.73	–	–	⋆–4.20
Zn	1.34	38.43	–	0.21–4.30

⋆below the detection limit.

6.7.1 Water Treatment Plants

Schematics of water treatment and wastewater treatment plants are shown in Fig. 6.30.[42] The water from a wastewater plant is fit to return to the natural waters and itself provides irrigation waters. It is also an appropriate input for the production of drinking water. The first stage of both steps is to remove significantly sized debris including rags, paper, and plastics with a mesh screen with 6 mm openings. Finer meshes are used to remove solids that would otherwise affect the operation of the plant. Grit grinding and removal will generally reduce particle sizes in the liquid stream to 100–250 μm. In wastewater treatment, by increasing the pH, metals can be precipitated as their hydroxides, and these removed by a sedimentation step, albeit leading to a sludge for disposal. Another approach shown in the water treatment pathway involves the addition of coagulating agents, typically aluminium or iron(III) sulphates ($M_2(SO_4)_3.nH_2O$) (*alums*), sometimes as mixed sulphates. These aid the separation of solids from the water stream through precipitation and/or flocculation and flotation. There is a considerable mass of the resulting solids, known as drinking water treatment residuals (DWTR), amounting to 3–5 % of the treated water. This itself has been investigated as an absorbent for heavy metal adsorption.[43] In the drinking water pathway a train of purification with filtration steps is followed. Firstly, in the rapid sand filtration, water is passed through a bed of carbon (anthracite) and silica sand. There follow ultrafiltration and reverse-osmosis filtration steps which will further reduce dissolved material.

For wastewater treatment there is a cycle including a membrane bioreactor which can digest much unwanted organic materials still in the water at this stage. This is followed generally by a series of absorption columns that provide the most commonly employed mode of retaining the heavy metal salts. The majority

[42] M. Enfrin, L. F. Dumee, J. Lee, 'Nano/microplastics in water and wastewater treatment processes—Origin, impact and potential solutions', *Water Res.*, 2019, **161**, 621–38.

[43] C. Shen, Y. Zhao, W. Li, Y. Yang, R. Liu, D. Morgen, 'Global profile of heavy metals and semimetals adsorption using drinking water treatment residual', *Chem. Eng. J.*, 2019, **372**, 1019–27.

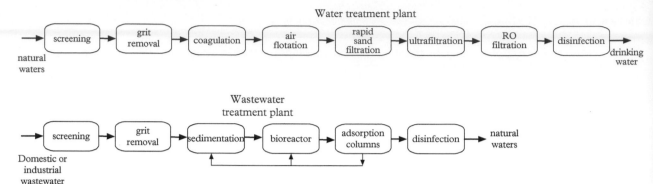

Figure 6.30 *Schematics of the steps in water and wastewater treatment plants (Based on Enfrin, Dumee, and Lee (n 42)).*

[44] G. Crini, E. Lichtfouse, L. D. Wilson, N. Morin-Crini, 'Conventional and non-conventional adsorbent for wastewater treatment', *Environ. Chem. Lett.*, 2019, **17**, 195–213.

Table 6.11 *UK Drinking water limits.*

Compounds	Maximum μg/L
benzo(a)pyrene	0.01
CH_2ClCH_2Cl	3.0
epichlorohydrin	0.10
Aldrin	0.03
other pesticides	0.10
total pesticides	0.50
polyaromatics (PAH)	0.10
C_2Cl_4 and C_2Cl_3H	10
total trihalomethanes	100
vinyl chloride	0.5

of operating plants use conventional, commercial absorbents based on activated carbons or silica gels.[44] An enormous variety of materials has been tested in the laboratory environment, but the large-scale installations tend to utilize materials of established engineering and performance properties. The materials employed are designed to provide large specific surface areas, typically 1,000 m^2 g^{-1} more more. The surfaces of activated carbons will generally have oxidized sites on the edge of graphite layers (Fig. 6.31), which can include phenols, ketones, and carboxylic acids separated by CH groups. Silica surfaces vary considerably with the pretreatment temperature. A classical view of a silica surface would be a quartz or cristabolite structure terminated by hydroxyl groups. These are non uniform. In Fig. 6.31, the left-hand silicone has an isolated silanol (Si–OH) group with its neighbour only having siloxane (Si–O–Si) bonds. The next two silicon sites show a *vicinal* silandiol. This allows a hydrogen bond link between the two silanols. This can stabilize the conjugate base and afford a more acidic site. The final site shown (on the right) is a *geminal* silanediol. In water there will be a layer of solvent molecules associated with these surface sites. We can see that, in different ways, activated carbons and silicas provide surface oxygen sites with a spread of Brønsted acidities and Lewis basicities. These reaction centres can also be used to further functionalize these surfaces. For example, the reagent HS(CH$_2$)$_3$Si(OMe)$_3$ can provide thiol groups on a silica surface.

6.7.2   Disinfection Methods

The final stage in both types of treatment plant is disinfection. The primary purpose is to remove pathogenic microorganisms which have passed through the upstream stages of the process train. This is generally carried out by oxidation processes and these can also remove other organic contaminants. The organic compounds limited by the UK Water Supply Regulations (2018) are listed in Table 6.11. They form three types: polyaromatics known to be carcinogens,

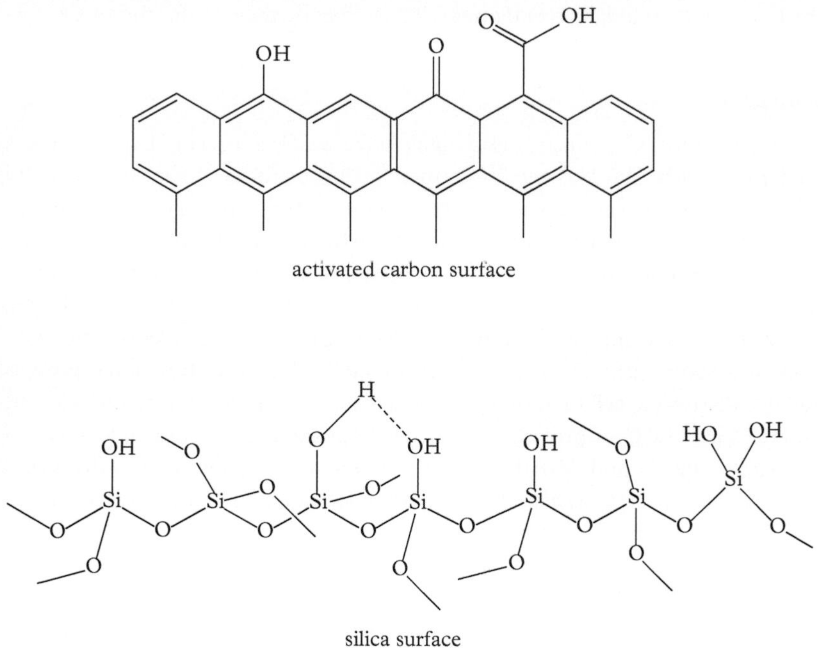

activated carbon surface

silica surface

Figure 6.31 *Sites on the surfaces of activated carbons and silicas.*

pesticides, and haloalkanes. The polyaromatics like benzo(a)pyrene, along with vinyl chloride and pesticides, can be adsorbed onto activated carbon filters.

Chorination

The long-established method of disinfection is by chlorination using Cl_2 gas. Cl_2 dissolves to a mole fraction of 1.88×10^{-3} at 298.15 K and so disinfectant baths can readily be permeated with it. Chlorine disproportionates in water to form $HO\text{-}Cl^I$ and HCl (Eq. 6.11). The hypochlorous acid dissociates between pH 7 and 8 (Eq. 6.12). The acid form is the more effective biocide and will remove bacteria, viruses, and protozoan cysts; the mechanisms appear to involve disruption of cell membranes and blocking enzymic action.

$$Cl_2 + H_2O = HOCl + H^+ \, Cl^- \qquad (6.11)$$
$$HOCl = H^+ + OCl^-. \qquad (6.12)$$

However, the environmental concerns about chlorine in the atmosphere and health concerns about haloalkanes have resulted in the move towards oxygen-based disinfectants. A mixture of trihalomethanes is generated by disinfecting with Cl_2 gas. These are regulated substances in the water supply (Table 6.11), largely due to their link to liver damage. Thus forming them at the end of a water-purification train is nonideal.

Ozonation

As a result ozone, O_3, which is also highly effective for removing the three types of microorganism, has become established in Europe. It has a short lifetime and so must be generated at the water-processing location by passing an electrode discharge through O_2. Its solubility in water (mole fraction 1.89×10^{-6} at 298.15 K) is much lower than that of Cl_2 and so mixing in the disinfectant bath is more challenging than for chlorination. The dose of O_3 is in the range of 5–15 mg L^{-1}, with contact time in the range of 15–50 min. Ozone reacts directly with multiple carbon–carbon bonds (ozonolysis) and with other functional groups, thus also disrupting cell membranes. It also reacts with water to generate radicals like $HO_2^{\bullet}$ and $OH^{\bullet}$ to provide further decomposition mechanisms. Ozone will also oxidize Fe(II) and Mn(II), allowing them to precipitate as $Fe(OH)_3$ and MnO_2, respectively. It will also oxidize trihalomethanes. The hydroxyl radical will also effect dehalogenation of haloethanes and ethenes.

Advanced Oxidation Processes

Disinfectant processes post-ozone treatment are called **advanced oxidation processes** (AOPs), and generally involve generation of hydroxyl radicals. This offers the possibility of improving the energy efficiency of the oxidizing agents over the electric discharge to form O_3 gas, and also improving the mixing of the oxidizing species. In Table 6.12, the electrode potentials of relevant oxidizing species are presented.[45] Hence ozone, hydrogen peroxide, and the hydroxyl radical provide stronger oxidants than encountered in chlorination, with $OH^{\bullet}$ being a stronger oxidant than O_3.

One enhancement of ozone treatment is to carry this out with UV irradiation. The light-induced reaction between O_3 and water affords hydrogen peroxide (Eq. 6.13). In the ensuing dark reactions, hydrogen peroxide can dissociate to give the hydroperoxide anion (Eq. 6.14). This in turn can react with ozone, and is oxidized to the radical (Eq. 6.15). The co-product radical anion, $O_3^{\bullet-}$, can transfer an oxygen anion to a proton generating $OH^{\bullet}$ (Eq. 6.16).

$$O_3 + H_2O + \text{UV} \rightarrow O_2 + H_2O_2 \tag{6.13}$$

$$H_2O_2 \rightarrow HO_2^- + H^+ \tag{6.14}$$

$$O_3 + HO_2^- \rightarrow O_3^{\bullet-} + HO_2^{\bullet} \tag{6.15}$$

$$O_3^{\bullet-} + H^+ \rightarrow O_2 + HO^{\bullet}. \tag{6.16}$$

The hydrogen peroxide can be added to the aqueous system in what is called **perozonation**. This has an advantage over a photochemical procedure when dealing with turbid systems that will scatter UV light.

[45] R. Hernandez, M. Zappi, J. Colucci, R. Jones, 'Comparing the performance of various oxidation processes for treatment of acetone contaminated water', *J. Hazard Mater.*, 2002, **92**, 33–50.

Table 6.12 *Thermodynamic oxidation potentials.*

Oxidant	Potential (V)
$OH^{\bullet}$	2.8
O_3	2.1
H_2O_2	1.8
HClO	1.5
Cl_2	1.4
O_2	1.2

An alternative method of generating hydroxyl radicals is using Fenton's reagent, invented by Henry John Horstman Fenton in Cambridge in the 1890s.[46] This reagent is typically formed by reaction of a solution of $FeSO_4$ with aqueous hydrogen peroxide (Equ. 6.17):

$$Fe^{2+} + H_2O_2 \rightarrow Fe^{3+} + OH^- + OH^\bullet. \tag{6.17}$$

The hydroxyl radical can then initiate chain reactions with organic pollutants. Both reagents can also terminate the chain by Eq. 6.18 and Eq. 6.19.

$$OH^\bullet + Fe^{2+} \rightarrow Fe^{3+} + OH^- \tag{6.18}$$

$$OH^\bullet + H_2O_2 \rightarrow H_2O + HO_2^\bullet. \tag{6.19}$$

Also the Fe(III) centres formed in the initiation step catalyse the decomposition of H_2O_2. Thus coordination of H_2O_2 to Fe(III) will increase its acidity, providing Eq. 6.20. Loss of the hydroperoxide radical reforms Fe^{2+} (Eq. 6.21). That in turn is oxidized by another ferric ion, causing the evolution of O_2 (Eq. 6.22).

$$Fe^{3+} + H_2O_2 \rightarrow [Fe - OOH]^{2+} + H^+ \tag{6.20}$$

$$Fe - OOH^{2+} \rightarrow Fe^{2+} + HO_2^\bullet \tag{6.21}$$

$$Fe^{3+} + HO_2^\bullet \rightarrow Fe^{2+} + O_2 + H^+. \tag{6.22}$$

The kinetics can be favourable for removal of organic pollutants since concentration ratios can make Eq. 6.23 faster than the chain-termination processes in 6.18 and 6.19. The organic radicals formed in Eq. 6.23 create further fast radical chain processes.

$$R - H + OH^\bullet \rightarrow R^\bullet + H_2O. \tag{6.23}$$

6.8 Other Organic Contaminants of Water

The regulations and guidelines covering irrigation and drinking water content do not, and cannot, cover all chemicals that might enter these supplies. Among the materials that have greatly expanded in use over recent decades, pharmaceuticals and plastics form important classes. The former compounds and formulations specifically have some biological effect and are designed to be administered to particular individuals. Thereafter they, and their metabolites, will generally be excreted and join water courses. The latter, generally, are considered (assumed) to be lacking any biological role. Collection and recycling systems cover only a proportion of used items. As considered in Section 5.2.3, degrading polymers follow differing pathways, largely dependent upon their backbone constitution. As a result there is a wide distribution of structures, chain lengths, and termination

[46] E. Neyens, J.Baeyens, 'A review of classic Fenton's peroxidation as an advanced oxidation technique', *J. Hazard Mater. B*, 2003, **98**, 33–50.

modes coexisting. Identifying these is a considerable analytical challenge. The section examines these two different classes of organic materials that have now entered our water supplies.

6.8.1 Pharmaceuticals

The situation with regard to pharmaceuticals in aquatic systems has been the subject of a very detailed assessment. It is estimated that there are $\approx$ 4,000 active pharmaceuticals available worldwide, with an annual consumption of $\approx$ 100,000 tonnes. Six of these comprise c.56% of this total, with non-steroidal anti-inflammatory drugs (NSAIDs) forming a substantial proportion. In order of decreasing consumption these are Metformin (Type 2 diabetes, 20%), Ibuprofen (NSAID, 12%), Metamizole (or Dipyrone: painkiller, 8%), Acetylsalicylic acid (Aspirin: NSAID, blood thinning, 7%), Acetaminophen (or Paracetamol: NSAID, 6%), and Iomeprol (X-ray contrast agent, 3%).[47] The structures of these compounds are shown in Fig. 6.32.

Over the last 40 years or so, pharmaceuticals have been identified in surface water, rivers, and wastewater at the level of μg L^{-1} or ng L^{-1}. Generally wastewater treatment plants were designed to include removal of relatively small, reactive organic molecules with an incoming concentration of mg L^{-1}, and outputs like those in Table 6.11. The tiny sample of structures in Fig. 6.32 demonstrates just a portion of the variety of chemicals within the pharmaceutical portfolio. The Australian government has introduced guidelines for pharmaceuticals

[47] M. Patel, R. Kumar, K. Kishor, T. Mlsna, C. U. Pittman, Jr., D. Mohan, 'Pharmaceuticals of Emerging Concern in Aquatic Systems: Chemistry, Occurrence, Effects and Removal Methods', *Chem. Rev.*, 2019, **119**, 3510–3673.

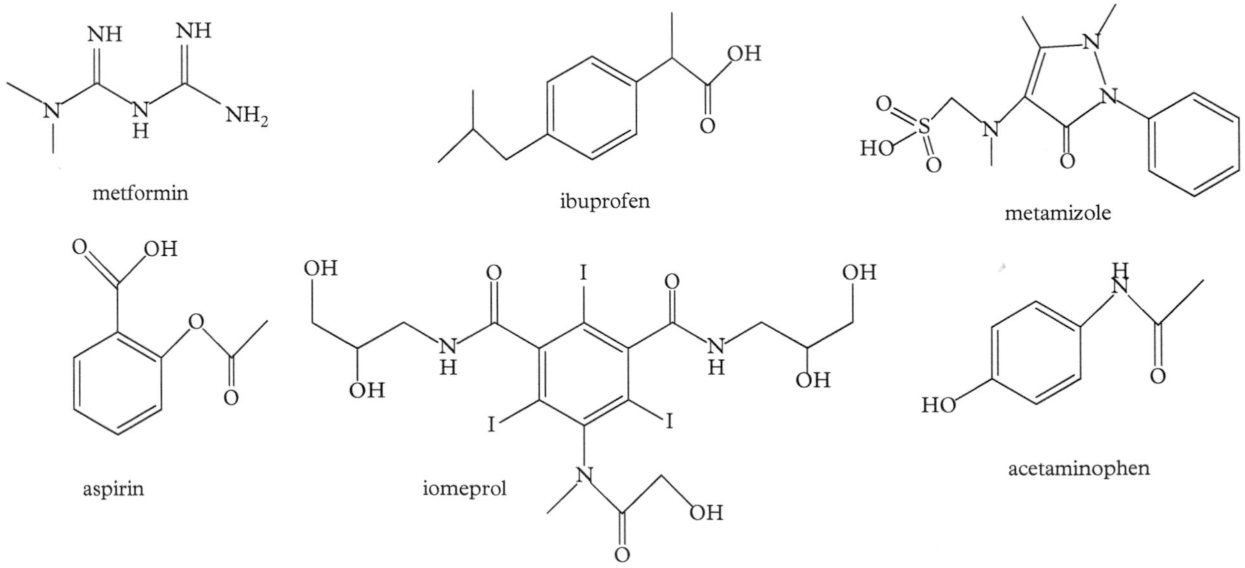

Figure 6.32 *Large-scale pharmaceuticals.*

starting from the lowest daily therapeutic dose (LDTD) for human use, from which surrogate(S)-acceptable daily intake (ADI) and drinking water guideline (DWG) were estimated.[48] These were all then reduced by three factors of 10 for safety (differences in human response, sensitive subgroups, and the lowest recommended doses above a no-effect level). In addition, cytotoxic drugs and hormones were given another safety factor of 10. From these, surrogate(S)-values of acceptable human daily intake (ADI) levels for pharmaceuticals in veterinary use and surrogate (S)-ADI values for those in human use and drinking water guideline (DWG) were derived.

A few examples of pharmaceuticals are listed in Table 6.13. Five of the six largest-scale pharmaceuticals are included in this table, with guidelines ranging from 29 to 525 μg L^{-1}. The excluded one is Iomeprol, which is designed not to have a pharmaceutical effect. Rather, it is the bearer of iodine. That has a high atomic number, and so very high X-ray absorption coefficients. Wastewater treatment plants are generally ineffective at removing this type of agent. Iomeprol has been identified in drinking water at a level of $\approx$ 1 ng L^{-1}, but this is unlikely to be a health risk. Three other pharmaceuticals are also included in Table 6.13: fluoxetine (Prozac), an antidepressant; diclofenac, an NSAID; and diazepam (Valium), the uses of which include treatment of anxiety (Fig. 6.33). All of these have considerably lower recommended limits for drinking water.

Observations can give wide ranges of reported values for these pharmaceuticals depending upon location. Taking ibuprofen as an example with a higher DWG value, many measurements place this at the ng L^{-1} range, but some locations are reported to be at much higher levels: 5 μg L^{-1} in influents to wastewater treatment plants, 20 μg L^{-1} in wastewater (Serbia), and 38 μg L^{-1} in a hospital effluent (Portugal). All of these values are within the guideline in Table 6.13. However,

[48] *National Water Quality Management Strategy, Australian Guidelines for Water Recycling: Managing Health and Environmental Risks (Phase 2), Augmentation of Drinking Water Supplies,* National Health and Medical Research Council, Canberra, Australia, 2008.

Table 6.13 *Australian guidelines for the limits of pharmaceuticals in drinking water (V = veterinary use).*

Pharmaceutical	LDTG mg d^{-1}	(S)-ADI μg kg^{-1}	DWG μg L^{-1}
aspirin	V	8.3	29
metamizole	V	150	525
acetaminophen	V	50	175
ibuprofen	800	11.4	400
metformin	500	7.1	250
fluoxetine	20	0.28	10
dichlofenac	V	0.5	1.8
diazepam	5	0.071	2.5

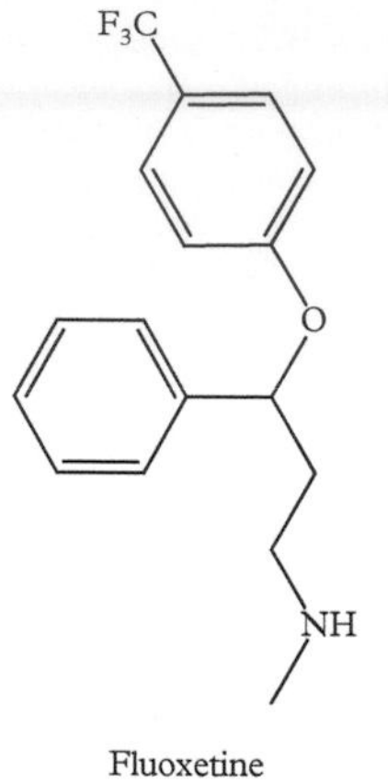

Fluoxetine

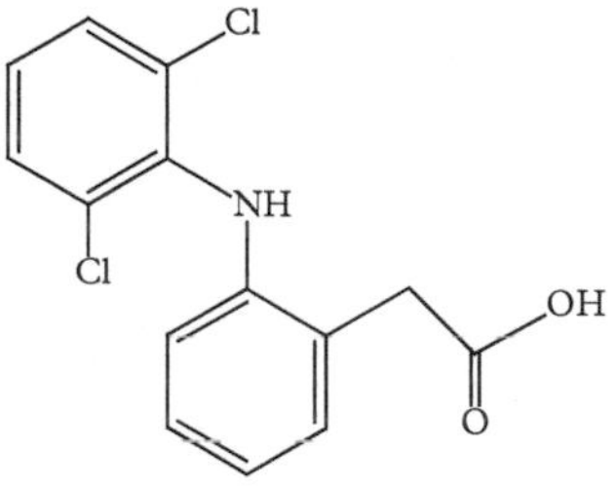

diclofenac

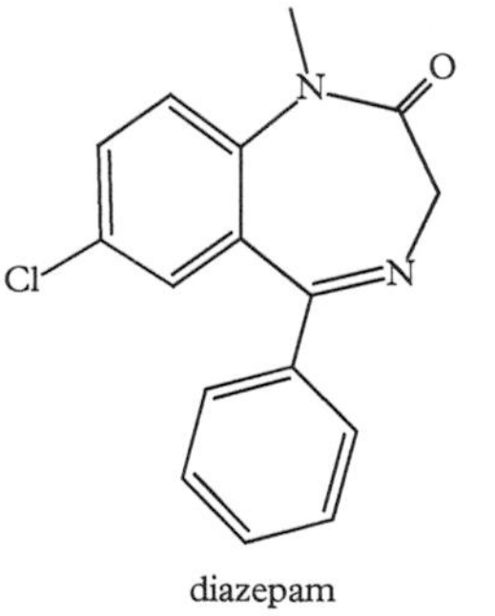

diazepam

Figure 6.33 *Three halogenated pharmaceuticals.*

if pharmaceutical manufacturing sites do not have appropriate waste treatment, these guidelines can be exceeded locally (e.g. 1.67 mg L^{-1} reported in Lahore, Pakistan). Considering the examples with the lower DWG values, diazepam and fluoxetine appear to be under 1 μg L^{-1} by a considerable margin even around wastewater treatment plants. However, for diclofenac, the guideline in Table 6.13 has indeed been exceeded in surface water (Kwa-Zulu Natal, RSA), in some wastewater treatment effluents (Greece, Cyprus), and greatly so in some industrial wastewater (Lahore).

This highlights the importance of the efficacy of treatment plants. Much can be removed by biological and chemical processes in the sedimentation/bioreactor stage. Aspirin and ibuprofen are degraded within 2–5 d, whilst diclofenac requires 5–15 d. Diazepam is unchanged in this stage of the treatment, even after 20 d. In general, it is considered that plants with ozonation remove > 90% of pharmaceuticals, leaving ng L^{-1}. Ozonation will remove diclofenac, with the secondary amine group being the susceptible site. Diazepam and ibuprofen, however, react slowly with O_3. AOPs do efficiently degrade pharmaceuticals in wastewaters and aquatic systems; however, they may be converted into toxic oxidation products. Given the cocktail of pharmaceuticals that might be present in a treatment plant, it requires a carefully judged mix of methods to ensure removal is optimized. Excepting for some poorly resourced locations, ambient concentrations in waters do not seem to be close to being health risks.

6.8.2 Plastics

The scheme of water treatment plants in Fig. 6.30 provides a basis too for discussing the implications of plastic waste on water treatment.[42] The initial screening will remove items which are millimetres in size or greater. As discussed in Section 5.2, types of polymer have differing linkages and this affects their rate and mode of degradation. Polyalkenes have no linking functional group and so degradation of the chain generally requires photolysis creating a free radical site (Fig. 6.34). In an aerobic environment the subsequent reactions of the radical with O_2 generates carbon, hydrogen, and hydroxyl radicals, setting up a set of chain reactions. The original polymer chain will undergo a C–C scission affording long-chain ketone and alkene. Polymers with functional groups as links (Fig. 5.22) will generate degrade by hydrolysis to reform the initial functional groups, but now as the terminals of polymer chains of lesser length than the original one. Hence for PET the terminations will be an alcohol and a carboxylic acid, and for a polyamide, these end groups will be a carboxylic acid and an amine.

This breakdown of polymer chains can be envisaged on beaches and water surfaces. Some is also mediated by bacteria, and physical stress can cause breakdown of plastics. This may often be due to interparticle and interchain binding, but severe stress can also engender high local temperatures. As a result, microplastics

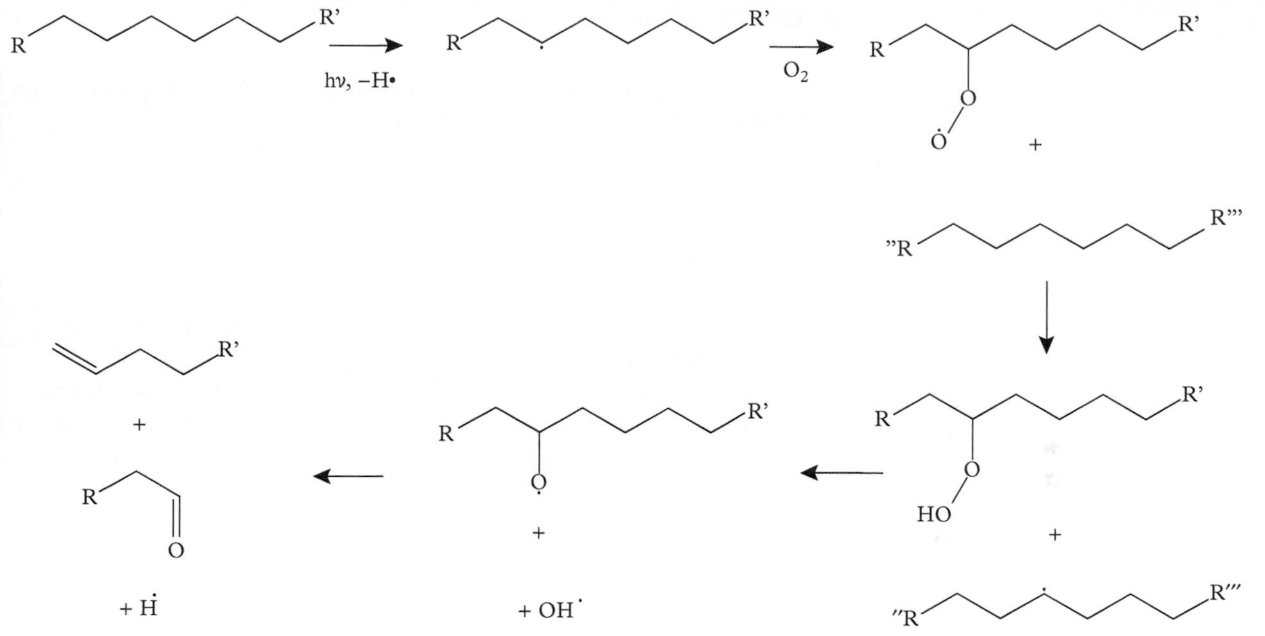

Figure 6.34 *Steps in polyethylene (PE) photodegradation.*

(MPs) have been identified in wastewater with particle sizes of 20–1,000 μm. Nanoplastics (NPs) with sizes down to 30 nm have been reported.[49] There are naturally occurring nanoparticles that have been found in a concentration range of $0.5–20 \times 10^8$ mL^{-1}. Studies on a variety of plastics in demineralized water at 30 °C under visible and UV (320–400 nm) light showed that the density of nanoparticles was considerably higher than the control. The distribution of particle sizes was polymer-dependent, with PLA (polylactic acid) producing a relatively high density of NPs but PET a lower mean particle size.

As a result of this, wastewater treatment plants can be assumed to be dealing with influents with MPs and NPs. There are many points within the treatment processes in which physical, chemical, biochemical, and photochemical mechanisms of change occur, with the expectation that will reduce particle sizes and polymer lengths through the procedures.[42] MPs and NPs may interfere with adsorption sites, and foul or abrade filtration membranes. This may impair the effectiveness of the treatment stream in removing other contaminants and so adaptation of plants to tackle this poorly understood stream of materials will be required. As noted in Section 5.2.3, there is over 200 Mt of plastic waste per year, and inevitably this will confront water treatment design in the years to come.

[49] S. Lambert, M. Wagner, 'Formation of microscopic particles during the degradation of different polymers', *Chemosphere*, 2016, **161**, 510–7.

6.9 Questions

1. The vapour pressure of water can be estimated by the Buck equation (P in kPa and T in °C):

$$P = 0.61121 \times e^{\left(18.678 - \frac{T}{234.5}\right)\left(\frac{T}{257.14 + T}\right)} \tag{6.24}$$

 (i) Given that the molar volume (V_m) at 273.15 K is 22.4 L, calculate the V_m values at 20, 25, and 30 °C.
 (ii) Calculate the vapour pressures (kPa) of water at these three temperatures.
 (iii) Calculate the vapour density (kg m^{-3}) of water at these temperatures.
 (iv) A simple model of two rain storms would be to imagine a volume of 1 km^3 of atmosphere dropping in temperature by 5 °C from 25 °C and from 30 °C. Calculate the volume of water held in this atmospheric volume at saturation vapour pressure at these temperatures. From this calculate the water released from the atmosphere after the 5 °C temperature drops and the depth of the rain if this were deposited over an area of 1 km^2.

2. The annual production of the major materials produced worldwide tabulated in 2016 and their virtual water (VW) values are shown in Table 6.14.

Table 6.14 *Large-scale materials.*

Material	Annual production (tonnes)	VW L kg^{-1}	VW L y^{-1}	VW km^3 y^{-1}	VW L d^{-1} pp
cement	7.5×10^8	37			
paper/card	3.6×10^8	1,000			
soda glass	8.1×10^7	18			
aluminium	4.4×10^4	1,200			
iron	2.3×10^9	50			
brick	5.0×10^7	5.5			
bamboo	1.4×10^9	700			
softwood	9.6×10^8	700			
polyethylene	6.8×10^7	58			
polypropylene	4.3×10^7	39			
polyester	4.0×10^7	200			
polystyrene	1.2×10^7	140			

(i) Complete Table 6.14 by calculating the total virtual water required for the production of the material in L and km^3, and also the daily consumption per person (take the world population as 7.73 billion).

(ii) Compare these values to those for the production of freshwater given in section 6.2. Consider the consequences of this comparison.

3. Table 6.15 includes information about the known land reserves[50] and seawater concentrations.

(i) Calculate the molar concentration of these elements in seawater.

(ii) Taking the volume of the oceans as 1.338×10^9 km^3, complete this table by calculating the estimated quantity of these elements in seawater and hence the ratio between that and the land reserves.

(iii) Comment on the prospects for mining these elements.

4. The standard electrode potentials of vanadium are as follows:
V^{2+}/V -1.175 V; V^{3+}/V^{2+} -0.255 V; $VO^{2+}, 2H^+/V^{3+}, H_2O$ 0.337 V; $VO_2^+, 2H^+/VO^+, H_2O$ 0.991 V.

(i) Construct a Frost diagram as described in Section 2.7 for vanadium (that is $n.E°$ versus oxidation state; n is the change in oxidation state of the electrode potential).

(ii) Two of these steps involve the consumption of protons. Write out the chemical equations for these steps.

(iii) Correct the potential for these steps for the pH of seawater (8.2) and amend the Frost diagram.

(iv) Comment on the observation that methods to selectively extract uranium from seawater experience competition from vanadium.

5. Molybdenum is present as different types of ores, for example as MoO_3 in oxidic environments and MoS_2 is sulphidic ones. Under oxidizing weathering conditions it can be present in groundwater at concentrations of the order μM in some regions. There is strong evidence that ruminant

[50] USGS National Minerals Information Center, Commodity Summaries, 2019.

Table 6.15 *Minerals on land and in the ocean.*

Element	AW	Reserves (tonnes)	Ocean mg L^{-1}	Ocean mol L^{-1}	Ocean (tonnes)	Ocean/land
sodium	22.99	1.0×10^9	10,800			
magesium	24.305	2.9×10^9	1,290			
potassium	39.098	4.8×10^9	392			
lithium	6.94	1.4×10^7	0.17			
molybdenum	95.45	1.7×10^4	0.01			
copper	63.546	8.3×10^9	0.0009			
gold	196.967	5.4×10^4	0.000011			

animals in these regions can suffer from diseases associated with copper deficiency. Indeed, treatments for this induced copper deficiency seek to ensure that the Cu:Mo molar ratio should be in excess of 2:1.

(i) What species may be responsible for the transport of molybdenum in groundwater?

(ii) What other molybdenum complex could be formed in the rumen of these animals?

(iii) How might this interact with copper?

Prospects for Planet Earth

7

7.1 Chapter 1: Planet Earth

The increase in world population was noted in Section 1.1, where its doubling during the writer's professional life was illustrated (Fig. 1.2). The expansion in the plot showed kinks in the increase post 1980. The trend from 1950 to 2020, as estimated by the United Nations,[1] (Fig. 7.1) shows a tendency to move from a concave to a convex curve, so indicating the prospect of avoiding runaway population growth. The annual change in world population, averaged over 5-year blocks, peaked around 1950 and again at over 2% in the 1965–70 quinquennium. It then fell to a plateau through 1975–80 to 1985–90. Since then there has been a further decline to $\approx 1.1\%$. Present estimates indicate a peak in world population of 11.2 billion might occur around 2100, with a prospect of stabilization thereafter.

In Section 1.1, a comparison in the change in human fertility and life expectancy was presented (Fig. 1.3). An improvement in health can be perceived from the increase in global average life expectancy from $c.52$ in 1960 to over 70 by 2015. Mean life expectancy is one of the figures that is grossly underestimated by most people![2] Peaks in average life fertility (i. e. live births per woman) of over 5 were reached around 1950 and again in the late 1960s, as also shown in Fig. 7.2. By 2018, the average fertility had dropped to 2.47; a value of 2.1 is considered to be required to give a constant population. Over the same time period changes in life conditions allowed the mortality rate of children under the age of 5 to be reduced from ≈ 22.5 to $\approx 4.0\%$. It does appear that this substantial reduction in the risk of losing a child can be clearly correlated with a reduction in children that are borne. This demographic transition has been demonstrated using much international data allied to imaginative graphics.[3,4,5] The place of a particular country, or region within a country, along these curves varies depending the state of development and living conditions. Factors such as the length of educational period of girls and female employment also show correlation with fertility—more so perhaps than with religious or political emphasis. Time series plots show that continent by continent this pattern predominates, starting with Europe. Africa is the last continent to be able to follow this demographic change. Aid and development are not at all dissociated from improving the prospects for planet Earth.

[1] United Nations Population Division, population.un.org.

[2] H. Rosling, O. Roslin, and A. Rosling Rönnlund, *Factfulness*, Hodder and Stroughton, London, 2018.

[3] https://www.gapminder.org/data.

[4] Robert Wilson, @countcarbon.

[5] https://www.ourworldindata.org.

Elements of a Sustainable World. John Evans, Oxford University Press (2020). © John Evans.
DOI: 10.1093/oso/9780198827832.003.0007

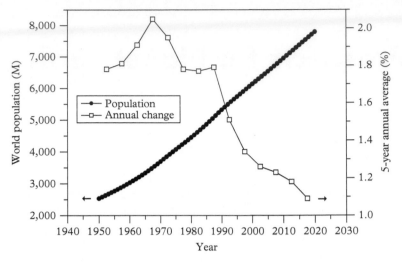

Figure 7.1 *World population estimates and annual changes (%) 1950–2020.*

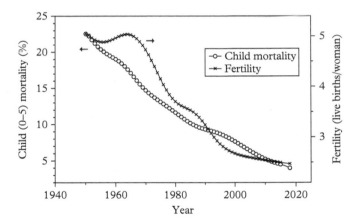

Figure 7.2 *Child mortality and live births per woman from 1950–2018.*

6 *Trends in global greenhouse gas emissions: 2018 report*, PBL Netherlands Environmental Assessment Agency, https://www.pbl.nl/en.

The annual emissions of CO_2 have risen from 22.7 Gt in 1990 to 37.1 Gt in 2017 (Fig. 7.3).[6] Per person, CO_2 emissions are estimated to have fallen from ≈ 4.3 t y^{-1} in 1990 to 4.1 t y^{-1} in the early part of that decade and remained stable for about 10 years. There was a significant rise through the next decade to ≈ 5.0 t y^{-1}, with the exception of a downward blip in 2009 following the onset of the financial crisis. There is a sense that this personal emission factor might seem to be relatively stable, and perhaps even have started to decrease. However, in the 12-month period from September 22 2018 the weekly average of the CO_2 concentration measured at the Mauna Loa Observatory increased by 2.7 ppm (to 408.3 ppm) and so the global emissions have certainly not peaked.

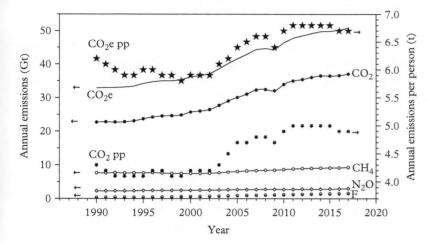

Figure 7.3 *Annual emissions (Gt) of CO_2, CH_4, N_2O, and fluorinated gases (labelled F) from 1990–2017. Also given are the total emissions in CO_2e (Gt) and the values of CO_2 and CO_2e per person (t).*

The emissions of CO_2 in 2010 by sector[5] are compared in Fig. 7.4.[7] As may be expected, it is the energy sector which dominates this (64%), followed in order by transport (17%), residential and commerce (11%), and agriculture and industry (each 8%). These values are partly offset by the net carbon capture estimated for the forestry sector, which is also at 8%. This would suggest that doubling the capture by forestry would have about the same effect as reducing the emissions by the energy sector by $\approx \frac{1}{8}$. Amelioration of CO_2 emissions can be significant across all of the sectors, but it is energy which dominates. The balance between agriculture and forestry offers interesting possibilities.

The other three types of emissions shown in Fig. 7.3 are all of gases with higher global warming (GWP) and temperature (GTP) potentials than CO_2 (Table 1.12); their emissions are expressed in terms of CO_2 equivalents (CO_2e). On that basis, emissions of methane (21%) and nitrous oxide (30%) have increased less than those of CO_2 (63%) over this time period. The patterns of emission sources for methane (Fig. 7.5) are quite distinct from those of CO_2. Energy is again important (37%), but agriculture (38%) is very similar in magnitude; emissions from the waste industry (17%) are also significant. For nitrous oxide emissions (Fig. 7.6), agriculture is the dominant source (72%), with energy (6%), industry (5%), and waste (4%) as more minor sources. Interestingly, transport is estimated to cause just 0.5% of the N_2O emissions. The importance of agricultural practices is highlighted by Figs 7.5 and 7.6.

The emissions of the fluorinated gases (in Fig. 7.3), which have very substantially higher GTP values than even N_2O, have increased to 450% of the 1990 value (from 0.353 to 1.544 Gt CO_2e). This percentage change can overestimate the effect of these gases. Unlike CH_4 and N_2O that had a baseline value arising

[7] H. Ritchie, M. Roser, *CO₂ and Greenhouse Gas Emissions*, 2019, published online at OurWorldInData.org.

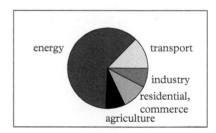

Figure 7.4 *Emissions of CO_2 in 2010 by sector.*

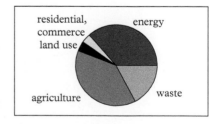

Figure 7.5 *Emissions of CH_4 in 2010 by sector.*

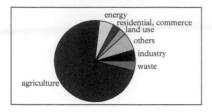

Figure 7.6 *Emissions of N₂O in 2010 by sector.*

[8] Oxford Environmental Institute, http://www.globalwarmingindex.org, using data from IPCC 2013, https://ipcc.ch/report/ar5/wg1/.

from natural and longstanding procedures in agriculture and waste processes, the fluorinated gases essentially started from a zero point until they were manufactured in the twentieth century. The overall result of this is that, although the curves of the CO_2e emissions are similar in shape to those of the dominant factor, CO_2, there is a divergence of slope. Additional emissions to the atmosphere equivalent to over 50 Gt of CO_2 are occurring. That there is a distinct increase due to the fluorinated gases is significant in the long term. Reversing the warming effects will have a longer time lag due to the very long residence times of the fluorinated contaminants (Table 1.10). They also have a potential to build up with little decay over millennia. Developing F-gas capture (for SF_6, NF_3, and perfluorocarbons (PFCs)) and their storage or chemical degradation may become a priority in future.

The variation in global temperature is often presented as temperature anomaly (ΔT) compared to the mean over a particular time period. In Fig. 7.7 a period before very large industrial age emissions has been employed (1850–79).[8] There is significant scatter in the observations but an upward curve is evident. Atmospheric modelling has been employed to discriminate between natural variations and the effect of human activity (anthropogenic effects). The changes due to natural effects do not form long-term changes in the baseline. Rather, they take effect for periods of years to decades and generally present a temperature-reducing factor of the order of 0.1–0.2°C. In contrast, the anthropogenic factors, dominated by the change in CO2e, form a longer-term upward trend. The total computed effects of changes of Earth's condition on its surface temperature are shown as a solid line. There are periods when natural effects dominated, specifically those close in time to the averaging period. There have also been periods when the effects were comparable, as in the 1960s. Now the dominant factor is evidently anthropogenic, and this is still on an upwards curve. Even if there is stabilization in the rate of

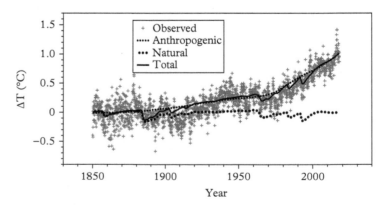

Figure 7.7 *Global temperature changes from 1850–2017 as a temperature anomaly (ΔT) compared to the mean from 1850–79. The computed effects from natural and anthropogenic variations are shown as solid and dashed lines.*

CO_2 emissions, the atmospheric concentration is increasing and this will result in a raising of surface temperature for a considerable period of time ahead.

7.2   Chapters 2 and 3: The Palette of Elements, and Earth

A selection of raw materials was listed in 2017 as having potentially critical supply issues for the European Union (Table 7.1).[9] Three materials had been removed from the 2014 list: chromium, coking coal, and magnesite. A factor in the assessment of criticality includes the substitution index. This represents the difficulty in substituting a material by another across the range of uses of the raw material. The scores range from 0 to 1, with 1 being the least substitutable. Two

[9] *On the 2017 list of Critical Raw Materials for the EU*, COM(2017) 490 final, European Commission, Brussels.

Table 7.1 *Elements and raw materials listed as Critical Raw Materials by the EU in 2017.*

Material	Substitution indexes (EI / SR*)	End-of-life recycling (%)
antimony	0.91 / 0.93	28
baryte	0.93 / 0.94	1
beryllium	0.99 / 0.99	0
bismuth	0.96 / 0.94	1
borate	1.0 / 1.0	0
cobalt	1.0 / 1.0	0
fluorspar	0.98 / 0.97	1
gallium	0.95 / 0.96	0
germanium	1.0 / 1.0	2
hafnium	0.93 / 0.97	1
helium	0.94 / 0.96	1
indium	0.94 / 0.97	0
magnesium	0.91 / 0.91	9
natural graphite	0.95 / 0.97	3
niobium	0.91 / 0.94	0.3
phosphate rock	1.0 / 1.0	17
phosphorus	0.91 / 0.91	0
scandium	0.91 / 0.95	0
silicon metal	0.99 / 0.99	0

Table 7.1 *continued*

Material	Substitution indexes (EI / SR*)	End-of-life recycling (%)
tantalum	0.94 / 0.95	1
tungsten	0.94 / 0.97	42
vanadium	0.91 / 0.94	44
platinum group metals	0.93 / 0.98	14
heavy rare earth elements	0.96 / 0.89	8
light rare earth elements	0.90 / 0.93	3

* EI = economic importance; SR = supply risk.

factors are listed and are used to correct the scores for economic importance (EI) and supply risk (SR). The final column presents the ratio of recycling of old scrap to EU demand, expressed as a percentage.

Unsurprisingly, there is considerable overlap with the minerals listed in Table 2.1 based upon the relative supply risk derived by the British Geological Society. The bulk of the raw materials given in Table 7.1 have zero or very low end-of-life recycling rates. Vanadium and tungsten are clear exceptions. Pathways to recycle platinum group metals (PGMs) are well established due to the dilute nature of the ores (Section 3.3.5). The high requirement for water in PGM extraction in regions of water stress is also a driver in increasing recycling rates (Section 6.4). It may be that the sequestering of PGMs in long-term products, such as jewellery, has a significant effect on the end-of-life recycling rate.

Employing the 'three Rs'—reduce, reuse, and recycle—is envisaged as an important component of a sustainable approach to engineering materials.[10] Good quality materials and engineering design can reduce the mass of an element required for a construction and extend its lifetime. These engineering solutions contribute substantially to lowering the emissions associated with the materials needed for a product.

A comparison between the EE and EC values for virgin (primary) and recycled (secondary) metals is presented in Table 7.2. For an element like lead, recycling provides reusable lead with about $\frac{1}{4}$ of the energy and carbon emissions required for processing from freshly mined galena (PbS). For more energetically intensive processes, such as the reduction of Al(III) to the metal, both the differences in the absolute values of EE and EC and the recycled/virgin ratio of $\approx \frac{1}{7}$ indicate even greater favourability for recycling. And in an example like palladium these ratios for EE ($\approx \frac{1}{30}$) and EC ($\approx \frac{1}{20}$) provide compelling reasons for the efforts spent in establishing the recovery systems. Considering Table 7.1, this can provide the basis of prioritizing further recycling units for those elements with poor retention. Metal-recovery sites at recycling centres can provide relatively

[10] J. M. Allwood and J. M. Cullen, *Sustainable Materials with both eyes open*, UIT Cambridge, Cambridge, 2012.

concentrated artificial mineral resources and their exploitation can have benefits in removing toxins from the environment and substituting from mineral extraction.

The demands for mineral resources are, in the main, increasing.[11] The four main regions of exploration in terms of numbers of sites are, in descending order, in Canada, Australia, Latin America, and Africa. About 51% of the exploration budgets in 2017 for mineral commodities were devoted to gold and silver, with the proportion devoted to the base metals (Co, Cu, Pb, Mo, Ni, Sn, Zn) the next-highest basket at 11%. Of this basket, $\approx$ 69% was devoted to copper exploration, 21% to lead and zinc together, and 10% to nickel. Diamond exploration amounts to $\approx$ 3% of this mineral exploration and PGMs to < 1%. Since 2015, Li_2CO_3 has undergone the largest price rise, driven by lithium batteries; cobalt exploration sites increased by $\approx$ 120% from 2016–17 because of the same driver.

The scope for extraction of elements from seawater was considered in Section 6.6. Two important parameters for viability were the absolute concentration of the element in seawater, and the abundance relative to that in the Earth's crust. Lithium (0.17 ppm, crust/sea 94) was seen as potentially viable, and uranium presently a stretch target (3.3 ppb, crust/sea 394). Of the elements listed in Table 7.1, those more favourable than lithium in terms of concentration are boron (4.45 ppm), fluorine (13 ppm), magnesium (1,290 ppm), and silicon (2.9 ppm). But the crustal abundances of most of these are substantially higher than those in seawater, so borate might be the most plausible option for extraction from seawater.

One evident use of the surface of land is in the production of food. Perhaps more than for many other processes there is a wide variety of environments and practices across the Earth which will have different energy requirements, emission levels, and water demands. Table 7.3 collects some of the estimates

[11] D. R. Wilburn, N. A. Karl, 'Exploration Review, Annual Report 2017', *Mining Engineering*, 2018, 28–50.

Table 7.2 *Embodied energy (EE) and carbon (EC) for virgin and recycled materials.*

Material	EE MJ/kg	EC KgCO$_2$/kg
Aluminium:		
Virgin	218	12.8
Recycled	29	1.81
Steel:		
Virgin	35.4	2.89
Recycled	9.46	0.47
Copper:		
Virgin	57	3.81
Recycled	16.5	0.84
Zinc:		
Virgin	72.0	4.18
Recycled	9.0	0.52
Lead:		
Virgin	27	2.0
Recycled	7.5	0.45
Palladium:		
Virgin	155,000	8,500
Recycled	5,400	426

Table 7.3 *EC* and VW* estimates for foodstuffs.*

Foodstuff	EC (kg CO$_2$e kg^{-1})	VW (L kg^{-1})
beef	27.0	16,000
cheese	13.5	5,290
pork	12.1	5,910
poultry	6.9–10.9	2,830
eggs	4.8	4,660
potatoes	2.9	160
rice	2.7	2,660
tofu	2.0	3,030
milk (2%)	1.9	865
tomatoes	1.1	261
lentils	0.9	7,060

* EC = embodied carbon; VW = virtual water.

Table 7.4 *Representative concentration pathways (RCPs).*

Pathway	Radiative forcing ($W\ m^{-2}$)	CO_2e (ppm)	Effects
RCP8.5	> 8.5 in 2100	> 1,370 in 2100	rising
RCP6.0	≈ 6.0 when stabilized	≈ 850 when stabilized	stabilized after 2100
RCP4.5	≈ 4.5 when stabilized	≈ 650 when stabilized	stabilized after 2100
RCP2.6	≈ 3.0 peak	≈ 490 peak	peaks before 2100 and declines

[12] K. Hamerschlag, K. Venkat, *Meat Eater's Guide to Climate Change and Health*, Environmental Working Group, Portland, Oregon, 2011.

[13] *Water: A shared responsibility*, The United Nations World Water Development Report 2, UNESCO and Berghahn Books, New York, 2006.

[14] R. H. Moss, J. A. Edmonds, K. A. Hibbard, M. R. Manning, S. K. Rose, D. P. van Vuuren, T. R. Carter, S. Emori, M. Kainuma, T. Kram, G. A. Meehl, J. F. B. Mitchell, N. Nakicenovic, K. Riahi, S. J. Smith, R. J. Stouffer, A. M. Thomson, J. P. Weyant, and T. J. Wilbanks, 'The next generation of scenarios for climate change research and assessment', *Nature*, 2010, **463**, 747–56.

[15] *Global Warming of 1.5°C*, IPCC special report, 2018, https://www.ipcc.ch/sr15/.

[16] A subsequent (2017) classification is a set of scenarios called shared socioeconomic pathways (SSPS).

made for the embodied carbon (EC)[12] and virtual water (VW)[13] of a selection of foodstuffs to provide a rough estimate for comparison with products with radically different purposes like metals, plastics, and structural materials (Tables 5.2, 6.6, and 7.2). Such comparisons can produce some rather eclectic coincidences like the EC values of rice and polyethylene and the VW values of tomatoes and nickel. However, it does provide some hints about the demands that we make with our daily choices.

7.3 Chapter 4: Air

A widely used classification of scenarios for the atmosphere is that of **representative concentration pathways** (RCPs).[14] The value associated with the RCP is the anthropogenic radiative forcing at the end of the twenty-first century in $W\ m^{-2}$. This can be related to the effects introduced in Section 1.7. The warming produced is measured against the global surface mean surface temperature (GSMT) over a reference period. The IPCC have adopted 1850–1900 as the reference, and temperature change at a given time is the average over a 30-year period around that. The target of 1.5°C warming is based on this definition.[15] Other reference times are used in publications which will give baseline shifts. For example, this '1.5°C' would correspond to a warming of 0.87°C from 1986–2005.

The envisaged results of these scenarios of the collective actions of humanity are given, in broad terms, in Table 7.4. RCP8.5 incorporates some mitigation strategies with a high estimate of emissions and population. It is not a 'do nothing' plan, and could be a consequence of positive feedback from global warming results (e. g. release of methane from permafrost regions). Refinements of this scenario (SSP8.5)[16] have been presented. These scenarios risk a warming of ≈ 4.9°C by 2100. Scenarios with lower population growth and more effective mitigation could achieve reductions in radiative forcing to 6.0–4.5 $W\ m^{-1}$, estimated to result in increases in GMST of 3.0–2.4°C. Clearly none of these scenarios model a recommended outcome of 1.5°C. It is thought that to achieve this, some pathway

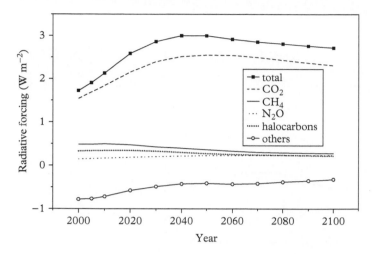

Figure 7.8 *Estimated radiative forcing up to 2100 according to RCP2.6.*

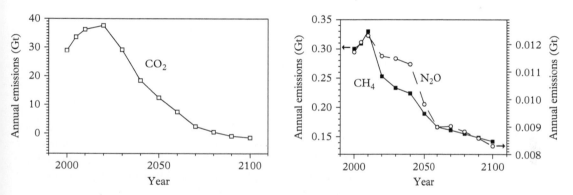

Figure 7.9 *The annual emissions envisaged under RCP2.6 to 2100.*

akin to scenario RCP2.6 would have to be attained.[17] That is shown in Fig. 7.8, using the RCP Database.[18] There is a peak in radiative forcing near 3 W m^{-2} in the 2040s, followed by a reduction to 2.6 W m^{-2} by the end of the century. The peak in radiative forcing from CO_2 emissions would be a decade later, and the earlier overall peak was largely derived from reductions from methane (residence time 12 y) by 2020 and halocarbons (some with residence times of tens of years) by 2030. N_2O (residence time 121 y) would not peak until 2090. Partly offsetting the reductions is a rise in the 'others' curve, which is a cleaning up of a combination of factors such as ozone emission and aerosols.

Attaining this pattern in atmospheric cleaning requires a pattern in annual emissions. Emissions matching this pathway of the major gases are presented in Fig. 7.9. This can be compared to the observed trends from 1990–2017 in Fig. 7.3. Emissions of CO_2 are required to drop sharply from 2020 and the continued pattern would result in a negative emission after 2080. Clearly this exceeds the

[17] D. P. van Vuuren, E. Stehfest, M. G. J. den Elzen, T. Kram, J. van Vliet, S. Deetman, M. Isaac, K. Klein Goldewijk, A. Hof, A. Mendoza Beltran, R. Oostenrijk, B. van Ruijven, 'RCP2.6: exploring the possibility to keep global mean temperature increase below 2°C', *Climate Change*, 2011, **109**, 95–116.

[18] https://iiasa.ac.at/web/home/ research/researchPrograms/ TransitionstoNewTechnologies/RCP. en.html.

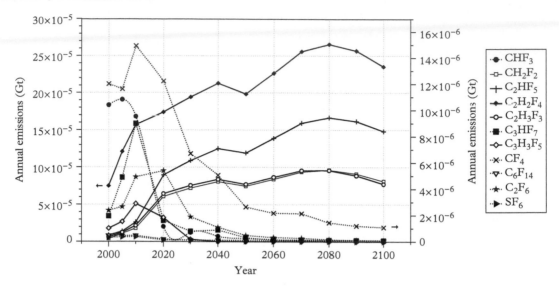

Figure 7.10 *Annual emissions of fluorinated gases envisaged under RCP2.6 (solid lines, left-hand scale; dotted lines, right-hand scale).*

reductions that might be envisaged for net zero emission and will require some carbon capture and storage. Emissions of methane and nitrous oxide were also required to turn down sharply in 2010. The relatively short residence time of CH_4 would then allow its concentration to fall within a decade and to track the proposed reduction of emission of 50% with a time lag. The overall reduction for N_2O is $\approx \frac{1}{3}$ but the time lag is much longer due to the residence time effect. However, it is evident from Fig. 7.8 that these reductions in the emissions of these gases had not occurred by 2017, and so the onset of the reductions in radiative forcing have been delayed.

It is evident too that the rise in the atmospheric concentration of fluorinated gases (Fig. 7.3) is envisaged to be reversed allowing their contribution to radiative forcing to reduce by $\approx \frac{1}{3}$ by 2100. This would reduce the proportion of their radiative forcing to the total from c.18% in 2010 to 8% in 2100. The appropriate emissions of fluorinated gases within RCP2.6 are plotted in Fig. 7.10. This inventory includes four perfluoro compounds: CF_4 (CFC-14), C_2F_6 (CFC-116), C_6F_{14} (FC-72), and SF_6. It is CF_4 which has the highest estimated emissions in 2010 (51,000 t). Given its exceptional estimated residence time (50,000 y) it seems appropriate to aim at a reduction in its emissions, here by an order of magnitude. All are envisaged to have much lowered emissions, presumably as a result of further restriction on their use and better containment. However, the major emissions in terms of mass are the hydrofluorocarbons (HFCs) seen as 'environmentally friendly' replacements of CFCs which led to the depletion of stratospheric ozone. All have large GWP and GWT factors, though. Emissions from $C_2H_2F_4$ (HFC-134a), C_2HF_5 (HFC-125), $C_2F_3H_3$ (HFC-143a), and CH_2F_2 (HFC-21) are expected to be the dominant ones with total emissions from them peaking over 600,000 t y^{-1} around 2080. The residence times of such gases vary considerably. HFC-23 (CHF_3) has a relatively long residence time (270 y) but is envisaged to have reduced emissions as restrictions on use are tightened. HFC-134a ($C_2H_2F_4$) has a much shorter residence time of 14 y. Restrictions in HFC applications as refrigerants, propellants, and foaming agents have been established in recent years so it is plausible that this prolonged increase in emissions may not transpire.

Further mitigation strategies are required to effect the sharp reduction in the emissions of the major greenhouse gases, CO_2, CH_4, and N_2O. For methane the key is to minimize emissions as it has a relatively short residence time in the atmosphere and so will naturally decay over a period of decades. But the spread of the emission types make this challenging (Fig. 1.5). In the energy field, safe capture of escaping gas in oil extraction and processing, avoiding flaring, could make a significant contribution. Likewise some inroads may also be achieved by increasing the proportion of organic waste that is digested under enclosed reactors. With a train of gas-separation processes, this can provide some 'green' natural gas for grid use. N_2O is challenging since the bulk of emissions come from agriculture, and it does have a longer residence time. Most of the agriculture emissions arise from soil management, and the degree to which organic and synthetic nitrogenous fertilizers are employed. Other mechanisms are through the management of manure and the burning of agricultural residues. N_2O is also emitted during wastewater treatment. Since these plants are in particular locations, methods of collection can be envisaged.

However, the main effect on radiative forcing will be to reduce the atmospheric concentration of CO_2. CO_2 is more persistent than N_2O in the atmosphere, so achieving a reduction in concentration requires a driven process. Under the RCP2.6 scenario this cannot be achieved solely by the reduction in emissions. As discussed in Section 4.1.3, there are seasonal changes in atmospheric contribution with photosynthesis rates. These are in the main reversible on an annual cycle and so do not provide much long-term storage of carbon. Trees, on the other hand, with lifetimes of tens to hundreds of years, can provide medium-term capture and storage of CO_2 largely as cellulose and hemicellulose (Section 5.3). Large-scale tree planting can provide some checking in the increase in atmospheric CO_2, which can afford some breathing space for longer-term solutions to be implemented. Methods of capture have been developed which can be implemented at known high-emissions sites such as cement works, power stations, and iron and steel production. Mineralization, as demonstrated by the CarbFix2 project in Iceland,[19] has the potential of fixing carbon for millennia.

[19] I. Gunnarsson, E. S. Aradóttir, E. H. Oelkers, D. E. Clark, M. T. Arnarson, B. Sigfússon, S. Ó. Snæbjörnadóttir, J. M. Matter, M. Stutte, B. M. Júlíusson, S. R. Gíslason, 'The rapid and cost-effective capture and subsurface mineral storage of carbon and sulphur at the CarbFix2 site', *Int. J. Greenh. Gas Con.*, 2018, **79**, 117–26.

7.4   Chapter 5: Fire

7.4.1   Plastic Waste

The plot in Fig. 5.26 of the production and waste generation of plastics shows some striking features. One is that (in 2015) almost all of the production ≈ 150 Mt

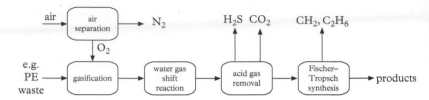

Figure 7.11 *Coupling of plastic waste disposal to Fischer–Tropsch synthesis.*

[20] R. Greyer, J. R. Jambeck, K. Lavender Law, 'Production, use and fate of all plastics ever made', *Sci. Adv.*, 2017, **3**, e1700782.

of packaging plastics rapidly became waste.[20] Crucial to the sustainable use of plastics is that life-cycle analyses, including the proper collection and treatment of the 'corpse', are established and enacted. It is not sustainable to have a production and delivery programme to be effected in a region that does not have the capacity to achieve wholesale collection and treatment. Much of the waste is polyethylene. In the UK at least, unlike for PET, there is no systematic retreatment process more subtle than combustion. The waste which is not discarded to the detriment of the environment can be incinerated as bagged waste to provide heat and energy. An alternative to full combustion is to give the carbon atoms a further life as commodity chemicals (Fig. 7.11), akin to the reforming of woody biomass.[21]

[21] I. S. Tagomori, P. R. R. Rochedo, A. Szklo, 'Techno-economic and georeferenced analysis of forestry residues-based Fischer-Tropsch diesel with carbon capture in Brazil, Biomass Bioenergy', 2019, **123**, 134–48.

$$3(-CH_2-) + H_2O + O_2 \rightarrow 3CO + 4H_2. \tag{7.1}$$

In that process, O_2 can be generated in an air-separation unit and fed, with steam, to a gasification unit containing polymer waste effecting the conversion to synthesis gas (Eq. 7.1). The preferred CO/H_2 can be provided using a water gas shift reaction. From this the acid gases can be separated, and ideally captured and stored. Fischer–Tropsch synthesis (Section 5.1.5) can be used as a means of forming commodity chemicals or their precursors of different chain-length ranges (e.g. liquids or waxes). The gases so formed may be used as fuel, or as precursors to polymer production.

Table 7.5 *Primary energy consumption in 2018 (10^6 toe).*

Fuel	Consumption 10^6 toe
oil	4,662.1
gas	3,309.4
coal	3,772.1
nuclear	611.3
hydroelectric	948.8
renewables	561.3
total	**13,864.9**

7.4.2 Prospects for Sustainable Energy Generation

The global energy consumption estimates for 2018[22] are presented in Table 7.5 and also in Fig. 7.12. The predominance of fossil fuel provision is evident. Of the major types of sources, renewables presently provide the smallest contribution. The overall growth rate for energy consumption was 2.9% in 2018, nearly double that recorded over the previous decade (1.5%). The growth rates for nuclear and hydroelectric power generation in 2018 were 2.4% and 3.1% respectively, similar to the overall rate. However, the growth rate for renewables was 14.5% in 2018, following a mean of 16.4% over the period of 2007–17. It can be expected that supply by renewables will exceed that from nuclear energy in the next year or two. The distribution of the major sources of renewable energy is given in Table 7.6. This is dominated by wind-based sources, with solar energy generation the

[22] *Statistical Review of World Energy*, 2019, BP, http://www.bp.com.

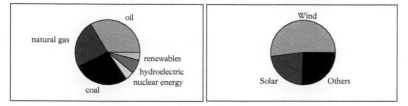

Figure 7.12 *Energy in 2018: consumption, left; renewables, right.*

second-largest factor, being close to the sum of contributions from many other methods including geothermal, biomass, and tidal. The growth rate for solar energy generation was very high in 2018. This was 28.9% in 2018, dominated by the growth in non-OECD countries (47.5%). These two methods—wind power and photovoltaics—can be envisaged to form the largest proportion of renewable energy generation in the coming decades. It is the elements key to those technologies that will come under increasing pressure, and research into substitutes may be crucial.

The world's primary energy consumption has grown markedly and is projected to continue to do so; it is essentially expected to have quadrupled from 1970 to 2040 (Fig. 7.13). In 2020, industrial consumption of energy, including the non-combusted use of fuels (e. g. petrochemicals), is estimated to be $\approx 50\%$ of the total. Both industrial and transport consumption are expected to reduce slightly as a proportion of the general increase. The latter can be ascribed to improved vehicle efficiency, η, which is simply defined as in Eq. 7.2; the remainder of the energy will otherwise be dissipated as heat. Table 7.7 provides a brief correlation of the types of motor used in transport. The petrol- (gasoline-) driven internal combustion engine (ICE) has an efficiency limited by the temperatures inside and outside the cylinders, and the efficiency range is $\approx \frac{1}{4}$ to $\approx \frac{1}{3}$; the majority of the energy is lost as heat. The efficiency is improved in both diesel and gas

Table 7.6 *Renewable energy generation in 2018 (TWh).*

Fuel	Generation TWh
wind	1,270.0
solar	584.6
geo, biomass, other	625.8
total	**2,480.4**

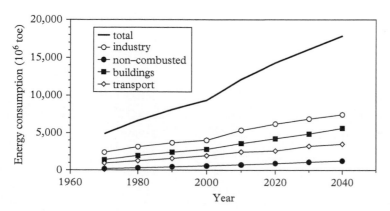

Figure 7.13 *Past and projected energy consumption by sector.*

Engine	Efficiency
petrol ICE	0.2–0.35
diesel ICE	≈ 0.45
gas turbine	0.46–0.61
EV motor	0.85–0.9

[23] *Greenhouse gas reporting: conversion factors 2019,* https://www.gov.uk/government/publications/greenhouse-gas-reporting-conversion-factors-2019

[24] 1 toe = 1.42 t hard coal = 0.805 t LNG.

turbine engines. None though approach the efficiency of an electric motor, which is close to 0.9. Thus this enhanced efficiency, and the lower demand on fossil fuels, provide a sound expectation for flattening of future demand for fossil fuels in transport.

$$\eta = \frac{work\ done}{energy\ absorbed}. \tag{7.2}$$

The data in Tables 7.5 and 7.6 provide a glimpse of the degree to which change is required to approach the CO_2e emission targets intrinsic to the IPCC scenario RCP2.6. A comparison of the average greenhouse gas emissions of a selection of fuels in the UK is presented in Table 7.8.[23] The general trend in reducing the GHG emissions for the carbonaceous fuels follows the expected trend for CH_x stoichiometries: $CH_4 < CH_{\approx 2} < C$. In 2019, the GHG emissions for electricity were similar to those for liquid fuels. For solid fuels, coal burnt domestically has high emissions, and for liquid fuel this is true for marine fuel oil. Both might be targets for replacement. For example, changing domestic coal could provide 38 or 47% reductions in emissions based on the fuel comparisons with LPG and natural gas, respectively.[24] If a quarter of the coal use were switched to fuel gases this would point to a reduction of ≈ 1.7 Gt CO_2e. An alternative scenario of doubling the consumption of renewables to replace oil consumption would afford a similar reduction. Within RCP2.6, the emissions of CO_2 required from 2020 to 2030 amount to 8.5 Gt, or about 0.9 Gt annually. The scenarios considered here provide some scaling for the changes required in energy use to move from the projections in Fig. 7.13 to the emissions proposed in Fig. 7.9.

Table 7.8 *Greenhouse gas emissions of a selection of fuels giving 1 kWh of energy (UK, 2019).*

Fuel	CO_2e kg	CO_2 kg	CH_4 g	N_2O g
electricity	0.2556	0.25358	0.65	1.37
natural gas	0.18385	0.18351	0.24	0.1
LPG	0.21447	0.21419	0.14	0.14
aviation turbine fuel	0.24776	0.24529	0.15	2.32
burning oil	0.24675	0.24553	0.61	0.61
diesel (mineral)	0.25267	0.24942	0.03	3.22
petrol (mineral)	0.24099	0.23961	0.72	0.66
marine fuel oil	0.26298	0.25918	0.11	3.69
coal (for electricity)	0.30561	0.30373	0.09	1.79
coal (domestic)	0.34473	0.31470	25.65	4.38

7.4.3   Prospects for Sustainable Energy Storage

Increased power generation by renewables which are dependent upon the sun and wind can cause significant variations between supply and demand. Hence, it will become increasingly important to be able to store energy and deliver a rapid response to demand changes. The classes of installations are shown in Table 7.9, with data from note 25.[25]

Physical Energy

The predominant methodology for energy storage is pumped hydro storage (PHS), otherwise called pumped storage hydroelectricity (PSH). It works by transfer of water between two storage regions of different altitude, generally between two dams, although this could involve an underground space. The flow of water from opened valves can provide high power levels, which globally afford $\approx$ 182 GW. It is anticipated that additional capacity will be created at a rate of $c.3$ GW per annum, largely in China.[26] The energy efficiency of these pumped processes is in the region of 70–75%, but the gain is being able to provide energy at high tariff times and to pump during periods of excess generation. The overall energy capacity is currently $\approx$ 9 TWh, compared with a global generation of $\approx$ 26,000 TWh.[27] The overall delivery of electrical energy will depend greatly upon the hours of use through the year; it is estimated that by 2023 the annual delivery will reach 146 TWh.[26] Smaller-scale developments involve cycling with compressed or liquified air. The release of the stored gases can then be used to drive turbines. An alternative energy storage principle is via thermal methods. A wide variety of methods have been employed, mostly for heating and cooling of processes and districts, with power generation as the third most used application.

Chemical and Electrical Energy

As is evident in Table 7.9, electrochemical means of static electricity storage are presently a distant second to PSH methods. The capacity of batteries was reported in 2018 to be 7 GWh.[26] However, installations of new static secondary battery facilities are expected to provide similar new power capacity over the period 2018–23: 22 GW and 26 GW for PSH.

   An important factor in the application of battery technology is its energy density (Section 5.7, Fig. 5.59). The energy density of batteries was seen to be orders of magnitude higher than for supercapacitors, with lithium-ion batteries providing relatively high values $c.0.6$ MJ kg^{-1} and 3 MJ L^{-1}. However, these values are much smaller than those afforded by the storage of chemical energy (Fig. 7.14). In terms of mass H_2 provides a very high energy density ($c.140$ MJ kg^{-1}), but in terms of volume, an elevated pressure (but not necessarily liquifying) is required to exceed the energy density of a lithium-ion battery. The effective mass of an energy-storage device must include the storage vessel. In Section 5.6, the electrochemical generation of H_2 gas by electrolysers was discussed together with its reconversion into electrochemical energy by fuel cells.

[25] DOE Global Energy Database, https://www.sandia.gov/ess-ssl/global-energy-storage-database/.

Table *7.9 Energy storage installations.*

Method	Power (MW)
pumped hydro	181,190
electrochemical	3,297
thermal	3,275
hydrogen	20
compressed air	7
liquid air	5

[26] *Will pumped storage hydropower expand more quickly than stationary battery storage?*, IEA 2019, https://www.iea.org/articles/will-pumped-storage-hydropower-expand-more-quickly-than-stationary-battery-storage.

[27] https://www.enerdata.net/publications/world-energy-statistics-supply-and-demand.html.

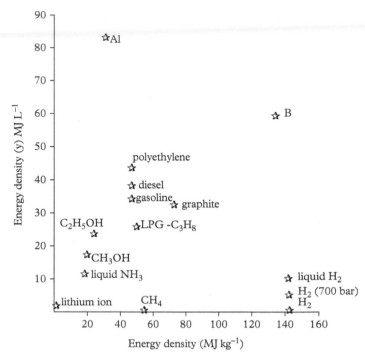

Figure 7.14 *Energy density by mass and volume of some energy-storage options.*

Other chemical methods for energy storage are under consideration, some being classed as thermal methods, as in Table 7.9.[28] Aluminium has a high volume-related energy density (84 MJ L^{-1}), but a much lower mass-related value (33 MJ kg^{-1}) as compared to H_2. The values for hydrocarbon fuel fall in an intermediate region ($\approx$ 35 MJ kg^{-1} and 25–45 MJ L^{-1}). Biodiesel and ethanol–diesel mixtures will afford energy densities between those of diesel and ethanol. Extremely high energy densities are held by nuclear fuels. For example, the energy density of uranium in a breeder reactor is in the region of 8×10^7 MJ kg^{-1} and 1.5×10^9 MJ L^{-1}. However, again the operational mass and volume are really provided by the nuclear reactor.

The storage method employed may well depend upon the application. Static storage can be envisaged as having components with different delivery rates. The fastest of these may be supercapacitors, with batteries providing intermediate rates and chemical storage providing higher-density, longer-term storage. In transport systems, energy density comes to the fore and the practical result will be a function of the mass of the transportation and the range. Here the high energy density of hydrogen can make it preferable to batteries as the alternative to (bio)diesel.

An interesting comparison of three alternative means of delivering 100% renewable electric vehicles has been presented (Table 7.10).[29] In a battery-driven vehicle the losses are relatively small in providing the fuel, largely due to the

Table 7.10 *Estimates of the energy losses of fuel types of car derived from renewable electricity.*

Process	Battery electric Direct charging	Fuel cell Hydrogen fuel	Conventional Hydrocarbon fuel
Renewable electricity to fuel			
power	100%	100%	100%
electrolysis for H_2		−22%	−22%
CO_2 capture + FTS			−44%
supply	−5%	−22%	
Fuel production efficiency	**95%**	**61%**	**44%**
Fuel to wheel			
charging	−5%		
battery charge efficiency	−5%		
H_2 to electricity		−46%	
inverter	−5%	−5%	
engine efficiency	−5%	−5%	−70%
Overall efficiency	**77%**	**30%**	**13%**

electrical supply and distribution system. More small losses are involved in the steps to driving the vehicle: charging equipment, battery charge efficiency, the inverter, and the engine efficiency. Overall, this affords a power-generation-to-power-to-the-wheel efficiency of ≈ 77%. For a fuel-cell-powered vehicle, it is considered that the losses are higher in the distribution system and those at the electrolysis step are also incurred. A further significant loss is likely to be incurred in the fuel-cell step to convert H_2 to electricity. Overall, the efficiency is estimated to be less than half that of the directly charged vehicle. The renewable fuel production steps are more complex for a conventional vehicle. H_2 is again to be generated by electrolysis, but the hydrocarbons are to be generated by CO_2 captured and then a Fischer–Tropsch synthesis (FTS) step. Finally the intrinsically lower efficiency of the ICE leads to a further lowering of the overall efficiency to less than half that of the fuel-cell vehicle. The high energy density per unit volume might still leave this as the most practical option for some applications.

7.5 Chapter 6: Water

The recent (2014) and projected (2040) demands for water are presented in Fig. 7.15. Demand for sources of water is expected to rise from ≈ 4,000 km^3 to 4,500 km^3 over that period.[30] The major sector of application is in agriculture, with

[30] *Leaving No One Behind*, the UN World Water Development Report 2019, UNESCO, Paris, 2019.

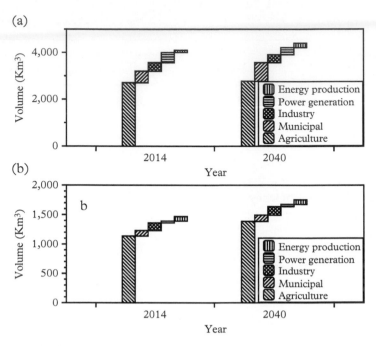

Figure 7.15 *(a) Annual water withdrawal from source and (b) annual consumption reported for 2014 and projected for 2040, shown by sector.*

municipal and power generation also demanding more than industry. The final sector, of lowest demand, is the production of primary energy, including fossil fuels and biofuels. The projected growth is substantially ascribed to increased municipal use, as population grows, and in primary energy generation, linked to increases in biofuel production. Power generation itself is predicted to have reduced water demand. However, much of the production demand is returned to the local source by drainage with some time lag. The consumption of water represents that lost to the local source, mainly by evaporation or transportation elsewhere. In 2014 the net consumption was $\approx \frac{1}{3}$ of demand, but this proportion is expected to increase to $\approx \frac{2}{5}$, with agriculture predicted to have the largest increase.

The maturing of desalination processes via forward and reverse osmosis processes (section 6.5) has provided routes to increasing freshwater supplies with much lower energy than evaporation techniques. Providing the power substantially by green means, by photovoltaics or off-shore wind farms depending upon the coastal location, seems a tractable approach. The recent development of reverse electro-dialysis (RED) as an electricity source in locations with both salt- and freshwater is an intriguing prospect, and in some locations this might also power desalination.

As shown in Fig. 6.7, there has been a reduction in the mass of the icecap of Antartica by about 2,500 Gt since 1992, which will have been converted into seawater. This has been shown to contribute to a rise in sea level of 7 mm, a trend which has increased in rate since 2007 (Fig. 7.16).[31] This is only one component in changes in global mean sea level (GMSL).[32] There has been a steady increase in GMSL since 1900 with an acceleration in the rate of change in the 1930s (Fig. 7.17). This is principally thought to be due to increased melting of glaciers, Greenland and Antarctic ice, but this became an influence for reduction through the 1940s. Consistent with Fig. 6.7, overall ice melting has become dominant again since the 1990s. A second factor is an effect from release from terrestrial water storage and this is thought to have been important during the acceleration period of the 1980s. The third factor is an interplay between atmospheric and ocean conditions. In the 1960s there was an increase in sea levels in the southern oceans

[31] The IMBIE team, 'Mass balance of the Antarctic ice sheet from 1992 to 2017', *Nature*, 2018, **558**, 219–22.

[32] S. Dangendorf, C. Hay, F. M. Calafat, M. Marcos, C. G. Piecuch, K. Berk, J. Jensen, 'Persistent acceleration in global sea-level rise since the 1960s', *Nat. Clim. Change*, 2019, **9**, 705–10.

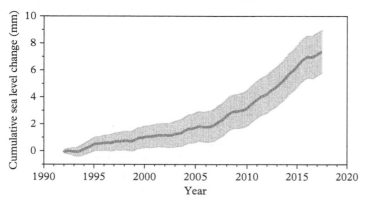

Figure 7.16 *Cumulative change in the sea level due to the melting of the Antarctic ice sheet from 1992 to 2017, with error estimates.*

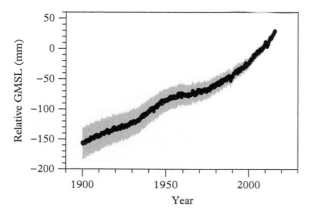

Figure 7.17 *Global mean sea level changes (GMSL) from 1900 to 2015 with error estimates.*

(Indo-Pacific and Atlantic) linked to an intensification and shift of the westerly winds. This appears to have caused an increase in heat uptake by the oceans and changes in the ocean circulation in the southern hemisphere enhancing ice melting in Antarctica and leading to a more rapid raising of GMSL. From 1993 to 2015 the annual increase in GMSL was estimated to have risen from ≈ 1.3 to ≈ 3.4 mm yr^{-1}.

It is estimated that if RCP2.6 is followed, the rise in GMSL by 2100 would be 260–550 mm. If this is not achieved then this rise will be higher; following RCP8.5 is expected to increase this to 520–980 mm. The Greenland icecap alone is estimated to have a sea-level equivalent of ≈ 7.4 m, and that of the Antarctic icecap is ≈ 58.3 m. Since about 10% of the world's population live within 10 m of the current sea level, even a fraction of these changes would cause massive disruption. Hence, if the RCP2.6 scenario can avoid a tipping point some very significant consequences might be avoidable. It is intended that this textbook might highlight the importance of this and also provide some pointers to the means of achieving it.

7.6   Questions

The questions in this chapter are deliberately of a different character to those in the chapters above. Here the emphasis is on evaluating the scientific evidence and the global context. From that assessments are required to choose the best course. Then this course is to be outlined. The resources provided stretch far beyond the text via the extensive referencing of sources. These questions can be considered as templates which can be adapted to other scenarios.

1. Imagine you are making a representation to the board of an international agency. Your interest is in population stabilization. Outline your views about the desirability of such a programme. How would you propose to the agency that they go about this?

2. Global mine production of cobalt is estimated to have been 120 kt in 2017, and is expected to rise. Secondary production of cobalt in the USA that year was 2.75 kt, and globally recycling was ≈ 10 kt. Consider the main uses of cobalt and assess whether demand for it is likely to increase in the next decade. What processes look to be good prospects for the extraction of cobalt from end-of-life products? Assess whether these can make a significant contribution to the supply of cobalt in the future.

3. Mindful of the EU's 2017 list of critical raw materials, consider the possibility of the extraction of boron compounds from seawater. Investigate the main uses of boron and its compounds. Assess if demand is likely to increase over the next decade. Examine the current sources of boron. Assess the viability of extraction of boron from seawater, considering the

forms will boron take, and how it might be isolated. What is your overall judgement about the future of the supply of boron?

4. Imagine now that you are a member of a national board with the responsibility of enacting policies to tackle climate change. One of your colleagues does not think this should be a matter of a national policy as they do not consider people's activities can influence the climate. You have a chance make the case in a 3-minute slot, and have one sheet of A4 available to you. Set out the case that you would make.

5. The same national board is considering whether to use RCP2.6 as a guideline for a national strategy. Assess any evident deviations from this pathway in current observations. What would be your priorities for steps aimed at achieving the target of limiting temperature rise to 1.5°C?

6. You have now been asked to advise on four energy proposals:
    (i)   an opencast mine to provide 7 M t of coal,
    (ii) extraction of up to 200 trillion cubic feet of shale (natural) gas,
    (iii) a wind farm of 22 turbines rated at 3 MW peak power, and
    (iv) a solar farm rated at 25 MW peak power affording 2.4×10^7 kWh annually.

    Assess the relative magnitude of these proposals in terms of energy generation, and comment on the contributions they might make towards RCP2.6. Which might you recommend for approval on this basis?

Index